MW00837699

SPANNING THE GILDED AGE

Hagley Library Studies in Business, Technology, and Politics
Richard R. John, Series Editor

SPANNING THE GILDED AGE

James Eads and the Great Steel Bridge

JOHN K. BROWN

Johns Hopkins University Press

Baltimore

Johns Hopkins University Press
2715 North Charles Street
Baltimore, Maryland 21218
www.press.jhu.edu

Library of Congress Cataloging-in-Publication Data

Names: Brown, John K., 1958– author.
Title: Spanning the Gilded Age : James Eads and the great steel bridge /
    John K. Brown.
Description: Baltimore : Johns Hopkins University Press, 2024. |
    Series: Hagley Library studies in business, technology, and politics |
    Includes bibliographical references and index.
Identifiers: LCCN 2023028038 | ISBN 9781421448626 (hardcover) |
    ISBN 9781421448633 (ebook)
Subjects: LCSH: Eads Bridge (Saint Louis, Mo.)—History. |
    United States—Social conditions—1865–1918.
Classification: LCC TG25.S15 B76 2024 | DDC 624.209778/66—
    dc23/eng/20230927
LC record available at https://lccn.loc.gov/2023028038

A catalog record for this book is available from the British Library.

*Special discounts are available for bulk purchases of this book. For more
information, please contact Special Sales at specialsales@jh.edu.*

Thanks to Sally and Gerrit
For Wendy

CONTENTS

PREFACE

THIS IS THE STORY of the St. Louis Bridge, known for more than a century as the Eads Bridge. Begun in 1867, completed in 1874, it was the first structure of any kind, anywhere in the world, built of steel. For nineteenth-century Americans, it rivaled the Brooklyn Bridge as a proud emblem of technological accomplishment and national greatness. An engineering marvel, the St. Louis Bridge broke records for the depth of its foundations and the length of its three steel spans, each a graceful arch. Yet this was the first bridge its promoter-designer had ever undertaken. To place the stone piers on solid bedrock beneath the churning Mississippi River, Captain James Eads adapted a range of daring innovations that allowed excavators to work, more or less safely, at depths of more than one hundred feet beneath the surface of the river. Pioneered in France, Eads's pneumatic caissons had their North American debut in building the St. Louis Bridge. Preparing for his own challenges in Brooklyn, Washington Roebling visited St. Louis to see how it all worked.

The financing of this demanding project required equally imaginative design skills. In nineteenth-century America, entrepreneurs, not governments, usually built this kind of infrastructure.* With construction scarcely begun, James Eads circulated a prospectus offering a 500 percent

---

* To clarify, entrepreneurs would create a company, secure a charter from a state government (in this case two states, Illinois and Missouri), amass financing, and contract with specialists to build a bridge. Upon completion, that company charged tolls and hoped for profits. This narrative uses a simplified name to identify the venture whose longtime name was the Illinois and St. Louis Bridge

On a bitterly cold day in January 1875, commercial photographer Robert Benecke ventured out onto the frozen Mississippi. In this upstream view, steamboats hibernated on the St. Louis levee, trapped in the ice. This perspective underscores why Eads embraced steel for his arches. That metal was stronger than iron, especially when bearing the compression loads of an arch. Photograph by Robert Benecke, MHS, VM89-000034.

return on investment. With that glittering prospect, the company sold equity shares to fifty-nine stockholders, nearly all bankers and merchants in St. Louis and New York, including J. Pierpont Morgan.[1] To complete the financing, Pierpont's father, the investment banker Junius Morgan, floated the bridge company's bonds on the London market. The Morgans cultivated their image as cautious men and safe bankers, so their enthusiasm for the bridge project is somewhat surprising given that it defined risk. It nearly defied fate. A record-breaking design to be built in a far-off city, employing a novel method to lay its foundations, using an untried metal for its arches, all projected by a steamboat man

---

Company. Throughout, *St. Louis Bridge* refers to the company, whereas *the St. Louis Bridge* (with the article) indicates the structure itself.

Studio portrait of James Eads, ca. 1874. The photographer captured his commanding presence. Photograph by A. Schelten, MHS, N 21961.

who had never designed a bridge in his life. This was not investment banking for the timid.

Before construction began, other giants of the Gilded Age came into the project. The president of the Pennsylvania Railroad, the world's largest corporation, assigned a young protégé, Andrew Carnegie, to link the bridge company to the railroad. The top two officers of the PRR also looked to Carnegie to shake out profitable insider deals for the trio, the kind of rich pickings that typically accompanied such a complex endeavor. When the bridge was nearly completed, the chief of the Army Corps of Engineers attempted to derail the project by declaring that it obstructed navigation illegally. At this eleventh hour, President Ulysses Grant stepped in to overrule the Corps and save the bridge.

The bridge's design, the methods used to build it, the financing plans, and the project's successful completion—even Grant's intercession—were largely the work of one remarkable man. Little remembered today, James Eads must rank among the most influential American engineers of the nineteenth century. His career is the second story told here. Born on the Indiana frontier in 1820, Eads quit school at 13 to strike out on his own in St. Louis, the leading city of the West. During the 1840s, he pioneered in salvaging sunken steamboats, searching the riverbed of the turbulent Mississippi in a diving bell to retrieve valuable cargoes. By the 1850s he had developed a suite of novel technologies to raise complete vessels laden with their freight from the river bottom. The work earned him a fortune and taught him valuable lessons about the Mississippi's treachery and power. In the early months of the Civil War, this charismatic innovator built the country's first ironclad warships, ahead of the USS *Monitor*. Unlike that famous vessel, Eads's ironclads won their first battles and many more thereafter. This wartime work bolstered his design skills in mechanical engineering, expertise he then applied in building the St. Louis Bridge.

Captain Eads had exceptional talent as a promoter, and few could resist his persuasive power. Carnegie saw him as a "man of real decided genius," grudgingly conceding that it was "impossible for most men not to be won over to his views."[2] Eads became a sophisticated player in the capitals of high finance, promoting his bridge company's stocks and bonds to the polished bankers of Wall Street and the City of London.

Despite his lack of formal education, James Eads made lasting contributions in many aspects of bridge design and construction. He had a talent for grasping and solving huge engineering challenges, a compulsion to see and think anew. Put off by his cocky self-confidence and his contempt for orthodoxy, many of his fellow engineers disliked Eads intensely. But the Fellows of the Royal Society of Arts in London awarded him their Albert Medal, honoring "distinguished merit in promoting the arts, manufactures, and commerce."

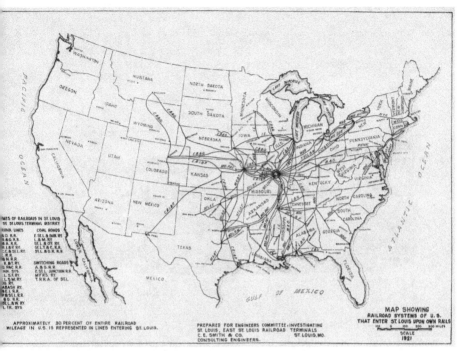

This 1921 map shows St. Louis as the point of convergence for a dense network of railroads, all reaching the city directly on their own rails. Only Chicago was a bigger hub. From its opening, the St. Louis Bridge linked carriers on both sides of the Mississippi. That connection then drew new lines to the city, connecting each with all. Known today as network effects, this skein of tracks also grew the city's population, regional industry, and the national economy. Map from *Report of Engineers Committee*, 14, MHS.

Even before his bridge had opened, Eads took up another daring, innovative project. By carving a deepwater channel through the Mississippi River Delta to the Gulf of Mexico, he remade New Orleans into a major port for oceangoing shipping.

While Eads focused on altering the flow of the Mississippi at its mouth, the St. Louis Bridge had only begun to transform his hometown, the midwestern rail map, and the commerce of the nation. This is the first account to examine its profound impact in the railway age, which remade the United States after 1865.[3]

While Captain Eads and his bridge were unique, this history of the project expands our understanding of a much broader story: how America

THE GREAT ST LOUIS BRIDGE.
ACROSS THE MISSISSIPPI RIVER.

Chromolithographs from the New York firm of Currier & Ives became popular icons of technological accomplishment, hung with pride in parlors across Gilded Age America. Widely seen, this image was also woefully inaccurate. The delineator flattened the graceful curve of the roadway and added statues that never actually stood atop the finished bridge. *The Great St. Louis Bridge,* Currier & Ives, 1874, MHS, N 22064.

became urban, industrial, and interconnected. Modern. Between 1864 and 1879, twenty-five long-span railway bridges were built across the three great rivers of the West: the Ohio, the Mississippi, and the Missouri. To place the St. Louis Bridge in its context, many of these bridges receive some consideration in this account. Nearly all were the work of profit-seeking entrepreneurs and investors.[4] The same was true of the big suspension bridges for roadway traffic built at Wheeling, Cincinnati, Niagara (road and rail), and Brooklyn. Countless prosaic roadway bridges in cities and towns across the country also were commercial ventures, put up by investors anticipating profits from patrons paying tolls to cross. Across nineteenth-century America, this private-sector approach built most of the urban waterworks that provided piped fresh water to growing cities.[5]

This halftone engraving from 1902 portrays much of note. Steamboats still docked on the levee and traveled the Mississippi, although the illustrator likely overstated their number. The city that Eads knew in the 1850s was just inland from the levee: the warehouses and merchant offices built to serve the steamboat trades. Further inland the modern business district towered, with office blocks for banks, insurance companies, wholesalers, and industrial corporations. The crowded bridge in the foreground gave rise to that prosperous city. *Bird's-Eye View*, printed by Woodward & Tiernan, MHS, N 13784.

It created the gas-lighting companies that finally ended the ancient tyranny of dark nights.[6] And it built the street railways that transformed market towns into sprawling industrial cities.[7] These vibrant industries all developed from the same fusion of entrepreneurship, investment, state action, and innovation that created the Eads Bridge. The men, motives, and coalitions that built this structure had counterparts across the country. For the most part, historians have overlooked the origins of these consequential ventures.[8] They built the Gilded Age.*

---

* Historians struggle for a suitable moniker to describe the decades between the Civil War (or Reconstruction) and the Progressive Era that originated ca. 1895. Social commentators of the 1920s originated the term Gilded Age to denounce the shallow vulgarities and frauds of the post–Civil War decades. But as this book describes, admirable people also accomplished much in the period that was

Until the 1890s or later, governments typically did not create these infrastructures directly, because they were restrained by inadequacies in revenues, administrative capacities, vision, or political will. But public authorities extensively shaped these ventures. As corporations, they originated with state charters. Many received franchises and bond guarantees from governments. Crucially, those bridges, waterworks, street railways, and gas companies were nearly all monopolies, protected by law from competitors thanks to city or state enactments. By shaping their markets in this way, government encouraged risk-taking promoters with private capital to build the infrastructures that grew and sustained American cities.[9]

When looking at the United States in the decades following the Civil War, we tend to focus on the squalor of its cities, the corruption of its politics, the unfulfilled promises of Reconstruction, and the fraudulent schemes of the men known to history as robber barons. An ample helping of fraud—at least by today's standards—shadows and complicates this narrative of James Eads and his bridge. Earlier biographies veiled those dealings behind hero worship.

This account reveals, for the first time, how this imaginative innovator and entrepreneur broke with engineering conventions to design an unprecedented and graceful structure, engineer its audacious financing, overcome a succession of challenges during its construction, and develop connecting infrastructures to grow and speed the commerce of the nation. Captain Eads's vision propelled St. Louis to the top rank of American cities by 1900. His bridge remains in daily use today, celebrating its 150th anniversary in 2024. Few protagonists of any kind have had such an influential role in American history.

---

essential and good, even heroic. I use the term simply as a convenient label for the years from 1865 to 1895, a transformational epoch that (like any chapter in human history) combined good and bad. For more on these debates, see John, "Gilders."

MONEY THEN AND NOW

COSTS, RATES, AND PRICES, including evolving estimates for the bridge, the tolls to pass over it, the par value of bonds to pay for it, and its final price tag, appear throughout this book. Contemporary values for all these figures and many more appear in the text. Today's readers may wish to know their modern equivalents. For example, the men who worked beneath the riverbed, digging away inside the pressurized caisson to place the stone piers onto bedrock, received a daily wage of $4 in 1870. How much would that sum be in today's dollars?

Economic historians have developed indices to approximate an answer. Those comparators reckon with the tides of inflation or deflation that have shaped money values to the present. The consumer price index (CPI), computed by the US Bureau of Labor Statistics, provides a starting point. Building on that foundation, historians have constructed data sets to project the CPI across time. For our sandhog of 1870, a good approximation of his daily pay in 2023 dollars is $95.

The historic CPI is available at www.measuringworth.com/ppowerus/, a resource that discusses the uses and limitations of comparators. They are imperfect tools. The value of a loaf of bread changed over time, as did that of an hour of manual labor, a ton of steel, or a US Treasury bond. Their *relative* worth (vis-à-vis one another) also evolved over time. So analysts use different comparators for different purposes. Using multiple price indices in this study would, however, overwhelm an already complicated issue. For that reason, the chapters that follow draw from the historic CPI exclusively. It adds a useful perspective.

The bridge's price tag in 1874 was $12.5 million in contemporary dollars. Using the historic CPI, that translates to $340.6 million in 2023 dollars. That figure appears to be reasonably accurate when compared with a similar project today. Just a few miles upstream from the Eads Bridge, the Merchants Bridge (1890) is undergoing a thorough rehabilitation at the time of this writing (2023). It will have three new, double-track trusses, each 520 feet long, and its foundations, piers, and abutments will be thoroughly rebuilt. The estimated cost for the Merchants Bridge rehabilitation is $219 million (in 2021 dollars). That number suggests that the CPI approximation for the Eads Bridge ($340.6 million in 2023 dollars) is about right, partly because the original cost of Captain Eads's structure included an upper (roadway) deck not found in the Merchants crossing. In another important distinction, structural steel was exotic and expensive circa 1873, while today's equivalent is mass produced at low prices with uniform qualities in a competitive market.

Because it would clutter the text excessively to give the modern equivalent for every dollar value in this book, and it would serve little purpose, a compromise seemed wise. In each chapter, comparative values typically appear just three times. All derive from the historic CPI found at www.measuringworth.com/uscompare/.

LEADING FIGURES

REFLECTING THE SIGNIFICANT roles they played in the creation of the St. Louis Bridge (SLB), these men appear repeatedly in the text:

James Andrews (1828–1897). Contractor for the masonry piers and abutments of the SLB. Completed excavations and stonework for the tunnel under downtown St. Louis.

Lucius Boomer (1826–1881). Proprietor of the American Bridge Company (Chicago), he built bridges in wood, iron, and steel for four decades. Promoted his own bridge for St. Louis.

B. Gratz Brown (1826–1885). Missouri senator and governor, candidate for vice president (1872), he spoke at dedication of the SLB, supplied Missouri red granite for the bridge.

Andrew Carnegie (1835–1919). Protégé of Tom Scott, chief salesman for Keystone Bridge. Negotiated bridge and bond deals (1866–74), stockholder in SLB (1869–77).

Octave Chanute (1832–1910). Chief engineer for Midwest railroads and a rail bridge at Kansas City (1867–69), he did pioneering investigations of caisson design and use in Europe.

Zerah Colburn (1833–1870). Editor of *Engineering* (London), he backed Eads's "noble arched viaduct."

Theodore Cooper (1839–1919). Steel inspector, then oversaw super-structure work for SLB.

Norman Cutter. Missouri state senator and original promoter of the bridge venture that Eads took over, he secured that company's Missouri charter (1864).

James Buchanan Eads (1820–1887). Chief engineer of SLB at a monthly salary of $1,000 ($327,000/year in 2023).

Henry Flad (1824–1898). First assistant engineer under Eads throughout the bridge project.

Daniel Garrison (1815–1896). A director and manager of the Missouri Pacific Railroad, president of Lucius Boomer's venture to build a bridge at St. Louis, competing with Eads's SLB.

Solon Humphreys (1821–1900). Partner in E. D. Morgan & Company (New York private bank), he was a receiver, then president of SLB. President of Wabash Railroad and ally of Jay Gould.

Robert Lenox Kennedy (1822–1887). President, Bank of Commerce (New York), early backer of Eads and SLB.

Jacob H. Linville (1825–1906). Designer of Steubenville Bridge (1864). Stockholder, chief designer, and president of Keystone Bridge Company. Critical of Eads's design for St. Louis.

William M. McPherson (1813–1872). Stockholder in SLB from 1868, then president (1871–72). Friend of Edgar Thomson and Tom Scott. Advocate of a transcontinental line from 1849.

Edwin Denison Morgan (1811–1883). New York governor, US senator, and principal of a private bank bearing his name. Stockholder in SLB.

J. Pierpont Morgan (1837–1913). Partner in Dabney, Morgan & Company (1864–71), then in Drexel, Morgan & Company. Invested in SLB stock (1869), served as a receiver (1875–79).

Junius Spencer Morgan (1813–1890). London investment banker, stockholder in SLB, he sold SLB's bonds on the London market, then oversaw its reorganization and lease to Gould.

William Nelson. Partner ca. 1840 in boat building with Calvin Case, he became Eads's partner in salvage business (1842). Built iron caissons for the SLB.

Charles Pfeifer (1843–1883). Assistant engineer at SLB, trained in Bavaria, expert in "the calculus," he worked on Coblentz Bridge before emigrating.

Simeon Post (1805–1872). Designer of a patented truss that Boomer advocated for St. Louis.

W. Milnor Roberts (1812–1881). Associate chief engineer of SLB (1868–70) while Eads recuperated in Europe.

Washington Roebling (1837–1926). Son of John Roebling, he led construction of the Brooklyn Bridge after his father's death (1869). Like Eads, an American pioneer in pneumatic caissons.

Thomas Alexander Scott (1823–1881). With Edgar Thomson, he led "Andrew Carnegie & Associates" in promoting midwestern railroads and bridges. Vice president of Pennsylvania Railroad.

William Sellers (1824–1905). Ran Midvale Steel Company (Philadelphia), which was successor to William Butcher in supplying SLB with steel.

Shaler Smith (1836–1886). With Benjamin Latrobe, he promoted a lift bridge for St. Louis (1867). He built the east approaches for the SLB (1874).

William Taussig (1826–1913). Born in Prague, he served SLB in diverse managerial roles (1867–96).

J. Edgar Thomson (1808–1874). Leader of Carnegie & Associates, president of Pennsylvania Railroad, he assisted SLB in many ways.

SPANNING THE GILDED AGE

Prologue

The Celebration

NOT MANY PEOPLE in St. Louis slept well the night of July 3, 1874, what with the muggy heat, boys setting off early fireworks, and the discomfort of dozing in chairs or on floors while relatives and guests filled every extra bed in town. At dawn on the Fourth, teams and carriages, horse-drawn floats and marchers began assembling in downtown streets for a grand parade. At 9:15 a.m. Grand Marshal Arthur Barret gave the signal, and the procession began to move. The hometown newspapers estimated the snaking throng at fourteen miles long, even as the heat reached 102 degrees by midday. Reporters from Chicago grudgingly conceded that the parade was "damnably long." The participants and upward of two hundred thousand spectators had come from near and far to celebrate the great feat of their age.[1]

In an era that loved its superlatives, the bridge across the Mississippi was both a proud civic boast and a great engineering triumph. The abutment on each shore and the bridge's two river piers, built of Missouri limestone and Maine granite, suggested the strength and permanence of a medieval castle. The stone piers extended down through the flowing Mississippi and its shifting sandy riverbed to land on the bedrock below. During construction, curious and well-connected friends of the project had taken giddy excursions to the riverbed, climbing down an iron staircase inside the stone pier. Once past the airlock, they found themselves standing on the bottom of the Mississippi, but safe and dry inside a riveted plate-iron caisson, a capacious inverted box. Elegant ladies made the otherworldly trip to watch as sweating laborers, mostly immigrants, excavated the sand—the scene darkly lit by smoky oil lamps.[2] While the sandhogs dug, skilled masons laid new courses of stone far above, building up the bridge pier atop the caisson. In a brute-force ballet, every piece of cut granite or limestone that they added pressed the caisson down, while growing the pier upwards.

Unfortunately, bedrock lay progressively deeper toward the eastern side of the river. Below depths of sixty feet, mysterious pains had struck down many of the sandhogs (or submarines, as they were also called) soon after they came up from the pressurized caissons. Fourteen had died by the end of 1870, a mournful toll for this great accomplishment.

The wonders of the piers mostly lay hidden below the river, but the superstructure was a self-evident marvel. The central arch spanned 520 feet; the two side spans each measured 502 feet. The essential structural members in each span were eight steel tubes, paired into four "arched ribs." The steel tubes and their iron braces provided the brawn to support the double-deck bridge. On the lower level, two railway tracks finally gave St. Louis a direct connection to the national railway network—more than sixty thousand miles of track, most of it east of the Mississippi River.[3] The 54-foot-wide upper deck accommodated four lanes of roadway traffic, a horse-drawn street railway, and two sidewalks, all providing unobstructed views of the river and the city.

The grand parade marched down Washington Avenue, the commercial heart of St. Louis, then promenaded through the downtown business district. Red, white, and blue bunting decorated nearly every building. Those traditional colors for the Fourth testified as well to the city's blood-soaked commitment to the Union during the late rebellion. Also prominently displayed on houses and stores, the black, white, and red bars of Germany signaled other pasts and allegiances.

Leaving downtown behind, the marchers ventured onto the bridge in exuberant delight. The procession was a pastiche of bridge company luminaries, St. Louis notables, and happy citizens, radiating a mix of patriotic fervor and booster humbug. After the police and the grand marshal, twenty-five junior aides rode in the pony corps, the youngsters dressed in short black coats, white pants, red cravats, and white straw hats. A detachment of thirty-eight mounted troopers from the US Cavalry escorted carriages bearing veterans of the War of 1812. Interspersed in the line of

Four arched ribs support each span. Each rib is made of two round steel tubes—*chords* in engineering parlance—placed twelve feet apart vertically, with extensive bracing that unites the two tubes to form a truss (best seen here in the rib on the right). The ribs land on wrought-iron skewbacks that transfer their loads into the masonry foundations. The skewbacks are held in place with steel anchor rods passing right through the stonework. Large hex nuts fasten the skewbacks to the rods. This 1990 image shows the result of long-deferred maintenance. Photograph by Tony Carosella, Historic American Engineering Record, Library of Congress, HAER MO, 96-SALU, 68.

march were seventeen German choral societies and a twenty-piece march-ing orchestra in Prussian uniforms. Fraternal groups such as the Druids and the Oddfellows took prominent positions. Local businesses filled out the procession, with the city's many brewers represented by twenty-two decorated wagons. Brunswick Billiards joined with two drays, each hauled by a matched pair of plumed horses.[4] In a practice typical of the era, many floats bore tradesmen demonstrating their skills and offering their wares.[5] Aboard one wagon, nine journeymen tin workers turned out 2,000 cups, all tossed to the crowds. A printer ran off free copies of the Declaration of Independence. A stove-maker's wagon carried several ovens baking cook-ies that were devoured by onlookers.

Only the very timid could fear that the crush of crowds threatened the bridge itself. The happy throng crossed to the Illinois side, where there was little to see besides rail yards, shanties, and oft-flooded muddy low-lands. No matter. From the bridge they gazed at the glory of their river and their city, hoping it was now poised to reclaim its rightful place atop the economy of the West.

Two days earlier, "thousands of citizens flocked to witness the interest-ing spectacle" of load testing the bridge. Fourteen locomotives, tenders fully laden with coal and water, had crossed back and forth in different combinations on the two tracks. The tests had culminated with all fourteen engines, coupled together, slowly traversing the three spans, their moving mass nearly seven hundred feet long and weighing seven hundred tons.[6] On the same day, the roadway deck had undergone its own test. A circus elephant had lumbered ponderously over the bridge, taken a sniff of Illi-nois, then ambled, without evident concern, back to Missouri.[7] Every-one understood that such wise beasts would never venture onto an unsafe structure.

While brass bands and pomp diverted spectators, less visible ele-ments of the new St. Louis Bridge project promised revolutionary con-sequences for the city and the region. After crossing the bridge, the double-track rail line entered a tunnel beneath Washington Avenue, ran eighteen blocks underneath the downtown business district, then emerged nearly a mile away in the Mill Creek Valley. At that point, the tracks linked to the westward route of the city's dominant railroad, the

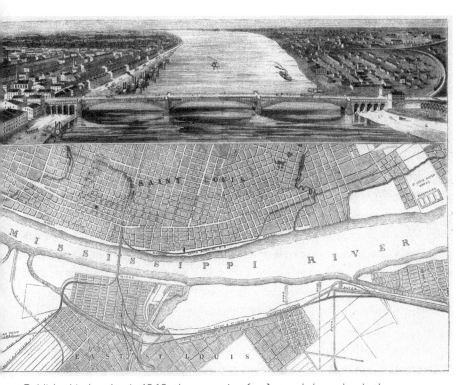

Published in London in 1868, the engraving (*top*) reveals how closely the completed bridge matched Eads's initial plans. The street map of St. Louis (*below*) shows the tunnel's proposed route from the bridge under downtown to a planned union depot. The engraving and the map depicted St. Louis accurately enough, but East St. Louis, Illinois (*at right in the engraving and at bottom in the street map*), was mostly mudflats and rail yards in 1868. The fictitious town shown here aimed to encourage would-be investors in Eads's venture. Illustration from *Engineering* (London), Sept. 25, 1868, 279.

Missouri Pacific.* The MoPac in turn carried travelers onward to connections reaching Denver and California. With their new bridge, St. Louisans planned to tap a growing share of the commerce crossing the continent. It would become a steel link to the expansive future of America. At

---

* In this era, railroads changed their names repeatedly, reflecting bankruptcies and mergers. To sidestep the resulting jumble, this account uses the most common name associated with a given carrier throughout.

the tunnel's western end, a large union depot would open in June 1875 to accommodate passenger trains of all the railroads that reached St. Louis, from every point on the compass. This joint facility would prove a great convenience to travelers, especially when transferring from one railroad or region to another, bolstering the city's position as a crossroads of the continent.

These new infrastructures—bridge, tunnel, and station—would surely attract new lines to St. Louis, ensuring its ascendence. True, a few American cities already boasted some of these transportation innovations. Carriers converging on Indianapolis had created the country's first union station in 1853. Still, they remained unusual into the 1890s.[8] Baltimore and Washington, DC, had their own downtown tunnels by 1874. But James Eads proposed more than isolated improvements. He sketched for St. Louis a fully integrated system to receive and move the freight and passengers of all area railroads. No other American city had achieved this kind of scale and coordination. Few would even attempt it until 1900 or later.

In the absence of organization and planning, railroads bedeviled and mangled other American cities. For example, Chicago had five main line terminals in 1874, each inflicting its own clanging, screeching miseries of shunting locomotives, backing trains, and smoky coach yards. Travelers passing through Chicago to make intercity connections often struggled from one station to the next, laden with baggage, unsure of their bearings, and cheated by unscrupulous cabmen. The city's rail anarchy tortured Chicago's residents. Noisome freight yards filled any available land, typically in the poorer quarters. Ponderous smoky trains ran in city streets, frequently snarling all traffic. Beginning in the 1850s, hundreds of Chicagoans had been killed or injured each year simply crossing the street.[9] Thanks to its newly unified rail infrastructure, centered on the bridge, tunnel, and Union Depot, St. Louisans could hope to avoid much of this mayhem.

While the grand parade proceeded happily past, the luminaries crowding the reviewing stand began to offer speeches befitting such an occasion. The engraved invitations for seats in the welcome shade of the stand had promised that President Ulysses Grant would attend and speak on this momentous occasion. But the organizing committee had been quite late in soliciting Grant's participation; as of June 22, his personal secretary still had not received the formal request. That delay may have been intentional,

Ornate invitations promised that Ulysses Grant would formally open the bridge, but as matters unfolded, the president enjoyed a relaxing holiday taking the ocean breezes on the Jersey Shore. Who had snubbed whom is unclear. Invitation to Mary L. Taussig, MHS, DO 3683.

as the president had many opponents in Republican circles in Missouri and across the country. As matters unfolded, Grant spent a relaxing Fourth at his cottage on the Jersey Shore, far from sweltering St. Louis.[10] His replacement on the reviewing stand was both an appropriate choice and a veiled snub of the president.

Like many Americans, B. Gratz Brown had undergone a difficult political evolution in the years after Appomattox.[11] A US senator and a Radical Republican in 1865, Brown had introduced the legislation authorizing this interstate crossing. He had given his initial remarks on the Senate floor on the same day that his fellow Radicals in the House of Representatives created their Joint Committee on Reconstruction, determined to forge an equitable democracy out of the defeated Confederacy. The Senate referred Brown's bill to committee on December 18, the effective date for the Thirteenth Amendment, abolishing slavery. The project to build a modern and free South—whose freedmen would reliably vote Republican—had its counterpart in the party's work to reorder the West.

Within a few years, Brown had broken with the congressional Radicals, fearing that their grip on power was upsetting all constitutional balance. Radicals dominated Congress and had driven the feckless president, Andrew Johnson, to impeachment and impotence. To bolster a national shift to the political center, in 1872 Gratz Brown ran for vice president on the Liberal Republican ticket, which challenged Grant for Republicans' allegiance.[12] Although resoundingly defeated at the polls, that effort revealed a serious breach in the party.

Through all that, Brown remained staunch in his support for the bridge. His 1865 Senate speech had offered a prescient if flowery portrait: "I want the structure when built to be one worthy of the great States it is to connect, and of such ample capacity as will permit the freight of all the railway lines that may hereafter center at the great distributing point of the continent. Let it be built, too, for the ages, of a material that shall defy time, and of a style that will be equally a triumph of art and a contribution to industrial development."[13]

The marchers and spectators on the roadway deck delighted in their new, expansive views of the city and the Mississippi. On the great rivers of the West, all the other long-span railway bridges were starkly utilitarian structures with wrought-iron trusses that dutifully carried their loads

The Mississippi River at flood stage in 1892 drew throngs of spectators to ob-serve the churning tempest. Floodwaters submerged much of East St. Louis, in the distance. Every day the roadway deck welcomed innumerable people to gawk and gaze. Baedeker's *United States* guide advised tourists to visit the bridge, "one of the lions of the city." Unknown photographer, MHS, N 02843.

while offering no sweeping panoramas nor any suggestion of aesthetic grace. By contrast, the roadway of the St. Louis Bridge was conceived as a grand public promenade, a Champs Elysée for a city determined to be great. In the eyes of St. Louisans, the bridge portended an auspicious future for the new commercial West at large.

Envious of Eads's creative and technical triumphs, many American engineers dismissed his bridge as an expensive anomaly—which it was. By contrast, the world's leading engineering journal offered high praise. The editor of the London weekly *Engineering* had watched the project closely from its inception. An American and an engineering prodigy himself, Zerah Colburn wrote in 1868: "No engineering work at present in progress, either in Europe or in India, can pretend to exceed in interest the noble steel arched viaduct designed, and already commenced, by Mr. James B. Eads."[14]

Following Gratz Brown to the podium, James Eads gave the keynote address on that hot July day. This wiry, intense man had generated all the bridge's major concepts, apparently single-handedly, shortly after his involvement in the project began in March 1867. His skeptical view of precedent had freed Eads to embrace radical innovations throughout his career. His decision to use structural steel in the bridge came two decades ahead of steel's widespread acceptance by the civil-engineering profession. This kind of fearless innovating was not cheap; the St. Louis Bridge had cost far more than a conventional iron-truss rail bridge. No matter to Eads.

In addition to his work in design, Eads served as the chief promoter of this complicated, seven-year project. He recruited the engineering and mathematical talent, lobbied political leaders, met with railroad chiefs, mastered the stock and bond markets, and toiled to raise the funds to build his masterwork. Merchants and bankers in St. Louis and New York City purchased shares enthusiastically, and their equity stakes had raised over $3 million for the venture ($78.6 million in 2023).[15] All hoped that lucrative profits would reward their investments.

Over in East St. Louis, Illinois, rail lines from the East crowded the riverbank, tantalizingly close to the city but blocked by the Mississippi. Until the bridge opened, ferries carried all freight and passengers across that 1,500-foot gap. This was a cumbersome and expensive makeshift in daylight. It became a perverse obstacle every night and during winter freezes, when the ferries stopped running. The new bridge company aimed to trounce the ferry operation with reasonable tolls that undercut its charges. Up on the roadway deck, the collectors would take 10¢ for a carriage, 5¢ for pedestrians. Sheep and hogs went across for 2.5¢ each (68¢ in 2023).[16] Freight and passenger traffic on the rail deck would also pay suitable tolls. The bridge company surely had a bright future, even if it counted the profits one hog at a time.

In his remarks, James Eads offered generous thanks to his talented team of associates. Barton Bates served on the bridge company's board of directors. The son of Lincoln's attorney general, Bates was president of the North Missouri Railroad, a line that aimed to profit by connecting its westward-reaching tracks to the bridge. Eads also thanked Milnor Roberts, the associate engineer on the project since July 1868; William Nelson, Eads's collaborator in building ironclads and the bridge caissons; and

William Taussig. Born in Prague, Taussig emigrated in 1847, arriving in St. Louis a year later. He had been a doctor, a county court judge, a leader of the Unionist and German émigré communities during the war, and a collector for the Internal Revenue Service. By 1869, the courtly, red-haired Taussig had joined the board of the bridge company. He would go on to serve as its general manager or president for three decades.[17] That work made him second only to Eads in shaping the company. Eads gratefully acknowledged Walter Katte, "the skillful engineer who swung [the] steel arches into place."[18] He studiously ignored Katte's employer, the Keystone Bridge Company. Keystone had fabricated most of the bridge's iron and steel components at its Pittsburgh shops, then erected the three arches and two decks above the swirling river. But Eads and Keystone had issues.

Although Eads singled out more than a dozen men by name, he offered no thanks nor acknowledgment to three others who had played influential roles in the genesis of the bridge: Andrew Carnegie and his two silent partners in "Andrew Carnegie & Associates," J. Edgar Thomson and Thomas Alexander Scott, respectively the president and vice president of the Pennsylvania Railroad. In March 1870 these men had contracted with Eads to form an alliance of bankers, railroads, and suppliers that originally aimed to complete the bridge by December 1, 1871. By the much-delayed opening day, each party to that deal believed with bitter conviction that the others had failed to meet their obligations.

No one had done more to shape the national railroad map of 1874 than J. Edgar Thomson, longtime president of the Pennsylvania Railroad, the nation's largest. In addition to his leading and largely hidden role within Carnegie & Associates, Thomson had generously assigned PRR engineering talent to the project, boosted its prospects to London financiers, and midwifed a subsidiary line into East St. Louis, Illinois, that was intended to funnel PRR traffic to and from the crossing. But Thomson died two months before the grand opening. His successor as president, Tom Scott, had taken a leadership role in Eads's bridge company from the start. One of twenty original stockholders, Scott had drawn personal profit from many parts of the project. Arguably, he was the most influential man in American railroading at that moment. Eads would not speak his name.

If Scott was a problem, his longtime right hand was a dark and unscrupulous figure in the eyes of Captain Eads. As the chief salesman for the

Keystone Bridge Company, Andrew Carnegie had wooed the St. Louis group starting in October 1867. In March of 1870, Carnegie and Scott had placed themselves and Thomson at the contractual center of the project. Over time, however, the demanding design had imposed burdensome delays on the construction schedule, raising costs for Keystone and wiping out much of the profit that Carnegie & Associates had calculated.

Dealings had reached a nadir on Sunday, April 19, 1874. Trains weren't quite running yet, but the city newspapers had promised that the bridge company would welcome visitors to its upper-level roadway deck that day. At the appointed hour, however, eager visitors who showed up at the bridge were turned away, blocked by armed guards under orders from Andrew Carnegie. The Keystone Bridge Company refused to surrender the structure until it received satisfactory agreements for payment in full.[19] Carnegie believed his action was fully justified, while Eads saw it as a betrayal.

In his speech on that hot day, James Eads did offer appreciative thanks to Junius Morgan. This American-born, London-based investment banker had already built a reputation on both sides of the Atlantic as a sober guardian of other people's money. With his US connections, Morgan seemed to be making America safe for British capital. In March 1870, he had agreed to provide much of the debt financing for St. Louis Bridge, marketing an initial stake of $4 million of the company's 7 percent gold bonds on the London market. Commonly used to finance new railroads, a bond on a bridge was unprecedented. As Morgan wrote to his New York partners, "It is novel, and yet perhaps none the worse for that."[20] Soon thereafter, Junius Morgan became the central financier in a venture that proved far riskier than his reputation for probity would ever admit. Besides the stockholders' equity investment, by opening day British bondholders had poured more than $10 million into the bridge and tunnel ($272.5 million in 2023), their money raised and vouchsafed by Junius Morgan.

The day's speeches and parade culminated that evening with a grand spectacle of fireworks over the river and across the bridge. At an advertised cost of ten thousand dollars, the "Pyrotechnic Programme" included seventy-eight named elements. Its planned finale was a "Railroad Phantom" of carefully timed and choreographed explosions on the bridge itself. In an edition published hours before nightfall, the *St. Louis Dispatch* described it as "a fiery train of cars, passing the whole length of the bridge, preceded by the Goddess of Liberty."[21] As matters played out, "that display fizzled completely," as did much of the show.[22]

That memorable day offered temporary relief from a great burden pressing down on the city and the nation. Nine months earlier, the leading investment bank on Wall Street, Jay Cooke & Company, had shuttered its offices in default. Cooke had financed the Union war effort, and his firm played a central role in the expansive postwar political economy that Republicans in and out of government had labored to create. Cooke's failure had touched off a Wall Street panic that quickly spread general economic depression across the country. Credit, land grants, fast-talking promoters, and sunny optimism had built the shaky structure of western railroading after 1865. Now defaults and general pessimism blew down that false-fronted house. Before the depression began to lift in 1878, half of all American railroads had fallen into receivership, unable to pay their bills, let alone earn a profit.[23] Up on the reviewing stand, General Manager Taussig smiled through the speeches and waved at the crowds. But his spirit was "heavy with the forebodings of impending bankruptcy."[24]

CHAPTER 1

Captain Eads

IN THE 1850S, St. Louis faced an existential challenge. Steamboats had transformed this fur-trading post near the confluence of the Mississippi, Missouri, and Illinois Rivers into the leading city of the West. But railroads had recently become the rising technology for transportation. Missouri's leaders eagerly embraced this wondrous innovation, hosting a railway convention in 1849 to boost St. Louis as the eastern terminus of a railroad to gold-rich California.[1] During the 1850s, Missouri's business leaders and politicians committed their capital to new railroads reaching westward across the state and to new lines that improved ties to the East.[2] And in 1855, St. Louis promoters chartered a company to build a 1,500-foot bridge over the river. They proposed a suspension design to carry railway traffic—a dubious combination, to say the least. Those men ran hard up against another obstacle, raising only $30,000 of the estimated $1.5 million cost.[3] The hesitancy of potential investors made sense, given the rudimentary state of American engineering and the inadequate structural materials available at the time: wood or the barely reliable output of the young iron industry. Even assuming that those problems could be solved, the nature of the Mississippi River at St. Louis presented its own profound challenges. Perhaps that helps explain why, in time, it was a riverman who stepped up to design and build a rail bridge at St. Louis.

The River and the Town

The Mississippi River and the identity of St. Louis have swirled together for three centuries. That link, however, obscures some realities of hydraulics and history. In many ways, St. Louis belongs to the Missouri River, which flows into the Mississippi just fifteen miles upstream of the city. Above that confluence, the average daily flow of the Mississippi in the mid-nineteenth century was 105,000 cubic feet of water per second. The Missouri contributed an average of 120,000 cubic feet per second to the combined river. Those were just averages; during a flood, the torrent passing St. Louis could swell fourfold. After the two rivers merge, their combined current can exceed 8.5 miles per hour at St. Louis, and the height of the river can vary as much as forty feet between the floods of June and the slack water of December.[4]

Occupied in prehistory by Cahokia Indians, the settlement was then a crossroads in an age when travelers over distances short or long went by canoe. From that central point, tribal people could access the Illinois River to head northward toward Lake Michigan, follow the Ohio River all the way to the Appalachian Mountains in modern Pennsylvania, or head southeast along the Cumberland and Tennessee Rivers. But the nearby confluence with the Missouri River set St. Louis apart, establishing its position as a strategic and commercial center. That stream and its tributaries (such as the Platte) formed an extensive arterial network across the Great Plains and into the foothills of the Rockies.

The indigenous peoples of the area represented a confluence of tribes. In the eighteenth century the Illinois dominated, while Shawnees, Miamis, and Kickapoos filtered in from the east. From the west came Missouris, Osages, and Iowas.[5] The Anishinaabe tribe gave the river its name, *Misi-ziibi*, a French rendering of the native phrase for "Great River." When French explorers arrived in the late 1600s, they immediately grasped the region's geopolitical importance, and the Missouri River became the highway to the West for fur traders and explorers.

Why French St. Louis grew up on the *west* bank of the Mississippi also reflected hydraulics. Lands to the north and east flooded frequently, a scant concern to the tribes, for whom the waters gave food and transport. For the French, however, the bluffs provided dry high ground, suited for permanent houses and a palisade. In 1764 the Frenchmen Pierre Laclède and his stepson Auguste Chouteau founded the settlement of St. Louis in an ironic aftermath to France's defeat in the Seven Years' War. Britain's victory had won her title to all lands east of the Mississippi, while territory to the west had become Spanish. For Laclède and Chouteau, distant and uncaring Spain promised far less trouble than perfidious Albion, so they sited their trading post on the west bank of the river. After Jefferson bought the territory for the United States, the lucrative fur trade continued, supplied mostly by trappers working the Missouri River valley all the way to the Rockies. Fur and the river more than justified statehood for Missouri in 1821.

Steamboats on the Western Rivers

In 1807 Robert Fulton's pioneering Hudson River steamboat, the *North River*, demonstrated proof of concept for this new transport marvel. After success in New York, he turned his focus to the West, exulting: "Everything is completely proved for the Mississippi, and the object is immense."[6] By that time a large and growing volume of western produce flowed down the Mississippi system to New Orleans, borne on flatboats, barges, and keel boats carried along by the currents. Their reliance on slow and primitive craft for these long and perilous journeys suggests the miserably limited markets available to western farmers seeking cash for their livestock and crops. The first generations of homesteaders, ranchers, merchants, and bankers in the vast interior of the continent would welcome anything that improved the volume and reliability of trade.

Steamboats accomplished both goals, since they made possible regular *upstream* travel and trade. Four years after his triumph on the Hudson, Fulton's first Mississippi steamer, the *New Orleans*, began service between its namesake city and Natchez, Mississippi, in 1811. In just two years, the 400-ton, 116-foot-long steamer earned clear profits sufficient to pay off all its construction and operating costs.[7] That kind of success quickly attracted new operators and ships. The technology seemed nearly ordained for the West. By 1824 more than 100 paddlewheelers steamed up and down the rivers of the Mississippi basin; by 1850 the fleet had reached 740.[8] Their routes stretched from Fort Benton, Montana, in the Northwest to New Orleans in the South, 3,500 river miles away.

St. Louis became the major hub for steamer routes and river trade. The lower Mississippi between St. Louis and New Orleans was the domain of

Steamboats crowding the levee testified to the commercial importance of St. Louis and its river in 1854. Engraving by Frederick Hawkins Piercy, 1854.

the big packet boats, which required river depths of six feet or more. Smaller vessels drawing three feet or less dominated the upstream trades from St. Louis to St. Paul, Minnesota. Those light and powerful craft also suited the challenging waters of the wild Missouri River. Steamboats on the Ohio River brought manufactured goods and iron from Pittsburgh and farm products from the Ohio valley to St. Louis, the trading hub of the West.

The Mississippi and its tributaries were the Interstate system of the 1840s, and the steamers were the eighteen-wheelers of their age. They also carried passengers in varying degrees of comfort, from abysmal to luxurious. But the boats seldom remained in service for long. Across the western rivers, they lasted four years on average, sunk by snags (submerged tree trunks), crushed by ice, holed by floating logs, broken open on sandbars, or lost in collisions, fires, and boiler explosions.[9] Young James Eads saw opportunity in all these losses.

A Career Made Afloat

The early life of James Eads is obscured by a scarcity of reliable information and an excess of the florid stories that nineteenth-century writers heaped on all great figures after their death. Eads was born on May 23, 1820, in Lawrenceburg, Indiana, a small village on the Ohio River, just downstream from Cincinnati. Mostly raw frontier, Indiana had gained statehood four years earlier. His parents, Thomas Eads and Nancy Buchanan, had grown up in Kentucky.[10] Thomas was a restless fellow of middling means and something of a wanderer. He was well connected politically, supporting the Whig Party, which favored government aid to develop the West. In 1833, he sent his young family by steamboat to St. Louis. Approaching the city levee, the boat caught fire, and Nancy Eads, James, and his two siblings barely escaped with their lives, losing their possessions to the blaze. Eight people died.[11]

Although real financial success eluded him, Thomas Eads amassed enough to build and operate the Hotel Berlin in Le Claire, Iowa, a river town on the Mississippi 250 miles upstream from St. Louis.[12] For steady income, the couple farmed their own fertile land at Le Claire, working the place themselves well into middle age. There, Thomas and his family enjoyed a middle-class life in a comfortable home they named Argyle Cottage.[13]

At the age of 14, James struck out on his own. Leaving his family and formal education behind, he clerked for five years in the St. Louis dry-goods store of Williams and During.[14] In his spare time, he read broadly in literature, engineering, and mathematics, freely borrowing books from his employer's extensive library.[15] His adult writings show fluid command of formal English, powerful logical abilities, thorough knowledge of the Bible, and facility with algebra and trigonometry.[16] He was raised as a Catholic; however, there is no indication that he carried that faith far into adulthood.

At 19, Jim Eads took his skills to the river, becoming the mud clerk on the steamer *Knickerbocker*.[17] His title signified that he was starting at the bottom, handling bills of lading and ensuring that cargos and passengers got on or off the vessel at the right landings. The work had its perks: travel, interactions with people of all sorts, steady meals and pay, and a private cabin. His career on the *Knickerbocker* ended in the late summer of 1839, a dry season with low water on the Mississippi. Near the mouth of the Ohio River, at Cairo, Illinois, the vessel ran up onto a snag, which punched "a large hole in her knuckle [bow planking] and bottom." These submerged tree trunks, lurking just below the surface, were a hazard on every western river. With sinking just moments away, the captain ordered the pilot to run the *Knickerbocker* hard aground in the shallows at the river's edge. That quick action saved the passengers and crew, but the ship and her cargo were lost. "River pirates" then made off with the light goods, while a valuable cargo of lead bars sank in the muddy waters at the mouth of the Ohio.[18] Eads shipped out on other vessels for three more years, then opened a new chapter in his life.

The loss of the *Knickerbocker* proved a formative experience for the young man, opening a new door. In 1842, Eads, William Nelson, and Calvin Case formed a partnership at St. Louis to enter the risky business of marine salvage. The firm combined real talents: a year earlier Case and Nelson had opened the first shipyard in the city, building and repairing steamers. To entice them to enter the salvage business, Eads had approached the pair with his own unique design for a "bell boat." Its twin hulls would impart stability, while a diving bell would give access to wrecks and cargos. Derricks would manage the bell and raise any valuable finds. This *Submarine No. 1* had no power plant, needing a tow to go anywhere. Eads proposed to handle operations on the river. Case and Nelson were swayed by this energetic and ambitious young man.

The salvage business reflected the harsh realities of steamboat operations. The wild western rivers themselves created everyday perils like snags and sandbars. Boat design added more jeopardy. A river steamer combined a lightly framed wooden hull (saving weight to achieve the shallow draft essential in dry seasons), powerful engines, and heavy iron boilers in which fires raged. To all that floating liability, captains added operational threats, especially in their passion to carry the largest possible loadings of cargos and passengers. Landings at unimproved riverbanks stressed hulls, while steam pressures mounted in the boilers during those stops. Under way, passengers and crews cheered up whenever an impromptu race beckoned, betting on the outcome. Wrecks, explosions, sinkings, and fires were the inevitable result. Recounting his travels by

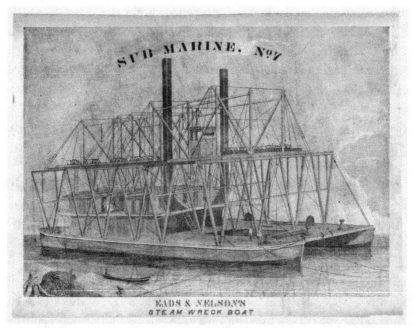

The wreck boats, or "submarines," designed by James Eads looked like no other craft on the western rivers. In 1857, Eads & Nelson commissioned *Submarine No. 7.* Its two steam boilers powered the engines that drove water pumps, lifting gear, and an air pump to supply the men in the diving bells. The twin-hulled design provided stability, while the forest of derricks lifted salvaged engines, boilers, and cargos out of wrecks. *Submarine No. 1* employed the same principles. Image from Taylor and Crooks, *Sketch,* MHS, N 13895.

steamer on America's western rivers in 1842, Charles Dickens concluded "that the wonder is, not that there should be so many fatal accidents, but that any journey should be safely made."[19]

In the summer of 1842, *Submarine No. 1* set out to harvest any valuable remnants from these calamities. Once anchored over a wreck site, his men lowered Eads to the river bottom in a diving bell of his own design to search for and retrieve any cargos that could be lifted, dried, and profitably sold. Engines, boilers, and capstans were additional valuable prizes, easily hawked in St. Louis. Eads and his fellow divers worked mostly by feel, given the poor visibility in the muddy water.[20] Sounding rods proved useful for finding iron machinery or cargos in casks. A diver communicated with his topside mates by tugging a signal cord. Apparently, the crew of this first submarine powered a hand-driven air pump, supplying fresh air to the bell.[21] It was difficult and dangerous work. Sudden death by drowning always threatened, as the divers worked at depths of up to fifty feet.[22] Daily prosaic hazards included hypothermia in winter and yellow fever in summer and fall, especially on the lower Mississippi.

As the boss, Eads shouldered further challenges: moving faster than competitors, finding the wreck he wanted, contracting with the owners or insurers if possible, hiring day labor—all before he could actually start to probe the silty riverbed for anything of value. Locating a boiler or a cargo of lead might take days or weeks of searching, walking the bottom as currents roiled everything submerged, while the topside crew on *Submarine No. 1* shifted her anchors to move the search into untrod territory. When they hit a worthwhile strike, more issues cropped up: sudden changes in the river water levels, equipment or anchors carried away by the floating driftwood that coursed past constantly, and the threat of injury or worse if salvage loads shifted as derricks swayed them aboard. Over the days and weeks, life afloat ground away at a man: immersion in cold and muddy water, visibility of a yard or less when overboard, perpetual damp when topside. Eads would spend the rest of his life tormented by bronchial complaints that laid him low for weeks at a time. The money might be good, or it might not—it was just a matter of chance.

After he had spent three years in salvage, romance started Eads down a different path. Martha Dillon grew up in the prosperous household of her father, Colonel Patrick Dillon. An Irish emigrant and an early settler in St. Louis, he had made a small fortune by selling dry goods and investing wisely in real estate. His first wife died in 1835, when Martha was 14. The colonel's second wife, Eliza Eads, was James's first cousin. That family connection brought the young people together. By the summer of 1844, Eads was courting Martha whenever his salvage work brought him to the city. When apart, the two wrote each other earnest letters of real love. In his first blush, James thanked her for a letter so "valuable" that he "would be willing to buy out a whole Post Office for the sake of getting more like it."[23] Though deeply smitten herself, Martha demonstrated proud independence, arguing that "you have never measured the length and breadth of my mind and consequently do not know what I am equal to."[24] James was intrigued, and by May 1845 the two young people had an understanding—and

Martha Dillon Eads had considerable presence. She challenged her father by her secret understanding with James. She also challenged Eads, within bounds. While courting, she conceded to James in a letter of May 18, 1845, that men's "minds are stronger, and their reasoning faculties far clearer and sounder than ours," but she said that this was "greatly owing" to differences in education. For their part, women were preeminent in "quickness of thought and vividness of fancy," and they outshone men with their ability to "arrive at a just conclusion by mere intuition." Letter 6, J&M. Unknown photographer, MHS, N 12177.

a problem. Colonel Dillon simply did not approve of James Eads, insisting, "I must see that a man knows how to make money." As a salvor, Eads could hope for wealth, but it would reflect those muddy origins. By contrast, the colonel was "anxious to unite my daughter to some one of the families of the highest standing in the country."[25] In Dillon's eyes, Eads fell well short of his plans. Refusing to accept that verdict, the couple agreed in July 1845 to marry before November. During the interval, Martha hoped to win over her father.

With that secret understanding between them, James decided it was time to see the world and acquire some polish. In late July, he embarked on a six-week tour that took him to Pittsburgh, Baltimore, Washington City, Philadelphia, New York City, Boston, and Montreal. On that trip the brilliant inventions of the age captured his imagination. After his first train ride, on which he sped at thirty miles per hour toward Baltimore, he wrote to Martha that "farms passed like thoughts in the brain." His verdict: "Truly these railroads are a great invention!"[26] The Patent Office in Washington mesmerized him, its glass cases displaying "models of every machine and implement you can possibly imagine, from a steam engine down to a baby's whistle."[27] In New York Harbor, Isambard Brunel's iron screw steamship, the *Great Britain*, was a monstrous "Leviathan of the Ocean."[28] His enthusiasm reflected his era, fascinated and transformed by mechanical invention.

The young man showed an earnest desire to improve himself. He loved seeing Baltimore's Washington Monument and its cathedral. There he stayed at the home of his uncle and namesake, James Buchanan.[29] En route from Troy, New York, to Montreal, he bought (and apparently read) an English translation of *Florentine Histories*, by Machiavelli.[30] At Québec City, he resolved to take up "fashionable" manners: dining late and never putting his knife in his mouth while eating.[31]

After a stop at Niagara Falls, James returned to St. Louis in late September. He and Martha were married on October 21, 1845, with her father still vehemently opposed. A letter he wrote her a few years later gives insight into Eads's mind. While on a trip for work, he thanked her for a recent letter, its tone "just as you would speak to the person if present. . . . With me it is different. I have given up trying to write as I converse and now labour to write as I think."[32]

Shortly after the wedding, Eads changed careers again. With the backing of his old salvage partners, Case and Nelson, the 25-year-old Eads opened the first glassworks west of the Mississippi. This venture as a factory proprietor might earn the approval of his new father-in-law. At least it promised a steadier income than the salvage work, without its threats to life and health. It was the kind of position that suited the person he intended to become: a settled family man and a player in the St. Louis business community.

Losing Money, Salvaging a Fortune, Enduring Tragedy

Young James Eads struggled to make a success of his glassworks. In this period (1846–47), Martha lived mostly with Eads's parents at Argyle Cottage, upriver in Iowa.[33] Eads himself spent little time on the farm, determined to master every detail of making and selling glassware, chiefly tumblers, bowls, and vases. Production began in a leased space in St. Louis, while Eads often traveled to Pittsburgh to hire skilled labor and buy essential materials such as fire clay and pearl ash. By the spring of 1847 he had eight men and twenty-one boys on his payroll. An April letter to Martha reported shipping twenty-five boxes of glassware to customers up the Missouri River and eighty boxes to New Orleans.[34] By August, employment and production had grown, gross revenues reached seven hundred dollars a week, and Eads had bought out William Nelson's interest in the firm, probably using promissory or debt notes.[35]

Against all the good news that he wrote home, the firm wasn't making a profit. Fearing that it would soon fail, Eads transferred the deed to Argyle Cottage to a trustee, blocking creditors from seizing that asset and evicting his family.[36] The close of 1847 marked the end of the glassworks.[37] It left Eads with no savings, debts of $25,000 ($942,500 in 2023), and a reproving father-in-law.[38] But his brief career as a glassmaker had long-lived consequences. Eads had learned about the challenges of mastering new technologies and wielding men. As he wrote to Martha, a manager must "have constant controul of his temper and be able to speak pleasantly to one man the next moment after having spoken in the harshest manner to another. Self must be left out of the matter entirely."[39] Eads grew even more committed to making money, the more the better. He had debts to pay and much to prove.

Wealth lay beneath the Mississippi, so early in 1848 James Eads went back on the river. With a loan from his creditors, he bought his way back into the salvage business, again in partnership with William Nelson. It was the same enervating, wet, and dangerous work. In a July 1848 letter to Martha, James described salvage operations on the steamer *White Rose,* which had burned at Cairo, Illinois, a week earlier and then sunk a mile downstream of that river town. He hoped to clear five hundred dollars on the operation.[40] Then he raced off to search for the wrecks of the *Clarksville* and the *Congress* before a competing salvor found them first.[41] The *Clarksville* eluded him despite five days of searching by probing and diving. The salvors did find the *Congress*, which was covered in ten feet of mud, so nothing of value could be lifted from the bottom.[42] His letters to Martha seldom conveyed a hint of danger, but she worried anyway, with good reason. Although Eads hired divers, she knew he did much of the underwater work himself. She wrote to him: "These horrible boats trouble me. What chance would you have, if an accident should happen while you were below, and you could not succeed in catching to the hook [in] the bell?"[43] Despite her worries, nothing could satisfy Eads's passion for riches. Stricken with gold fever, he seriously considered joining the Forty-Niners out in California before deciding that the river offered better bets.[44] Eads now searched for better ways to do the work.

Innovations and enlarged scale made the difference. Eads & Nelson launched *Submarine No. 2* in 1848, *No. 3* in 1849, and *No. 4* in 1851, each more capable than its predecessor.[45] With four boats out searching, Eads and his fellow captains found the wrecks before sand and mud filled the hulls, before competitors snatched the jobs. Equipped with their own propulsion systems and thus no longer reliant on towboats, the new *Submarines* hurried to job sites. Their power plants drove air pumps to sustain the men down in the diving bells, steam windlasses to lift the spoils, and new raw-water pumps that enabled them to salvage entire sunken vessels. Eads & Nelson had purchased exclusive rights throughout the Mississippi River to use the patented Gywnne Centrifugal Pump. With these steam-driven pumps aboard, the *Submarines* anchored above a wreck, sent divers down to patch the holes they could reach, then pumped out the sunken hull so quickly that further repairs and temporary patching became comparatively easy once the vessel floated again. In just seven years (1851–58) the firm raised fifty steamers in this fashion, their cargos salvaged and many of the boats returned to service.[46] Innovations developed to raise wrecks from the Mississippi's depths would later prove essential to placing the stone piers of the St. Louis Bridge on bedrock beneath the river's sandy bottom.

Martha saw little of her husband for years; work and his scramble for money pulled him away. She pleaded especially for him to come home when she was pregnant. At that time and place, the late stages of confinement portended terrible hazards that could culminate in a joyful delivery, difficult illness, or death for mothers and babies. On the eve of her second delivery, attended by James's sister Sue, Martha wrote him, "While I long for your presence as the greatest earthly good, I yet resign myself to the will of Providence."[47] His son was born on August 14, 1848. Three weeks later, the joyous news that both Martha and the baby had survived childbirth reached Eads, deep in his work in Louisiana. He did not return to see James Jr. until October; the baby would die seven months later.[48]

In August 1852, Eads again went out on the river, confessing to Martha that "it is almost totally inexcusable" given that she was "as ill as when I left you." His letter reported good news: her doctor believed "you will be very much benefitted by the water cure."[49] Reunited, the couple left in September for their first vacation in memory, a trip to the resort and spa at Brattleboro, Vermont. With her health much improved, they started for home in October, traveling by trains and river steamers. Business called Eads away briefly to Louisville. Martha wrote to him on October 6 from Lexington, Kentucky, closing a chatty letter with "the joyful and earnest hope that no accident may happen to either of us" before Eads rejoined her a day later. She signed it "your own loving Matty."[50]

Perhaps she had contracted cholera before boarding the *Dr. Franklin II* at Louisville. By the time they reached Indiana, Martha was badly stricken by this water-borne bacterial infection, which was all too common at the time. She died on October 12, still on the Mississippi, not quite home.[51] Eads was crushed. Martha had written him perceptively in 1847: "Believe me James, I have never thought that your almost utter inability to wait on me proceeded from want of thought or love."[52] The work had always drawn him; now he went back to the river, leaving his two young children in the care of his sister-in-law at Argyle Cottage.

Flush Times for St. Louis

The triumphs and tragedies of Eads's life played out against the fortunes of the city where he made his home. The leading metropolis of the American West, St. Louis suffered two devastating blows in 1849. On May 18, a night watchman discovered a small fire inside a lady's cabin on the steamer *White Cloud*, moored on the levee. Firefighters arrived quickly even as high winds carried embers to the adjacent *Edward Bates*. When the burning *Bates* drifted into other boats, the fire spread to the levee, where it ignited cargos ideally suited to fuel an inferno: hemp, cordwood, and barrels of bacon and lard. A thousand men fought the blaze, saving most of the city, but the riverfront commercial district was a blackened ruin, with 430 buildings lost to the flames. The death toll was comparatively light. Twenty-three steamers had burned, and insured losses were estimated at $5 million.[53] This bad luck for the city became a stroke of good fortune for Eads. By year's end, Eads & Nelson had upwards of fifty men at work removing the levee wrecks under a contract from the city.[54]

At the same time, a slow-motion killer was stalking the region. Cholera spread as springtime temperatures hastened bacteria growth, especially in the wells and privies of daily life. In July 1849, an epidemic of suffering overwhelmed St Louis. By the time it tapered off in the fall, 10 percent of the city's sixty thousand residents were dead.[55]

Despite these miseries, St Louisans mostly enjoyed boom times in this era. America's short, sharp war with Mexico (1846–48) was disastrous for that country but wonderful for St. Louis. Army troops and money poured into the city, which became the embarkation point for expeditions to capture distant Santa Fe and Las Vegas. (Soldiers traveled by river steamer to Missouri's western border, where the Santa Fe Trail began.) Shortly after the war, the California gold rush jolted St. Louis with men and money. Steamboats crowded the levee, most of them engaged in the Missouri River trade that took prospectors and their gear out to the Rockies.[56] When that river was at a favorable stage (not in flood or too low), steamers routinely traveled the two thousand river miles to Fort Benton. Near the center of modern-day Montana, that army post was less than two hundred miles from the Continental Divide.

By the 1850s, St. Louis had sophistication and wealth. In these years, residents and visitors delighted in comparing the city to long-established eastern centers. Commercial prosperity had largely eclipsed its fur and frontier past. A writer for the *Atlantic Monthly* decided that "old-fashioned, square, roomy brick mansions . . . the brick pavements . . . the prodigious extent of the city for its population, the general quiet and neatness all call to mind comfortable Philadelphia."[57] In 1855, a resident of the city proudly enumerated its "churches, schools, hotels, steamboats, newspapers, and other institutions of civilized life." With its paved streets, gas lighting services, and eighteen miles of public sewers, St. Louis was in Richard Elliott's view the equal of any eastern city.[58] Its distinguishing qualities were notable too. Alongside its French roots and cultural flavor, German emigrants made up one-third of its 1850 population. Reflecting those origins, half the residents were Catholic. Roughly 4 percent were enslaved persons. In a city where thirty dealers bought and sold human beings, comparisons to Quaker Philadelphia do not stand scrutiny.[59] St. Louis served as the economic and cultural capital of the entire Mississippi watershed, reaching from Montana and Pittsburgh to the Gulf of Mexico. And thanks to the river, the city's wholesalers supplied the plantations and booming cotton economy of the Deep South.

Vast regions newly acquired as a result of the victory over Mexico were a blank slate for most Americans in the 1850s. The promise of those western lands—and California gold in particular—motivated St. Louis merchants, bankers, and politicians to convene a great railway convention in October 1849. Delegates from fourteen states ate oysters by the bushel, drank German beer and Kentucky whiskey, and clamored for a massive public works project to build a railroad linking St. Louis to California.[60]

The idea was prescient, if premature. In 1850 the United States had nine thousand miles of railway lines, all east of the Mississippi. Missouri entrepreneurs soon changed that, breaking ground for the Missouri Pacific Railroad on July 4, 1851, in St. Louis.[61] The MoPac's promoters had good friends in Jefferson City, the state capital. Actual results on the ground, however, proved disappointing, veering into disastrous. By November 1855 the Missouri Pacific had laid just 125 miles of main line from St. Louis to Jefferson City. To celebrate the completion of that division, on November 1, 1855, the railroad loaded a thirteen-car special with its officers, stockholders, directors, and other business leaders. At milepost 88, the locomotive overwhelmed the road's new wooden truss bridge over the Gasconade River. Half the train and the bridge itself crashed into the river, then the wreckage caught fire. The catastrophe killed thirty-one, maimed many more, and crushed the vitality of the MoPac for years to come.[62]

That failure offered grim evidence that it might be decades before a railway bridge could be built over the Mississippi.

Fresh Starts

James Eads remarried in 1854. Eunice Eads, the widow of his cousin Elijah, would prove a capable mother to his two surviving daughters by Martha. His new wife also was a suitable mistress of his prosperous St. Louis household. The couple took their honeymoon in Europe that year. Over the next three decades, Eads returned to the Old World six times to improve his health, search out novel technologies, raise funds, or promote new engineering work. The child of the frontier would transform himself into a polished and professional citizen of the world.

For the present, the salvage business preoccupied him. He focused on deals, not diving. In 1855, Eads & Nelson added five vessels to their fleet.[63] Two years later, the partnership became a corporation, the Western River Improvement and Wrecking Company. Eads made a persuasive pitch to investors, and seventy business leaders bought shares in the company, with a total par value of $250,000 ($8.9 million in 2023). Eads and Nelson were now wealthy men, members of the St. Louis business elite. The cash infusion also funded construction of *Submarine No. 7*. Outfitted with innovative gear, much of it designed by Eads, the new salvage steamer had two Gwynne pumps, capable of discharging 2,000 barrels of water per minute. On its first job, *Submarine No. 7* raised the *Switzerland* days after that large steamer sank in fourteen feet of water near Natchez with 900 tons of cargo aboard.[64] Success on this scale boosted the firm's reputation up and down the rivers.

Building on those innovations in organization and technology, the Western River Company developed an ambitious new approach in marketing, forming contractual relationships with "nearly all the prominent Insurance Companies throughout the Union." Fifty-seven underwriters granted salvage rights to the Western River Company for all their insured losses on the Missouri, the Mississippi, and most of their tributaries. The deals gave the company a near monopoly on this work, speeding salvage operations while it received a standard percentage payout on the value of the recovered goods and vessels. According to one approving commentator, the company "saves yearly vast amounts of property," thus lowering in-surance rates and freight costs for all goods moving on the rivers.[65]

One investor in the Western River Company, Emerson Gould, offered a more skeptical view. Late in life, Gould credited Eads with a genius touch in finance, organization, and mechanical innovation but condemned his sharp financial practices. Gould believed that Eads had puffed the value of Western's stock just as "the tide of its success was about to turn." For Gould, this investment proved "the means of wrecking" the fortunes of its investors, suggesting that they had paid far too much for their shares.[66] Credit reports on the company indicate that its business was failing late in the decade and its assets were mortgaged.[67] It seems that James Eads had a talent for making money, then moving on to new ventures at just the right moment—for himself at least.[68] At age 37, he retired from active busi-ness affairs in the wrecking company. Reflecting his years on the river, his friends, associates, and the press would address him as Captain Eads for the rest of his life.

Retirement did not check Eads's boundless energy. He remained pres-ident of the Western River Improvement and Wrecking Company and ran the Citizen's Railway Company, a horse-drawn trolley line that helped Eads and other investors develop new residential areas of the city. In 1857, he bought "Compton Hill," a large mansion with lovely grounds on the city's south side. Eads settled into this opulence with Eunice and five children, her three daughters and his two.[69] The census taker in 1860 counted seven servants in the house: a carriage driver, a gardener, a cook, two maids, a washerwoman, and a seamstress. None were Black (slave or free); most were Irish emigrants. Eads gave his net worth as $25,000 in personal property and $375,000 in real estate ($13.3 million in 2023).[70] It was too bad that Matty and Colonel Dillon had not lived long enough to bask in Captain Eads's worldly success.

The Wiggins Ferry

During the prosperous 1850s, residents of St. Louis lived with a daily par-adox. Railroads were a well-established fact in the East and promised to open the exciting next chapter in western settlement. In the meantime, the Wiggins Ferry Company ruled the commerce of the city itself. In 1797, James Piggott began a ferry service across the Mississippi at St. Louis. Two decades later, Piggott's heirs sold out to Samuel Wiggins. In 1819, Wiggins acquired from the Illinois legislature the right to purchase a mile of Mississippi waterfront in St. Clair County Illinois, opposite the little fur-trading settlement of St. Louis, Missouri. Then called Illinoistown, it was barely a hamlet in any conventional sense, just a patch of boggy ground frequently immersed by floodwaters. In subsequent years, Wiggins and his successors wrested further value out of the Illinois legislature. In 1829, the lawmakers awarded to Wiggins the exclusive right to run a Mississippi ferryboat service from St. Clair County, Illinois. His new steamboat (replacing a horse-powered ferry) had one destination: the growing village on the opposite shore. In 1849, the Illinois legislature designated Illinoistown as the terminus for all new railroads traversing the southern portion of the state. And in 1853, the legislature granted "the right of perpetual succession" to the Wiggins company, blocking forever any attempt by other companies to operate competing boats.[71] With another legislative gift—incorporation in 1853—the Wiggins Ferry Company acquired unlimited life, authorization to raise $1 million in capital, clear title to fifteen hundred acres of land, and other attractive perks.[72] Some shares of its new stock probably became quiet gifts to legislative friends.

With all its legal advantages and assets, the Wiggins Ferry Company played important roles in the economic life of St. Louis and the region. The men atop the company, mostly citizens of St. Louis, dominated the city. A lawyer, Lewis Bogy, served as Wiggins's president for eighteen years (until 1871) while developing railroads and iron mines in Missouri. By 1872, Bogy headed the St. Louis City Council; a year later he was a US senator.[73] Other stockholders included Joseph Brown, the mayor of St. Louis in 1872; William Aspinwall, the leading magnate behind steamship and railroad lines linking the East with California via Nicaragua; descendants of Jean-Pierre Chouteau, a founder of St. Louis; and Samuel Wiggins.[74] These men wielded real power. Sometimes their self-interest served St. Louis, but always at a price.

With its monopoly on trans-river traffic, the Wiggins Ferry Company welcomed the railway age for the legions of passengers and mountains of freight that the trains brought to Illinoistown and its ferries. By 1856, new railroad lines connected that hamlet to Chicago, Illinois, and Terre Haute,

Indiana. A year later, the Ohio & Mississippi inaugurated connecting rail services from the East Coast via Cincinnati to the Wiggins boat transfers to St. Louis.[75] And a number of short lines drew the rich farm produce of southern Illinois to the ferry docks. As the landowner, wharf proprietor, and ferry operator, the Wiggins company dictated terms to the carriers and extracted rents from all.

Average citizens paid tribute too. In 1868, a St. Louis–area farmer spent twenty cents to ship a barrel of flour across the river to Illinois. The same farmer could ship the same barrel to New Orleans for twelve cents. Despite its exorbitant rates, in 1866 Wiggins's boats carried half a million passengers and nearly a million tons of freight across the river.[76] That year, the Illinois legislature changed the map, renaming Illinoistown as East St. Louis (the name used hereafter). The change reflected the facts on the ground: the Missouri city gave East St. Louis, Illinois, its raison d'être.

Aside from tracks in the mud and a haphazard assortment of rough wooden buildings, East St. Louis was little more than a ferry landing.[77] Passengers departing St. Louis to catch an eastbound train in Illinois endured a sequence of miseries. The railroads offered an omnibus service with scheduled departures from the Planters Hotel. After passengers had boarded, a four-horse team set off at a gallop, hauling the heavy wooden omnibus over roughly paved streets, thoroughly shaking the passengers and luggage inside. That was only a curtain-raiser for the main event. "When the driver came to the steep [downward slope] of the levee he first slowed up, then, bracing for a final effort, he galloped the horses over the apron [a ramp] to the ferry-boat. For this catastrophe every experienced passenger prepared himself with resignation, largely modified by anxiety. He clinched his gripsacks firmly . . . and held on like grim death to whatever could steady him when the final thump came."[78] An observer described how "huge coal vans with four mules attached come jolting down the slope . . . animals on the run." People destined for a train had to wait anxiously until the boat filled with freight wagons and their teams. Demand for space often outran supply. A full load aboard the ferry might include four omnibuses, fifteen coal vans, and five merchant wagons, with foot traffic wedged "among the motley crowd of trucks and drays."[79]

After the drama of loading, the ferry ride could prove a pleasant five-minute diversion—if the boats were running. Railroads operated all night,

The original caption for this 1871 engraving read, "The crowd awaiting transporta-tion across the stream has always been of the most cosmopolitan and motley character." The curious structures with derricks out in the river are the pontoon boats at work building the two piers of Eads's bridge. Wood engraving by J.M.D. from King, *Great South*, 220.

to the approval of most Americans in this go-ahead age, but Wiggins ferries ran only in daylight.[80] Winter brought ice floes to threaten the boats, while hard freezes shut service down entirely. No ferries ran for twenty-seven days during the winter of 1865–66; the suspensions lasted longer in the next two winters.[81] If the ice appeared reliable, the hardy or the desperate walked across the Mississippi, while the whole city suf-fered isolation and mounting prices during winter's grip. The ferry com-pany created great wealth for its owners, and year-round inconveniences or worse for the 205,000 residents of St. Louis in 1866.

Obstacles to the Future

Every Wiggins passenger could imagine the benefits to come from a St. Louis bridge, while America's leading civil engineers envisioned the fame that would accompany that commission. Charles Ellet, the nation's best-known builder of suspension bridges before Roebling, in 1839 proposed a roadway bridge with an audacious center span measuring twelve hundred feet. The city council balked at the estimated price, over $730,000 ($24.3 million in 2023).[82]

The advent of railroads on both sides of the river during the 1850s heightened public interest in a bridge, as well as the skepticism that one could ever be built. By the 1850s, Ellet and John Roebling had proven the capabilities of wire suspension bridges for common roadway traffic—pedestrians, horse-drawn carriages, omnibuses, even herds of cattle. The massive dynamic loads imposed by a thirty-ton steam locomotive with its train of loaded freight cars were another matter entirely.

In 1855, Roebling completed his suspension bridge over the Niagara River, near the falls. That impressive structure carried railway trains on an upper deck and common roadway traffic on its lower level. Swaying in the wind and undulating with every passing train, the bridge scared passengers, riders, and pedestrians out of their wits. As Mark Twain described the experience, "You drive over to Suspension Bridge and divide your misery between the chances of smashing down two hundred feet into the river below, and the chances of having a railway-train overhead smashing down onto you."[83] The Niagara example confirmed the skepticism of most engineers in Britain and the United States: suspension bridges simply could not handle railway loadings.* No less an authority than Robert Stephenson, the British progenitor of the modern railway, shared that view.[84]

---

* We associate suspension bridges with the Roeblings and their Brooklyn Bridge (begun in 1869, finished in 1883). The type was comparatively rare in the nineteenth century, however. Roebling's Niagara Bridge was the only significant suspension bridge in railway service in North America. With periodic renovations, it remained in use until 1896. Growing locomotive weights finally led to its replacement by an arch bridge much like the St. Louis crossing. The lighter dynamic loads of cars and trucks made suspension types well suited to twentieth-century needs.

John Roebling's Niagara Suspension Bridge, as seen from the Canadian side of the famous gorge. Railway trains used its upper deck. The carriage and team in the foreground had just crossed on the lower-level roadway. Overhead suspension cables carried the loads, while the trusses that connected the two decks stiffened the structure, allowing trains to cross safely—although Mark Twain had his doubts. Photograph by William Notman, 1869, McCord Stewart Museum, Montreal, I-37302.1.

Such a massive infrastructure project faced daunting obstacles in the United States of the 1850s: inadequate capacities in structural ironwork, a weak St. Louis government, weaker capital markets to provide the funds, an engineering profession barely into adolescence, and substantial professional and public skepticism that such a bridge was even feasible. After visiting St. Louis in 1850, Ralph Waldo Emerson wrote: "They think the river here unconquerable and that man must follow and not dictate to it. If you drive piles into it to construct a dam or pier, you only stir the bottom which dissolves like sugar and is all gone."[85]

Political divisions pulling the nation apart grew into another obstacle to any bridge at St. Louis. During the 1850s, the mounting deadlock between South and North blocked St. Louisans' passionate desire to become the entrepot of commerce serving the West with new railroads to the Pacific. By 1855, the US Army had laboriously surveyed four potential routes for railway lines to cross the plains and reach the West Coast. The central route would have started in St. Louis, while the two alternatives further south (at the 35th and 32nd parallels) also would have boosted Eads's hometown. Deepening bitterness in Congress prevented any decision or compromise. Representatives from northern states blocked the southern projects, fearing they would lead to the expansion of slavery. Southern politicians would only consider a Pacific railroad that benefitted their region.[86]

The fault lines of slavery and abolitionism passed directly through Missouri.[87] Under the Compromise of 1820, Missouri became the northernmost state to sanction the ownership and exploitation of human property. In 1857, the US Supreme Court struck down that compromise, finding in the case of *Dred Scott v. Sandford* that Congress lacked the power to ban slavery in *any* federal territory. This radical finding overthrew law and precedent dating from the Northwest Ordinance of 1787, passed by the country's founding generation to govern Ohio. Instead of settling the slavery question, the court had made civil war far more likely.

By then, Missouri's western border with Kansas was aflame in guerrilla warfare, with settlers and partisans from both sides of the border fighting over whether slavery or freedom would rule in the Kansas territory. Upon the death of Martha's father in 1851, Eads had inherited a slave, Mary, from Colonel Dillon's estate.[88] For a time, he hired her out as a cook, but he feared "we will have trouble with Mary at any place we may hire her if we let her have her way at all."[89] Given those sentiments, it seems likely that Mary left the lives of James and Martha soon thereafter. A slave was no help to Martha at Argyle Cottage, for Iowa was proudly a free state. There is little reason to believe that Eads manumitted the woman, given his enthusiasm for money. His views on abolitionism are unknown. He *was* a committed Republican, advocating for free labor, a strong national government, and a powerful federal role in western development. Like most white men of his time, class, and place, business mattered far more to Eads than abstract notions about the rights of the dispossessed.

Eads may have welcomed war when it came. Few could imagine its grievous cost in lives; certainly Americans had a choice to make. Missouri as a whole was sympathetic to the South, with substantial slaveholding in portions of the state. On the other hand, St. Louisans, especially the German-born third of the population, lined up solidly for the Union. Many of these emigrants had already fought for political freedom during the European revolutions of 1848. After losing in the subsequent counterrevolution, they stood resolute for freedom and union in their adopted country.

On January 1, 1861, emigrants bulked out the crowd of two thousand that thronged to the St. Louis County Courthouse, drawn there by a sheriff's advertisement for a sale of probated property. When the sheriff tried to auction off those seven enslaved people, the crowd roared out the impossibly low bid of three dollars. For two hours the mob shouted its disapproval, drowning out any bidders. Finally, the sheriff gave up.[90] The question had been called. St. Louis voted for freedom.

With the April 1861 attack on Fort Sumter, the fighting that had started on the border of western Missouri and eastern Kansas broke the Union itself. The war would have contradictory consequences for any bridge at St. Louis. No progress was possible during the conflict. For one thing, investors would not put their capital into such a difficult venture while the survival of the nation itself lay in doubt. But wartime needs would drive wide-ranging innovations in engineering, metalworking, and finance. Eads participated directly in some of those developments. When peace came, he would seize on new capacities in iron and steel production, develop his own novel contributions, and take up the once impossible challenge.

Days after Fort Sumter fell, James Eads wrote to his old friend Edward Bates, now the attorney general, arguing that "vigorous action must be taken to defeat the South." While many northerners favored conciliation, blockade, or limited war, Captain Eads sought outright victory and restoration of the Union. Bates replied: "Be not surprised if you are called here suddenly, by telegram. If called, come instantly. In a certain contingency, it will be necessary to have here the aid of the most thorough knowledge of our river and the use of steam upon it."[91]

CHAPTER 2

Advances from War

IN COUNTLESS WAYS impossible to foresee, the Civil War became a bloody calamity for most Americans. Straddling the divides between North and South, East and West, slave and free, Missouri was wracked by the conflict. Profiling the state in 1867, a writer for the *Atlantic Magazine* wrote that "the whole population was at war." Lawless disorder and Confederate guerrilla raids were chronic threats across southern Missouri, driving thousands of rural folk to the comparative safety of St. Louis. Many sympathized with the Confederacy, but sheltered under the safety and order ensured by the Union army. The city was never seriously threatened by Confederate attacks. As the chief supply base for the Union's western theater, it was stoutly garrisoned, protected by the cannons of ten new earthen forts. From August 1861, martial law ruled the city.[1] The influx of Union blue, however, scarcely compensated for the end of its river commerce with the South. During the rebellion, the volume of commercial trading at St. Louis fell by two-thirds.[2] Amid the exigencies of war, any bridge over the river there was simply impossible.

Despite the conflict, many foundations for that structure originated in the war years. Like most St. Louisans, James Eads was a Lincoln supporter and a Union man. His wartime work—building the country's first ironclad warships—proved crucial to victory. That experience bolstered his political contacts, acquainted him with the special qualities of steel, introduced him to Tom Scott of the Pennsylvania Railroad, and refined his design skills in mechanical engineering. He would call on all that knowledge and those connections after 1865 in meeting the challenges of creating a bridge for St. Louis.

Another project six hundred miles from the city originated during the war years and shaped later developments in St. Louis. Completed in 1864, the Steubenville Bridge over the Ohio River near Pittsburgh established the paradigm that would guide the design of most long-span bridges for the next half century. A new firm, the Keystone Bridge Company, created in 1862 to design, fabricate, and erect the Steubenville crossing, would move aggressively after the war to put up similar pin-connected iron truss bridges across the western rivers. Eads would choose a radically different design path. Even so, Keystone would shape his work in St. Louis, supporting and challenging his plans simultaneously.

The top officers of the Pennsylvania Railroad, the nation's largest carrier, created the Keystone company. Edgar Thomson and Tom Scott also turned their eyes westward during the war, looking to build new railroads from the Mississippi River to the Pacific Ocean. Although little actual construction began during the fighting, their visions of new western lines solidified their interest in St. Louis and their ties to James Eads. Driven by the emergency, President Lincoln, the US Congress, and the leadership of the Union army forced key innovations that shaped northern railroads broadly during and after the conflict.

A New Gamble on the River

Responding to that urgent telegram from Attorney General Bates, James Eads arrived in Washington City little more than a week after the garrison at Fort Sumter surrendered. Bates was telling all who would listen that the Union should build a fleet of steam-powered gunboats to take control of the Mississippi River. At a cabinet meeting called by the president, Eads assured these men that civilian vessels, including his own *Submarines*, could be quickly converted into a powerful force of armed gunboats able to capture or block all Confederate traffic on the river, the main highway binding the rebel states. To protect the ships against enemy shore fire (and to save time), Eads proposed initially to clad their deckhouses and power plants with bales of pressed cotton.[3] Soon thereafter, he became a strong advocate for building a new fleet of powerfully armed steam gunboats clad in iron armor plate.

That warships protected with iron would play a role in the American war required no special insight in 1861. During the 1850s, Great Britain and France had developed prototypes, moving from unpowered floating batteries in the Crimean War (1854–56) to fully rigged oceangoing steamers with iron hulls and heavy cannon. Britain's 9,000-ton, 400-foot HMS *Warrior* (1860) was both a powerful example and a useless precedent for the American conflict. The Confederacy could not hope to make her thick iron plates. The Union could not fight in the bays and rivers of the South with a ship that needed twenty-seven feet of water simply to float. Unique conditions called for special knowledge, as Edward Bates had foreseen.

After meeting with Lincoln's cabinet, Bates brought his friend Eads, an expert on the western rivers, together with the top men in the navy. Those blue-water sailors had no experience on the brown, shallow, and treacherous rivers of the West. Furthermore, many had spent most of their careers aboard traditional sailing vessels. Eads then collaborated with procurement officers to develop the broad concepts for a radically novel warship. An experienced ship designer, Samuel Pook, labored over the summer drafting design plans for "an armored light-draft river war steamer." In August the War Department sent the drawings out for bids from shipbuilders. On August 7, Eads closed a contract to build seven of these 500-ton warships, each 175 feet long, with a draft of just 6 feet. He committed to a price of $89,600 each ($3.2 million in 2023).[4]

Eads had never built a warship; he did not even own a shipyard or a machine shop. Yet the contract specified delivery of the first ironclad in just sixty-five days, and it levied financial penalties if deliveries lagged. His mix of audacity, persuasion, and charisma won the contract. Now he had to produce these gunboats, each with five boilers, twin high-pressure engines, and wooden hulls—all protected by seventy-five tons of armor plate.[5] At least, the navy's Bureau of Ordnance would supply their powerful guns, thirteen for each ship.

In taking up this new challenge, Eads would draw upon his unusual business and engineering work on and under the Mississippi. Creating the *Submarines* had imparted lessons in designing and building river steamers. Operating them had translated into years of learning river currents and conditions. His experience in repairing and selling salvaged engines, boilers, and boats had given him connections to shipyards and engine builders up and down the Mississippi and Ohio Rivers. The boldness of his leap itself proved an advantage. The steamboat trades on the western rivers were largely moribund in the summer of 1861, their business wounded by the war and its uncertainties. Shipyards had little work; engine and boiler shops were largely silent. Thousands of skilled tradesmen had been laid off. This was considerable latent capacity if managerial skill, organization, and money could put it to work.

Charles Boynton's *History of the Navy during the Rebellion* describes what happened next. "Telegraphic orders issued from Washington" marshaled men and resources. Woodcutters from Minnesota to Kentucky went into forests to fell standing trees, which were immediately transformed into the dimensional lumber needed for hull frames and planking. Plate mills in Ohio, Kentucky, and St. Louis began rolling the iron armor, 2.5 inches thick, to cover and protect the wooden hulls. "Nearly all of the largest machine shops and foundries in St. Louis . . . were at once set to work day and night." In all, the fleet required fourteen engines, thirty-five boilers, auxiliary engines and pumps, high-pressure piping, and hundreds of other parts to build integrated marine powerplants. Needing so many components, Eads used telegraphic ordering to enlist metalworking factories in Cincinnati and Pittsburgh. "Within two weeks not less than four thousand men were engaged."[6]

As the work advanced, barges delivered materials and components to a shipyard that Eads had leased in Carondelet, just south of the city. That yard's prewar operator had taken his shipbuilding skills to the Confederacy.[7] The Carondelet yard constructed four ships in Eads's contract; he subcontracted three more to the Mound City Dockyards, on the Ohio River just above Cairo, Illinois. The first, the USS *St. Louis*, was launched on October 12, 1861, just two days past the deadline. Although fitted with engines and boilers, she was not quite finished under the contract terms. The two yards completed the other six vessels by December, a delay partly owing to design changes mandated by the navy. The delays and alterations resulted in a final cost double the contract price. Nonetheless, Eads had accomplished a truly noteworthy feat, building seven utterly novel steam-driven warships ready for their guns in just one hundred days. That result reflected the man. Back on a cold December day in 1849 he had

Two gunboats under construction at Eads's leased yard just south of St. Louis, ca. September 1861. The ship in the foreground had her five boilers installed. Iron armor plating would soon cover the oak framing shown on each side. Those casements were angled to deflect enemy shot. Construction of the ship in the background was further advanced. Unknown photographer, *City-class Armored Gunboats*, US National Archives, 165-C-702.

written to Mattie while working a salvage job on the river, *"Drive on* is my motto."[8]

Flag Officer Andrew Foote formally accepted the vessels into government service on January 15, 1862. Army hearings investigated and excused the cost overruns, largely resulting from postcontract changes that included an additional forty-five tons of armor on each vessel.[9] Called City-class gunboats, the seven ships all bore the names of midwestern river towns: *St. Louis, Carondelet, Louisville, Pittsburgh, Cairo, Cincinnati*, and *Mound City.* Because their sternwheels were protected by a hump of iron plating, the ships were known affectionately as "Pook's Turtles."

A proud man, James Eads wrote to President Lincoln in September 1863, sending along a photograph of the USS *St. Louis,* "the first iron-

James Eads had every reason to take pride in the innovations and accomplishments embodied in the seven City-class gunboats that he contracted to build in 1861. They fought in every Mississippi River campaign of the war, bearing the brunt of that fighting. In the first such contest, the Battle of Lucas Bend (Jan. 11, 1862), the USS *St. Louis* engaged Confederate vessels and forced their retreat. This image shows the nearly identical USS *Cairo*. Unknown photographer, *USS* Cairo *on the Mississippi in 1862*, US Naval History and Heritage Command, NH 61568.

clad built in America, being the first of eight constructed by me during the Fall of 1861." In writing to the president, Captain Eads bestowed on himself an accolade largely denied him then and since. In February 1862, a month before John Ericsson's USS *Monitor* fought to a draw in Virginia, Eads's innovative ships gave the Union its first significant victories of the war, leading the attacks that captured Forts Henry and Donelson. The combined army and navy assaults overwhelmed those Confederate strongholds and led quickly to Union control of the Tennessee and Cumberland Rivers, the fall of Nashville, and the effective end of Kentucky's participation in the Confederacy. Those victories made a national hero of Ulysses Grant and proved fundamental to the Union's eventual victory in the war. Eads's claim to Lincoln was accurate:

the *St. Louis* was "the first ironclad that *ever engaged a naval force in the world.*"[10]

Iron warships may appear far removed from a steel bridge, although Eads's wartime work provided essential foundations. Fearlessly, perhaps even recklessly, he sized up and took on novel challenges in innovation. Marshaling his talents in finance and organization—and his charismatic leadership of men—he then achieved all he had set out to do.

Making Good Money

After his City-class successes, Captain Eads took an important role in designing and building the four-ship Milwaukee class. Three of those ships had Eads's new steam-powered turret, "a marvel of the times and far more complex" than the Ericsson turret of the USS *Monitor.* He devised steam-powered methods to elevate and lower the guns, run them out for firing, cushion their recoil, and operate the gun ports. The modern authority on these vessels, Donald Canney, declares that "all in all, for 1864, this was a technological *tour de force.*"[11] His work in cannon design earned him his first patents (1864) and brought him into contact with naval officers experimenting in steel, a novel material in most industrial contexts in America.[12]

The work earned Eads considerable profits. His declared income for 1864 totaled $60,719 ($1.2 million in 2023), an astronomical sum. Just nineteen residents of St. Louis, a city of two hundred thousand or more, reported higher totals to the new Internal Revenue Service that year. As a temporary wartime measure, the 10 percent tax on incomes over $10,000 proved to be a popular demonstration of patriotism for most Americans, even among the wealthy who paid it. And when the *Missouri Democrat* published its list of "The Wealthy Among Us," everybody knew who ranked atop the city's commercial elite.[13]

The intense work of design and construction (his Carondelet yard also built two ironclads in the Milwaukee class) broke the Captain's health. March 1864 brought recurring fevers, followed by some kind of surgery, a month of recuperation, then a trip to Europe in the summer with Eunice. He carried an appointment as a special agent of the navy, charged with visiting "the different Navy Yards and principal iron-works of Europe."[14] Sojourns to Paris and Geneva did little to improve his health.

In August he wrote from Geneva to his friend Gustavus Vasa Fox, the assistant secretary of the navy: "This state of things naturally exerted a depressing influence on my spirits and I felt disinclined to do anything but think of new methods of operating your big guns. In fact I began to think ironclads and such matters are with me a kind of monomania. I think too, like some poets who produce their most brilliant strains just before they go crazy, I am most successful as the disease increases in its intensity."[15] Just what ailed him mystified doctors in the United States and Europe, so no diagnosis seems warranted or possible 160 years later. The syndrome recurred in later years: hard work, a bout of "nervous collapse," a recuperative voyage to Europe, recovery there, then return to the New World and new work armed with new ideas. He came home in December, drifting a bit in search of challenges, the war nearly won.

Married and middle-aged when the conflict began, James Eads was not expected by anyone to put on a uniform for glory or country. In any case, his talents well served the Union. On the other hand, Pierpont Morgan (aged 24 in April 1861), Andrew Carnegie (25), and Jay Gould (25) could have entered the armed forces. After the war, these three men would shape the bridge at St. Louis. During the conflict, they had other priorities. Each bought an exemption from the draft (conscription began in 1863) by paying a fee to procure a substitute for army service. Many affluent men made the same choice. In the general view, this was a practical decision, not a moral issue. Everyone knew that any day and in every way wealth or poverty shaped life's outcomes.

Andrew Carnegie took a wandering path during the war. Back in 1853, this Scots emigrant had had the break of a lifetime when, at age 18, he became the personal secretary and telegraph operator for a rising star at the Pennsylvania Railroad. In December 1859, Tom Scott became a PRR vice president, and Carnegie succeeded him as superintendent of the road's Pittsburgh division. He had just turned 24, and he remained in that post until April 1865. His wartime work on the PRR was essential to that company and to the Union war effort. Confederate cavalry frequently tore up the Baltimore & Ohio main line, an east-west thoroughfare that, with connections, tied Washington City to St. Louis. When that artery was severed, many of the B&O's burdens in freight and passengers came over to the PRR, adding to its own wartime work of moving men, munitions, and the civilian economy. All this meant towering challenges for the superintendent of the Pittsburgh division.[16]

Still, there was much more to Carnegie's war than working for the railroad. Like Eads, he profited handsomely from the conflict, becoming a partner or stockholder in a dizzying array of businesses: express companies and telegraph lines, horsecar and sleeping car lines, two coal companies and an oil company, an ironworks and a factory making iron bridges. It is hard to imagine how a young middle manager on the PRR could create or find all these opportunities. He didn't. Although Carnegie and his biographers tended to obscure the details, other men were the prime movers behind all these start-ups. Carnegie's mentor, Tom Scott, was often the chief instigator. Together with Scott's boss, PRR president Edgar Thomson, Scott and Carnegie pursued a surefire path to wealth. According to Carnegie's best biographer, David Nasaw, their "standard operating procedure" consisted in "capitalizing on insider information to invest in companies that were about to be enriched by lucrative contracts."[17]

The creation of the Keystone Bridge Company in 1862 well illustrates how these men used their PRR connections to create wealth. The new firm essentially stripped the railroad of its in-house design and building capacities for new bridges. John Piper had headed the Altoona shops of the PRR, Aaron Shiffler was the line's general bridge supervisor, while another partner in the new company, Jacob Linville, was the PRR's chief bridge designer.[18] Whoever dreamed up the idea, President Thomson and Vice President Scott were the principals behind the new firm. "Scott's Andy" (as Thomson called him) was present at the creation, but it surely distorts history to call Keystone "another Carnegie company."[19] After the war, his work for Keystone focused on sales, a role well suited to his talents.

Because of its partners' abilities and their insider connections, Keystone was immediately lucrative. The PRR needed new bridges for its expansions as well as many iron replacements for the wooden spans in use across the system. An 1871 ledger of Keystone stockholders reveals how far the shadow of the PRR fell across the company.[20] Of its sixty-nine shareholders, at least twenty were managers or directors of the railroad. The ledger's overall portrait: Thomson and Scott had created Keystone to build bridges and to throw profits to themselves and their professional associates.[21] Paying handsome annual dividends, Keystone bound these men together and tied them to the railroad. All could subscribe to the comforting notion that what was good for Keystone Bridge was good for the PRR. Perhaps for America as well. By harvesting investing opportunities across the United States, it became a very good war for Thomson, Scott, and Carnegie.

The Steubenville Bridge

From its first job, the Keystone Bridge Company, located in Pittsburgh, took a leading role in creating a new industry: fabricating and erecting long-span iron bridges to standard truss designs, crossings to carry railroads across the great rivers of the West. To improve the PRR's western connections, in 1862 Edgar Thomson launched a project to bridge the Ohio River at Steubenville, Ohio, forty miles west of Pittsburgh.[22] His determination to bolster traffic on the PRR provided the immediate spur to create Keystone and build this bridge.* At that time, no other railroad crossed the Ohio at any point in its 981-mile course from Pittsburgh to the Mississippi at Cairo, Illinois. The engineering challenges were simply too forbidding.

Besides the river itself, the Commonwealth of Virginia presented further obstacles. Thomson's proposed line needed to run through Virginia's narrow panhandle on the Pennsylvania state border. For years, Virginia politicians had blocked this route, seeking to protect the western business of the Baltimore & Ohio Railroad. Much of that carrier's main line passed through the Old Dominion. By 1862, however, Virginia had removed itself

---

* The historian John Kouwenhoven believed that Thomson created Keystone in 1862 to build the Steubenville Bridge. Beyond the aligned timing of the new company and the new project, Kouwenhoven noted that the PRR's charter (from the Commonwealth of Pennsylvania) provided no authorization for the railroad to manufacture iron bridges for sale, particularly for the Ohio corporation that would own and operate the bridge. As plausibly, Thomson may have established Keystone to limit overall liability in case the new crossing proved a failure. Outright collapse was a clear threat during construction or in service. No evidence supports Peter Krass's picturesque assertion that the threat of wartime sabotage of wooden bridges drove Thomson and Scott to create Keystone. *Carnegie*, 67–68. Carnegie's *Autobiography* helped mislead Krass by its claim that Keystone specialized in iron bridges (116). According to its 1874 catalogue, since its founding Keystone had built iron bridges equaling 64,000 feet in overall length, while its output of wooden spans totaled 47,600 feet (pp. 34–42).

from the United States and its Congress, so the PRR had cause for optimism.

On July 5, 1862, the US Senate debated authorizing construction of the Steubenville Bridge, an interstate crossing. The cloudy legal status of rebel Virginia made federal imprimatur especially important. Pennsylvania's own senator, Edgar Cowan, proved no friend of the project. The original bill to charter this bridge called for a 200-foot-long span above the main navigational channel of the Ohio River. Advocates for the steamboat interests then pushed for a longer channel span of 270 feet. The bill's sponsor, Republican senator Benjamin Wade of Ohio, grudgingly accepted that change despite fears that the mandate was impossible to meet, given the state of the art in railway bridges. Committee testimony had indicated that the longest continuous span then in service in the United States was 250 feet, a composite wood and iron truss across the Connecticut River.

Such limitations did not concern Senator Cowan. He wanted to kill the project outright, and his chosen weapon was an amendment to stretch the requirement for the channel span length to 300 feet. Taking the Senate floor, Cowan asked rhetorically who wanted the bridge—the country, the people?

> No, sir, a corporation, and one of the class which I say is swallowing up to-day more of the political privileges of the people and exercising a more dangerous and deleterious influence over the politics of the country than all other interests and all other mischiefs combined. The very same men, sir, that are charged with purchasing Legislatures wholesale, charged with buying them for a price like oxen in the shambles [slaughterhouse], are here and were here to get this bridge fixed to suit their particular notions.[23]

July 1862 was a dark moment for the republic. Yet this Republican senator denounced his own state's dominant corporation as a threat to political freedoms greater than "all other mischiefs combined." Tom Scott tried to be discreet when bribing state and national legislators, but the senators knew full well the root of Cowan's enmity.[24] On July 14, Congress authorized the Steubenville Bridge, requiring Cowan's 300-foot clear span, positioned 90 feet above the river surface.[25] It was a rare defeat for the carrier. Now the railroad needed to stretch the reach of engineering against the mark set in law.

The new Keystone Bridge Company took up the challenge. Jacob Linville, still on the PRR staff as well as the chief designer for Keystone, set to work on the drawings for Steubenville. He enlarged the proven Whipple-Murphy iron truss to heroic proportions for a channel span of 320 feet, with seven adjoining spans of 232 feet each.[26] Tension members, including the bottom chords, were in wrought iron, while parts in compression (including the top chords and posts) were of cast iron.*

With Thomson's authorization, Linville took the unprecedented step of ordering a testing machine to ascertain and verify component strength.[27] Keystone could now test to destruction any part of the bridge. Pulling eye bars apart (a tension force) and crushing channel beams (compression) sounds like primitive stuff, but this machine marked an important step toward *engineered* bridges. With it, Keystone developed reliable data on the tensile and compressive strengths of the major bridge components. Testing to destruction gave Keystone confidence in its design as well as the ability to demand quality improvements from its iron suppliers. Thomson's young firm completed the Steubenville Bridge in 1864, exceeding the congressional mandate for the channel span, opening a new era in long-span iron bridges, and ensuring its own place at the forefront of the new industry. And its interest in a bridge for St. Louis.

Rivals, Railroads, and the War

While the Civil War set the scene for a bridge at St. Louis, the burdens of discord over slavery began to fall on the city years before the assault on Fort Sumter. In 1854, Democratic senator Stephen Douglas pushed the Kansas-Nebraska Act through Congress. Because the politicians in Washington City could not resolve the slavery question, Douglas's bill gave the choice to the people of those territories lying to the west of Missouri. "Popular sovereignty" soon resulted in "Bleeding Kansas" as nightriders, bushwackers, jayhawks, and border ruffians fought with knives, guns, and

---

* The chords of a truss bridge are the main structural members, running horizontally. Braces, posts, or web members connect the top and bottom chords to form the truss.

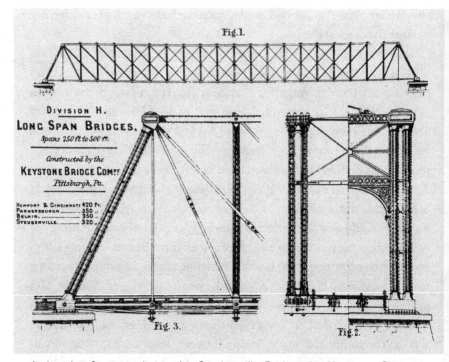

A decade after completing the Steubenville Bridge, the Keystone Company published these engravings in its catalog, justly proud of its record-breaking accomplishment. The firm scaled up this proven design for later crossings over the Ohio River at Bellaire, Ohio (1871), Parkersburg, West Virginia (1871), and Cincinnati, Ohio (1872). Success begat success, while curbing risks and costs and speeding construction. Image from Keystone Bridge Company, *Catalogue*, plate 6.

intimidation—partisans for slavery against advocates for free soil and free labor. The violence scared away peaceable folk, regardless of their politics, so in-migration to Kansas and western Missouri fell sharply. The editor of the *St. Louis Intelligencer* described the consequences for his city, 250 miles to the east, as follows: "St. Louis is retarded in a most woeful way. Our railroads creep at snails' pace. We build ten miles while other western states build one hundred. In every department of life we feel the paralysis. Instead of bounding forward . . . we sit with dull eyes and heavy spirits, and listen to the tick of the death-watch."[28]

By contrast, Chicago boomed across the 1850s. Eleven "great railroads" reached out from the city by 1856 to serve the new farmers flooding into

The Rock Island Bridge as originally built. The city of Davenport would grow up on the pastoral Iowa shore shown here. The drawspan at right pivoted to allow the passage of steamboats, while the two rafts shown in the river passed easily under the fixed spans. To the right of the draw, two more fixed spans (not shown) reached Illinois. Built largely of cut timber, all the spans were patented Howe trusses. Image from the Putnam Museum, Davenport, Iowa, reproduced with permission.

Illinois and Iowa.[29] In senses both figurative and practical, these carriers transformed the grassland prairies into farms growing wheat and corn. In that year, an average of one hundred trains a day served the city.[30] Increasingly the farmers of western Illinois, Iowa, and Missouri who once had shipped their crops by Mississippi River steamers to St. Louis now sent their goods to Chicago by rail. With farming products dominating the freight of those railroads, they became known as granger lines.

In another lift for Chicago (and a blow to St. Louis), in 1856 a granger line from the Windy City began running trains across the first rail bridge over the Mississippi River. Serving the Chicago and Rock Island Railroad, that single-track crossing was rudimentary, built cheaply for one purpose: to push rails into the farmlands of Iowa.[31] Its five Howe trusses, each 250 feet long, broke no records.[32] Patented a decade earlier by a Massachusetts farmer, this simple design relied mostly on wooden structural members, using iron rods just for the vertical ties to carry the tension loads. Such composite bridges were the state of the art in the 1850s.[33] To allow river steamers to pass under this low bridge, it had a pivoting drawspan that swung open for boats as needed, then closed when trains approached. A Chicago firm, the American Bridge Company, built the trusses and the drawspan.[34] In all, the bridge and its approaches were estimated to cost $260,000 ($9.5 million in 2023), a lot of money for anything in the prairie West but a pittance compared with the costs of later bridges.[35]

The independent company that projected and owned the Rock Island Bridge was chartered by the state of Illinois; it did not bother to secure authorization from Iowa's legislature. The Railroad Bridge Company pressed ahead without any federal sanction. With its Corps of Topographical Engineers, the War Department was the logical agency to exercise federal oversight over the navigable waterways of the interior West. But the secretary of war, Jefferson Davis, opposed the Rock Island Bridge in every forum and in any way possible. He would have lobbied hard against a congressional charter.[36] In his eyes, any advance westward by northern rails posed an intolerable threat to his southern allegiances.

In all, the Rock Island Bridge was a rudimentary effort, but a significant harbinger. St. Louis rivermen howled when the bridge went up, claiming it obstructed navigation. It did. The piers for the new bridge created cross-currents and eddies in the two channels through the draw, posing real difficulties for the pilots of river steamers. Just two weeks after the bridge opened, a nearly new 240-foot steamboat, the *Effie Afton*, collided with the structure, ramming in succession the piers on either side of the open draw. The collisions upset heating stoves aboard the steamer, its upper deck caught fire, and the flames leapt to the wooden truss bridge directly overhead. Soon the truss collapsed onto the burning steamer, which sank immediately.[37]

Consequences large and small resulted from these dramatic events. The leading salvage firm on the Mississippi, Eads & Nelson, owned the rights to the wreck and its cargo. Their local agent sent all recoverable freight downriver to St. Louis and then disposed of the wreck itself.[38] American Bridge rebuilt the wood and iron composite superstructure. And a rising Illinois lawyer, Abraham Lincoln, helped defend the Railroad Bridge Company against a suit for damages, filed by the *Effie Afton*'s owners, who claimed that the bridge had obstructed her safe passage. More broadly, the collision that day at Rock Island was a turning point in a struggle between river and rail commerce, St. Louis and Chicago, the past and the future.[39]

In January 1865, another granger line, the Northwestern, completed a bridge over the Mississippi at Clinton, Iowa, just thirty-five miles upstream from the crossing at Rock Island.[40] Iowa farmers now had a second rail connection to Chicago and the East. This structure demonstrated more so-phistication, with a pivoting drawspan (made entirely of iron) constructed to a patented design known as a Bollman truss. A steam engine opened the draw for river traffic and closed it for trains.[41] Beyond the immediate chal-lenge it posed to river commerce, the Clinton Bridge conveyed an ominous future for St. Louis. Its construction had required just a year, thanks to a range of innovations in the design and manufacture of iron bridges. With these factory techniques, the comparatively placid and shallow upper Mis-sissippi would scarcely hinder bridge construction or slow the advance of Chicago's railroads.[42] The river at St. Louis was another matter entirely, even before the US Congress heightened the challenge there.

Wartime Railroad Policies

As with ironclads, it had required no great insight to predict that railroads would matter in the Civil War. By 1860 the United States had more rail-road mileage—30,626 miles—than all the nations of Europe combined. During the 1850s new construction had added roughly 2,000 miles of track annually. The war crippled that pace. In 1863, a decisive year on the battlefields, new lines totaled just 574 miles.[43] And wartime disruptions touched much of the railroad map in the North and the South. Even so, the conflict laid some foundations in the North that profoundly shaped rail-road developments after 1865, especially in the West.

The demanding challenges of transforming separate railroads into coordinated regional systems had barely received the notice of railroad managers in 1861. The war forced the issue. In the fall of 1861, the Union's top western commander, John C. Frémont, ordered construction of a new union depot at St. Louis to hasten troop movements through the city.[44] Grudgingly, St. Louis's three carriers built the required track connections, only to abandon this coordination and convenience after the war.[45] In the East, the wartime emergency finally forced carriers there to cooperate in providing through passenger service between New York City and Washington City.[46] Travelers still encountered delays at the Hudson and Susquehanna Rivers, transferring to ferries because those broad waterways lacked bridges. Even so, the coordinated service was a great improvement.

Federal support for the proposed transcontinental railroad had been a central Republican promise in the election of 1860. An 1862 statute gave President Lincoln the duty to choose the gauge (the distance between the rails) for that new line. He picked five feet, the norm for California's railroads. In an act of enduring significance, Congress in 1863 overruled the president, mandating 4 feet, 8.5 inches. The world's first railway engineer, Britain's George Stephenson, had selected that gauge decades earlier, and many American carriers, especially in the East, made the same choice. Even so, the country's railroads operated with an array, or more accurately a disarray, of gauges at the time. Acting for the nation, Congress believed the Stephenson standard would be essential in war and eventually beneficial to all railroads. In this view, a common gauge would allow freight or passenger cars to move with scarcely a pause from one railroad to the next. Real gains in efficiency beckoned, especially in moving cargos. Still, interchanging cars across connecting lines (regardless of gauge) was quite rare in 1863.

The Senate debate on the transcontinental gauge question revealed conceptual, practical, and geographic obstacles to the interchange of freight cars between different railroads. For example, Senator James McDougall (D-CA) argued that "the running of strange cars over the roads is something that proper business interests will not permit." Furthermore, he argued, no single company "could conduct their business if they distribute their cars all over . . . the United States." Senator John Henderson (R-MO) believed that a single common gauge for east and west had no value "since the Missouri river is not bridged; and I do not think it is possible to bridge it." Moreover, he doubted that the lower

Mississippi—from its confluence with the Missouri River (just above St. Louis) to the Gulf of Mexico—would *ever* have a railroad bridge.[47] The fast currents, crushing ice jams, and sandy riverbeds seemed likely to overwhelm or undermine any creation of human ingenuity or pride.

Transcontinental Dreams

All the ventures to bridge the western rivers are best understood as elements in a vast project to build out the country's railroad lines, integrating them into networks that would bind and grow the US economy in the decades following the war. At the time, that process resembled figurative chess matches: played by shifting rosters of opponents—shaped by money, geography, and government—without rules and unfolding on a continental scale. Those contests began during the war.

On July 1, 1862, the federal Army of the Potomac finally scored a tactical victory at Malvern Hill, Virginia, after seven days fighting and losing to the Army of Northern Virginia and its new commander, Robert E. Lee. The lone victory could not erase the truth: Union forces under General George McClellan had failed to take Richmond.

Also on July 1, President Lincoln signed into law the Pacific Railway Act. That statute sought to encourage railroad men and investors to step forward to build a 2,000-mile line across the grassland prairies and the Rocky Mountains to California. The law too proved to be a failure. Despite its offers of generous federal land grants and bonds, those rewards amounted to little against the specter of risks either overwhelming or unknowable. Even so, the 1862 statute laid some foundations that endured. This grand project to bolster the Union would start on the Missouri River, at a point somewhere on the western border of Iowa or Missouri. It would terminate on the Sacramento River in California. A new company, the Union Pacific (UP), would begin construction on the plains, heading west, while the established Central Pacific would pierce the Sierra Nevada range, building eastward from California.

The 1862 statute also offered an imaginative solution to a fraught question: where to locate the eastern starting point of the new transcontinental line? That issue had deadlocked the US Congress throughout the 1850s. In a classic political compromise, the 1862 act chose to put it in the middle of nowhere. The UP segment of the transcontinental line was to start its

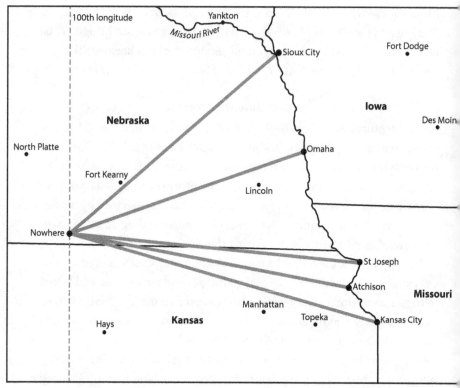

The Pacific Railway Act of 1862 fixed a site south of Fort Kearny, Nebraska, as the starting point for federal land grants to build the westward-reaching tracks of the Union Pacific. To connect that point in Nowhere to the East, the statute proposed more land grants to foster construction of the five spokes shown here. All would terminate at Missouri River ports: Sioux City, Omaha, St. Joseph, Atchison, and Kansas City. At that time, only St. Joseph had an operating railroad connection to the East. Townsfolk at the four other villages aspired to such grandeur. Map by Jade Myers.

westward route at a point on the 100th meridian of longitude in the Nebraska Territory, north of the Republican River valley and south of the Platte valley.[48] In 1862 that land sustained prairie grass, buffalo, Plains Indians, and little else. A decade later its first white settlers would organize their local government as Frontier County.

Subsequent clauses in the 1862 statute illustrated the motives behind this origin at a place we will call Nowhere. Turning to the east from that spot, five spokes were to radiate to different points on the Missouri River, all real places with white settlement. Once at the river, three routes would then connect with railroads in Missouri; two aimed to make Iowa connections. All five of those eastbound spokes from Nowhere would receive the same land grants that the bill proposed for the westbound stem of the UP, the tracks heading to California that would originate in Nowhere.[49] In this fashion, the Republican Congress planned to bypass the prewar deadlock and shower developmental largesse across Nebraska, Kansas, Missouri, and Iowa. Land rich in farming potential would gain settlers and rail connections to markets. Furthermore, if all this came to pass, Congress would satisfy the voters and merchants of St. Louis *and* Chicago. For one question preoccupied their residents and partisans: which western cities would claim the benefits to flow from the new cross-continental railroad?

Despite all the aid Congress stuffed into the 1862 statute, financiers did not step forward with the private-sector funds needed to build these visions. With the odds for a Union victory improving by 1864, the legislators tried again. More to the point, midwestern politicians and railway promoters fell on Washington City, determined to extract a Pacific railroad bill that would boost their own ventures. The Kansas Pacific Railroad looked like a real contender.[50] By December 1864 the road had begun operating trains over forty miles of track westward from Kansas City, Kansas. With ferry connections across the Missouri River, that city in turn gained a rail link eastward to St. Louis in March 1865, after the completion of the Missouri Pacific line across that state.

Meanwhile the Union Pacific barely existed on paper, even as it glowed in the dreams of Nebraska politicians and Omaha real-estate speculators. The UP would not lay its first westward rails from that Missouri River town until July 1865.[51] In the meantime, its Washington lobbyists produced an ace. In March 1864, President Lincoln issued an executive order fixing Council Bluffs, Iowa, as the transcontinental terminus of the Union Pacific.[52] Omaha and the UP's boosters were just across the river. Lincoln's decision mattered, boosting the advocates of an axis of lines from Chicago to Omaha to the Pacific. Thus, it challenged proponents of the rail route from St. Louis to Kansas City and onward across the plains and mountains. But Lincoln would not have the final word. Congress would have its say, while facts on the ground mattered most of all.

To ensure that they both received generous federal aid, the KP and the UP temporarily put their rivalry aside and launched a joint lobbying effort. The Pacific Railroad Act enacted on July 2, 1864, amended the 1862 bill by doubling the acreage of federal lands granted to the companies building the line. It also effectively doubled the number of bonds the carriers could issue as construction advanced. The statute named the UP as the main stem of the transcontinental route to the Rockies, while granting "all of the benefits . . . conditions and restrictions of this act" to "any company to construct its [own] road and telegraph line from the Missouri River" to link up with the main stem of the UP.[53] This time congressional largesse transformed dreams into a race to lay rails, a contest of competing crews from the KP and the UP. Although imperfectly worded, the statute implied that the Kansas Pacific could become the main link between east and west if its tracks reached the 100th meridian before the UP's did.

The Kansas Pacific had originated with an 1855 charter of a predecessor road. By 1864, the company had tracks, trains, and truly scandalous owner-managers. Few grieved when the road's chief engineer shot its grafting promoter, Samuel Hallett, in the back, killing him on a dusty Kansas road in July 1864.[54] The exit of both men certainly improved the company's prospects. A year later, they looked even better after Edgar Thomson and Tom Scott took a controlling interest in the road.[55] At that moment, the KP was a strong contender to reach the 100th meridian before the UP. The distances requiring new tracks were reasonable enough at 260 miles for the KP and just 220 for the UP.[56] And the KP was actually a railroad in August 1865. Meanwhile, every rail, car, and locomotive destined that summer for the UP had to be loaded aboard steamboats at St. Louis, then laboriously transferred up the Missouri River to Omaha.

In their Kansas Pacific play, Thomson and Scott headed an investor group known as "the Philadelphia interests." The group included other top officers of the PRR, the proprietors of iron and locomotive works in Philadelphia, and principals of four investment banks—including Jay Cooke and Anthony Drexel. According to the historian James Ward, this "tight little circle of capitalists . . . owned investments across the nation."[57] Starting in 1865, the Philadelphia group moved to dominate the construction of railways and bridges throughout the trans-Mississippi West. Their Kansas Pacific thrust against the UP might boost the fortunes of the PRR as well. If everything fell into place, goods from across the Pacific Rim would flow over the transcontinental rails onto the Kansas Pacific and the Missouri Pacific, pass through St. Louis, then enter the Pennsylvania Railroad system heading eastward.

The Philadelphia interests bought their way into the KP by funneling a million dollars into its stocks and bonds, matching another million put up by St. Louis investors. That group included James Eads.[58] Eads had met Tom Scott during the war, when Scott was serving as assistant secretary of war, advancing the Union armies with his skills in railroad and telegraph operations. In that post, Colonel Scott had undertaken an inspection tour of the West in February 1862, meeting Captain Eads while inspecting his new ironclads at Cairo, Illinois.[59] The two men renewed their friendship when the PRR vice president visited St. Louis to inspect the KP in August 1865. At that point, Eads was a St. Louis business leader, known for his peerless Unionist credentials, innovative ironclads, and investing savvy. Not one of those radical Republicans agitating for equality for the freed slaves, he appeared to be the kind of man you could trust with your money. The first evidence of *his* interest in a St. Louis bridge dates from April 1866, an outgrowth of his railroad investments. Across that year, Eads deepened his ties to the Kansas Pacific, writing to a friend that he was "pretty largely interested" in the road.[60] Through this KP connection, Eads bolstered his ties to the Philadelphians.

Despite all its advantages, the KP lost its race to Nowhere in the fall of 1866; the UP reached the 100th meridian first. According to his biographer, Edgar Thomson "badly underestimated the resources and determination" of the UP's leaders. Nonetheless, the KP men remained buoyant. With new land grants from Congress (awarded in July), their line would now aim for Denver. The road's backers in St. Louis and Philadelphia had fresh hopes of securing another charter and further grants to extend their tracks to California along a route south of the combined UP and CP. Pushing the bill through Congress could cost upwards of $500,000 in bribes and free stock, but surely the country was big enough for another transcontinental line.[61]

As these wartime projects accelerated in 1866 with the peace, Tom Scott dreamed of the wealth he would extract from a third transcontinental venture, the Atlantic & Pacific, chartered that year with generous land grants from Congress.[62] James Eads too was seduced by that vision, writing a friend that "the A&P is a gigantic enterprise with such magnificent gov't aid that it cannot fail in good hands to be made immensely profitable."[63] The A&P promised to build its line from Missouri to California along a route south of the KP's proposed track. While debating further subsidies for the KP, Congress authorized an interstate bridge at St. Louis on July 25, 1866. Once built, that crossing could become a key link tying the PRR, the KP, and the A&P.

CHAPTER 3

Conventional or Radical

IN JULY 1867 CAPTAIN EADS unveiled his plans for a St. Louis bridge after months of work and consultations.[1] The drawings drew inspiration from an 1864 structure over the Rhine at Coblentz, Prussia. Eads had seen that three-arch rail bridge during a river voyage with Eunice that year.[2] Its 308-foot spans in wrought iron carried a double-track railroad. Coblentz amounted to a suggestion, however, rather than a full antecedent. Bare facts reveal that the Captain's proposal that July was audacious, even reckless. He had no experience in civil engineering, and his plans had no precedent in North America. Yet he proposed to enlarge and adapt for St. Louis a bridge that he had seen on his summer vacation. He did not lack nerve.

Even as the Eads design had its public debut, a national leader in the business of building bridges had already begun a St. Louis project. Lucius Boomer's career in the industry had started two decades earlier, and in 1856 his American Bridge Company had added iron bridges to its product line.[3] Boomer had undertaken design work and site investigations at St. Louis many months ahead of Eads. In August 1867, he unveiled plans for a largely conventional structure, proposing to use a patented design for iron trusses already well proven in bridges across the Midwest.

And with that St. Louis had a contest: the long-established Boomer advancing a proven design, and the challenger Eads, a neophyte in civil engineering, promoting a radical approach. The site and the river posed

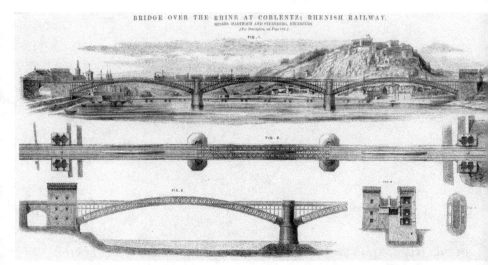

The Coblentz Bridge in Germany inspired Eads's plans for St. Louis. The two bridges shared many elements, yet the differences were striking too. When working in the comparatively shallow and placid Rhine, the builders used coffer-dams to hold back the river while laying the stone piers. Temporary falsework supported each arch during assembly. When locomotive weights outgrew its safe loadings, railway service moved in 1879 to a new crossing nearby. The adjacent town is today known as Koblenz. Engraving from "The Coblentz Railway Bridge," *Engineering* (London), June 7, 1867, opposite p. 586.

knotty problems for both projects. The elements that would determine which venture went ahead all lay in the future. The recent past suggested that it would not be Eads's. In June the Captain had sent his preliminary drawings to Jacob Linville at Keystone, seeking his input and improvements. After a thorough review, Linville declared that Eads's plans for a steel bridge with three arches were "entirely unsafe and impractical."[4]

Varieties in Play

In the first two years after the war, engineers and promoters offered five different plans to move trains, people, and roadway traffic across the river at St. Louis. The outliers need not concern us here, although the variety of designs testifies to a truth too easily forgotten.[5] Then as now, good engineering was fundamentally a creative endeavor. This bustling activity

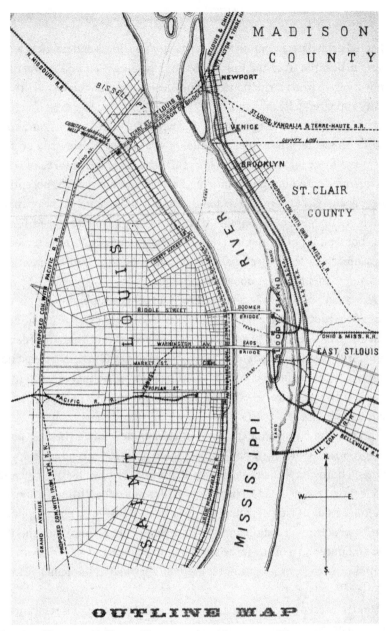

This detail from an 1867 map illustrates the rising competition among promoters to build a rail bridge for St. Louis. The Baltimore Bridge Company (Smith, Latrobe & Co.) advocated a crossing at the northern location shown here. Arousing little interest among railroads or financiers, that proposal quickly faded, leaving the field clear for Lucius Boomer or James Eads. The map also shows their plans for bridges into downtown St. Louis, as well as the ferry routes and railroads operating or planned on both sides of the river. Detail from a map published in *Alton & St. Charles County and the St. Louis and Madison County Bridge Companies Consolidated* (n.p., [1868]), author's collection.

reflected grim determination among the city's business and political leadership to best the river and link up to the railway age. Thanks to wartime innovations in iron production and truss design, that long-deferred dream finally appeared within reach.

The diversity in designs, however, reflected a deep division about *how* to cross the Mississippi. Engineers who focused primarily on the river—the force and scour of its currents, the destructive force of winter ice—advocated designs requiring few piers, which meant accepting longer spans. Others saw the cost and strength of the spans as the primary design challenge. Shorter spans were less expensive and posed fewer risks but would put more piers into the Mississippi. That in turn meant more obstacles to the free passage of currents, snags, and winter ice floes, not to mention steamboats and barges.

The conceptual variety also reflected a reality shaping many big urban infrastructure projects in nineteenth-century America. Most were built by profit-seeking promoters. City governments seldom ordered these large custom bridges, and they would not pay for them, at least not directly. Local business elites adamantly opposed tax increases to fund general infrastructure, so competition among promoters shaped design goals and results.[6]

In 1865, the city engineer, Truman Homer, developed the first postwar proposal. Prudently, he simply copied a design already proven in Great Britain: Robert Stephenson's Britannia Bridge (1850). It would have carried trains inside a large box girder more than fifteen hundred feet long, made mostly of iron plates and riveted I-beams. The simplicity of this structure—a long iron box—had some appeal, although the design amounted to a brute-force approach to bridging. It never advanced beyond drawings, as no promoter stepped forward to implement Homer's plans.[7]

Homer's design and accompanying report did clarify issues relevant for any project in St. Louis. He envisioned a crossing that combined a single-track railroad right-of-way (inside the box) with two lanes of roadway traffic outside the box, on the same level. Atop the box girder, pedestrians would enjoy a promenade deck. All were practical choices, and the promenade was a lovely amenity.

Perspective view of proposed Bridge across the Mississippi River at St. Louis

In Truman Homer's plans, trains would cross the river inside a box girder. Road-way traffic would travel in lanes outside the box, on each side. His design offered little clearance for steamers on the Mississippi, as the low piers saved costs. Here a solitary steamboat slips under the bridge thanks to its hinged smokestacks. Detail of a drawing by Thomas Schrader, 1865, MHS, F 11390.

Every other element in Homer's plans revealed a maze of problems. The drawings show a comparatively low, fixed bridge to save on construction costs.[8] Steamboat pilots and owners would surely have howled at that obstruction. Its location would have aroused more opposition. Homer's bridge aimed for the heart of downtown, putting all its rail traffic onto city streets. The drawings failed to show the inevitable mayhem among carts, carriages, pedestrians, locomotives, bolting horses, clanking trains, and shrill whistles, all coated in soot and smoke. The report predicted that the crossing would carry upwards of 240 trains a day. That traffic would have strained the capacity of a single-track bridge, while the trundling processions of freight and passenger cars in arterial streets would have overwhelmed the patience of downtown merchants and shoppers. Estimated at $3.3 million, these problematic improvements would have proven costly.[9]

The US Congress and the Western Bridges

In the summer of 1866, a decade after he failed to block the Rock Island Bridge, Jefferson Davis languished in the US military prison at Fort

Monroe while Congress debated legislation authorizing *eight* new bridges for railroads over the Mississippi. The war had erased most doubts about expansive federal powers to shape and grow the economy. The bill did include provisions to safeguard river commerce against undue obstructions by the new bridges. Even so, the contest between river and rail had turned into a rout.

The Omnibus Bridge Act grew out of the granger passion to transform prairie grasslands into wheat fields. It also reflected broad strategies of the country's most powerful corporations and politicians. The eastern trunk lines (or their subsidiaries and feeders)—the New York Central, Erie, Pennsylvania, and Baltimore & Ohio—began on the Atlantic coast and ended at the Mississippi River. The embryonic transcontinental railroads—the Union Pacific, the Kansas Pacific, or the Atlantic & Pacific—were projected to run westward from the Missouri River. The railroads between those two natural barriers drew far-off potentates into battles to control the region and its routes. Bridges over the two watercourses would be a primary weapon.[10]

Enacted into law on July 25, 1866, the Omnibus Bridge Act was the comprehensive legislation promised by its name. The statute authorized two new bridges over the Mississippi in its northern reaches, at Winona and Prairie du Chien in Wisconsin.[11] Those crossings would benefit Milwaukee, a port to the east on Lake Michigan. Three bridges connecting Iowa with Illinois—at Dubuque, Burlington, and Keokuk—received federal sanction. And the law provided for three new bridges connecting Missouri and Illinois. The crossings at Quincy and Hannibal would funnel the farming and ranching products of northern Missouri and even Kansas toward eastern markets. In all, the statute authorized five new bridges at latitudes and locations that effectively benefitted Chicago, with just one to boost St. Louis. Salting that wound, the carriers using those bridges were all tapping farming regions that had been tributary hinterlands to St. Louis before the war, when river steamers carried the region's freight and passengers. Then and since, boosters and historians have argued that railroads lifted Chicago's fortunes at the expense of St. Louis.[12] While that analysis has merit, matters would have unfolded quite differently if the Mississippi River at and south of St. Louis had

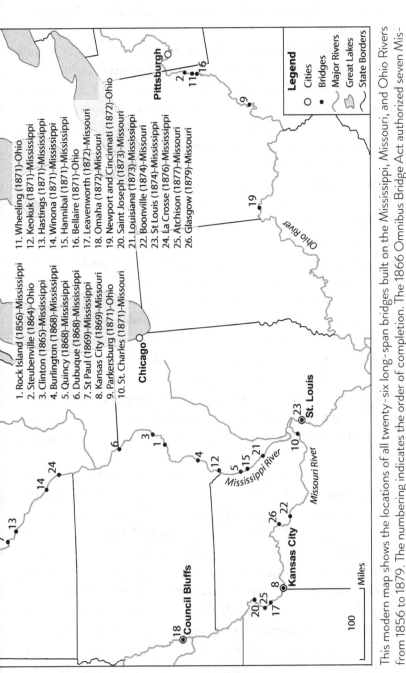

1. Rock Island (1856)-Mississippi
2. Steubenville (1864)-Ohio
3. Clinton (1865)-Mississippi
4. Burlington (1868)-Mississippi
5. Quincy (1868)-Mississippi
6. Dubuque (1868)-Mississippi
7. St Paul (1869)-Mississippi
8. Kansas City (1869)-Missouri
9. Parkersburg (1871)-Ohio
10. St. Charles (1871)-Missouri

11. Wheeling (1871)-Ohio
12. Keokuk (1871)-Mississippi
13. Hastings (1871)-Mississippi
14. Winona (1871)-Mississippi
15. Hannibal (1871)-Mississippi
16. Bellaire (1871)-Ohio
17. Leavenworth (1872)-Missouri
18. Omaha (1872)-Missouri
19. Newport and Cincinnati (1872)-Ohio
20. Saint Joseph (1873)-Missouri
21. Louisiana (1873)-Mississippi
22. Boonville (1874)-Missouri
23. St Louis (1874)-Mississippi
24. La Crosse (1876)-Mississippi
25. Atchison (1877)-Missouri
26. Glasgow (1879)-Missouri

Pittsburgh

Chicago

St. Louis

Council Bluffs

Kansas City

Mississippi River

Missouri River

Ohio River

100 Miles

Legend

○ Cities
● Bridges
〜 Major Rivers
Great Lakes
State Borders

This modern map shows the locations of all twenty-six long-span bridges built on the Mississippi, Missouri, and Ohio Rivers from 1856 to 1879. The numbering indicates the order of completion. The 1866 Omnibus Bridge Act authorized seven Mississippi bridges shown here: Burlington (4), Quincy (5), Dubuque (6), Keokuk (12), Winona (14), Hannibal (15), and St. Louis (23). The map reveals how thoroughly geography and technology combined to boost the fortunes of Chicago. No bridges were built south of St. Louis until the crossing at Memphis in 1892. Map by Jade Myers.

been as easy to bridge as it was between Dubuque and Keokuk (see illustration on p. 67).[13]

The new bridges all served railroads, but few were built or owned directly by the carriers. Six of the eight Mississippi spans authorized in 1866 were the work of independent bridge companies.[14] Such a firm would secure an authorizing charter, raise the capital, and operate the bridge. It contracted with a far-off manufacturer like American Bridge or Keystone to make and erect the superstructure. Local worthies, railway officers, New York bankers, and European investors financed these independent bridge companies. The promoters anticipated generous profits from levying a toll per car or from lease deals with railways using the span. By avoiding direct ownership, the railroads sidestepped a major capital burden.[15]

The independents all followed the same business plan. They would quickly build a utilitarian crossing, then extract tribute from the local, regional, and continental trade that everyone involved confidently projected after the opening day. This business model all but required cheap bridges, manufactured and erected in a hurry. Like the western railways they served, these structures exemplified the "good enough" philosophy dominating American engineering of the era. Their promoters had no aesthetic considerations, no eye to longevity, little regard for maintenance costs or problems, and scant concern that growing traffic and heavier loads would render initial designs obsolete within decades.

Congress and geography then raised another obstacle at St. Louis. Striking a technological compromise between rail and river traffic, the seven new bridges over the upper Mississippi could use the same kind of pivoting drawspans that had been employed at Rock Island and Clinton. The legislature prohibited such a compromise at St. Louis. Any bridge there needed high fixed spans. The Omnibus Bridge Act required at least fifty feet of vertical clearance to allow river steamers with their high smokestacks to pass unimpeded. This meant that Truman Homer's plans were out. (The drawspan bridges were allowed a height of just twelve feet.) In a second provision to protect river interests, Congress required that any bridge at St. Louis would have at least two spans that provided 350 feet of horizontal clearance between the piers or abutments. Or the legislature would accept a single channel span giving 500 feet of navigable width for river traffic.[16]

To put these terms in perspective, on the date that President Andrew Johnson signed the statute, the *record* span for a rail bridge anywhere in North America was 320 feet at the two-year-old Steubenville Bridge.[17] In protecting the interests of steamboat operators, Congress created a difficult challenge for every promoter at St. Louis. Even if such a bridge proved technically feasible, these provisions ensured that it would cost far more than the seven upstream crossings authorized in the act. To carry the unprecedented spans, the river piers would need exceptional strength and depth. The time required for construction would prove burdensome as well. Furthermore, the act explicitly banned a suspension bridge, deemed unsuitable for railway loadings. In these clauses, the statute demonstrated a Congress flexing new muscles, wielding the law to force innovation. It was unclear, however, whether any bridge builder in 1866 could meet the congressional requirements for St. Louis.

Those requirements had a curious genesis. They were drafted initially by a special committee of the St. Louis Chamber of Commerce. Its charge: to craft recommendations for Congress to shape a rail bridge, while protecting the city's river commerce. In its final guidance, the chamber requested a truly heroic (or perhaps impossible) congressional mandate for "one span of 600 feet or two spans of 450 feet each."[18] On its face, that language looked like an outrageous attempt by the rivermen to block *any* bridge over their Mississippi at St. Louis. Appearances can deceive, however, for the committee chair was Captain James Eads. Deeply committed to western railroad ventures by the spring of 1866, Eads had begun to plan the great steel bridge that would link those lines to the nation's growing rail network. At that time, the source, cost, and quality of its steel were all questions without answers.

Rival Railroads and a Bridge at St. Louis

While any bridge at St. Louis would prove difficult to build, railroads on both sides of the river had powerful reasons to cheer those projects. By 1865 St. Louis was the nodal point for six lines, three terminating on each side of the Mississippi.[19] Three would play key roles in the contest between Lucius Boomer and James Eads to build a bridge. On the east side of the river, the St. Louis & Alton ran to Terre Haute, Indiana.[20] On the west side, the Missouri Pacific had opened its line across the state to Kansas City by 1865. The North Missouri would complete its own line to that city by 1868.[21] Thanks to that competition, the managers and directors of the MoPac chose to back Lucius Boomer's bridge, while the North Missouri linked its destiny to James Eads.

Chartered in 1852, the North Missouri aimed initially to become a commercial alternative to the Missouri River, paralleling that waterway into Iowa farming territory. The line suffered badly from Confederate guerrilla raids in the summer of 1864. The expense of rebuilding, coupled with its strong prospects after the war, made it a ripe target for a takeover bid. In June 1867, a syndicate that included James Eads, St. Louis colleagues, and New York bankers acquired "ultimate control" of the company by purchasing at steep discounts its unsold first mortgage bonds.[22] Captain Eads had become a player in the high-finance gambits of Wall Street. Once its tracks reached Kansas City, the North Missouri would connect with the Kansas Pacific, controlled by the Philadelphians. From its eastern end in St. Louis, the line could funnel traffic across a new bridge and onward to the Pennsylvania Railroad.

As those developments played out in the West, Edgar Thomson and Tom Scott labored to improve the PRR's connections to St. Louis. The city's longtime commercial prominence made it a compelling destination. Back in 1852, President Thomson had sketched his ultimate vision for his young railroad in a bond prospectus aimed at London financiers. Connecting with western lines (not yet built), the PRR would take the freight and passenger traffic of both Chicago and St. Louis and deliver it to the Atlantic Seaboard.[23] For 1852, this was a bold bet on the expansive future of the American West.

By 1867 those connections existed, including the St. Louis & Alton. But Thomson could not take them for granted, especially because his PRR only held a half interest in that carrier. Worse, the other half had come under the control of promoters associated with the New York Central. Archrival to the PRR, the Central sought to improve its own links to St. Louis. Each of those two powerful trunk lines wanted the revenues that traffic to—and through—St. Louis would provide. Equally important, each wanted to block its main competitor from taking that business.

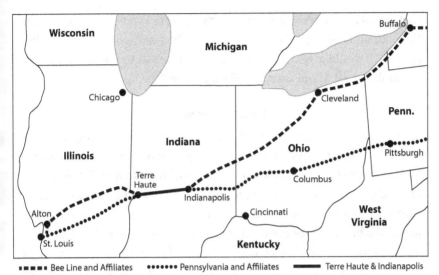

This map shows how a small link had huge strategic importance. The Bee Line promoters wanted the Terre Haute & Indianapolis to complete their proposed route from Cleveland to St. Louis. Edgar Thomson blocked their efforts by making that short line part of the Pennsylvania Railroad's route from Pittsburgh to St. Louis. The PRR had also acquired a half interest in the St. Louis & Alton, connecting Terre Haute and St. Louis (shown here, it passed through Alton). Because the Bee Line promoters had the other half, the PRR then financed a new line, the St. Louis, Vandalia & Terre Haute into East St. Louis (also shown here as a PRR affiliate). Creating through routes out of short lines often proved a messy business. Map by Jade Myers.

In the summer of 1867, this contest focused on two assets: an existing short line in Indiana and the bridges proposed for St. Louis. A consortium with ties to the Lake Shore line and the New York Central planned to stitch a route from Cleveland to St. Louis. To that end, these Bee Line promoters tried in June 1867 to take control over the Terre Haute & Indianapolis Railroad.[24] A leader of the Bee Line group was Amasa Stone Jr., the brother of Lucius Boomer's longtime partner in his bridge-building business.[25] Amasa was a strong, if silent, force behind Boomer's interest in building a St. Louis Bridge. That fact alone provided sufficient motive for Thomson and his Philadelphia friends to back James Eads. Those men could foresee many other upsides in this alliance.

St. Louis as a destination and the Bee Line as a rival were so important to Thomson that he also aided construction of a direct line, the Vandalia, that would run between East St. Louis and Terre Haute, which were also the endpoints of the St. Louis & Alton.[26] That new route (it opened in July 1870) would give the PRR untrammeled access to the leading city of the Midwest—if Thomson could block Boomer's bridge.

Boomer's Plans

Lucius Boomer's proposed bridge for St. Louis has been misrepresented, and his motives impugned, in nearly every account written about James Eads since 1947. That year Florence Dorsey published a fanciful biography of the Captain. To bolster his greatness, she presented Boomer as a villain. Dorsey claimed that Boomer was a partisan of Chicago who "never meant to build a bridge, but only to make sure that none would be put up at St. Louis for a generation, at least."[27] Accounts published since Dorsey's fable repeat this implausible fiction.[28] By contrast, this study takes Boomer and his bridge seriously. In the first place, his was a sophisticated proposal, a real option at one time for St. Louis, and in certain respects a better choice than the Eads design. Furthermore, when placed alongside Boomer's alternative, the St. Louis Bridge comes into focus, a structure with pros and cons rather than the sui generis product of unique genius.

In 1867 Lucius Boomer ranked among the most successful bridge builders in the country. For more than thirty years, he produced spans of all types in wood, iron, and eventually steel for common roadways and railroads across the country. He entered the business in 1846, benefitting from close family ties to William Howe, patentee of the proven wood-iron-composite Howe truss. Four years later, Boomer relocated his American Bridge Company to the growing town of Chicago. The firm did impressive work. Its 1854 bridge for the Illinois Central at Peru, Illinois, used a series of Howe trusses to reach thirty-five hundred feet across the Illinois River valley. The double-deck bridge (a railway deck above a roadway) required a million board-feet of lumber, its trusses and planking "all worked up in Chicago." At that time, American Bridge had seventy-five truss bridges under contract, including the Rock Island Bridge, which would meet its match in the *Effie Afton*. By 1856, the company reported revenues exceeding $1.2 million ($44 million in 2023) and a payroll of 650 men and boys.[29] Boomer's core accomplishment lay in transforming bridge construction from a craft, done largely in the field, into a factory product.

Lucius Boomer entered the drama at St. Louis in November 1866 at the invitation of Norman Cutter. A Missouri state senator, Cutter had secured a charter for a St. Louis bridge company from the Missouri legislature in February 1864. The venture required an Illinois charter as well, which it received a year later. Boomer was a welcome addition to Cutter's St. Louis promoters. A man who built bridges was just what they needed. Eads had not yet entered the picture.

A flurry of action in the Illinois and Missouri legislatures between December 1866 and March 1867 resulted in a *second* company chartered to build a St. Louis bridge. Later accounts cast Boomer as the architect of this anarchy, but all were written by Eads partisans. The confusing thicket of accusations is complicated and inconclusive. We know that Lucius Boomer was sent to the Illinois legislature by the Cutter company, instructed to seek improvements to its original and imperfect charter. On February 21, 1867, he came away with a new, much more supple authorization for a new company.[30] His motive is unknown, but one explanation seems straightforward. Cutter's venture had already promised the contract for its superstructure to Boomer. Boomer probably wanted to boost his compensation by taking the promoter's role for himself in this newly chartered company. That would leave Cutter and his colleagues in the lurch, holding the imperfect charters of the older company.

Crucially, the Illinois legislature included in the new charter a clause giving Boomer's new firm a twenty-five-year monopoly on the right to operate a bridge from St. Clair County, Illinois, over the Mississippi. By its practical effect the clause blocked construction of any other rail bridge into downtown St. Louis. These kinds of legislative grants had a long history and a legitimate purpose.[31] They encouraged and remunerated the investors who put up the capital to build these public infrastructures. And they blocked late-arriving "green-mailers" from projecting rival crossings solely to extract payoffs from the first chartered venture.

There should be little doubt that Boomer wanted very much to build *his* bridge at St. Louis. That feat would buy incalculable prestige for American Bridge, already a leader in the industry. By controlling the crossing at St. Louis, Boomer would anchor Amasa Stone's plans to create a new trunk line connection via the New York Central to the East. In November 1866, Boomer had taken the preparatory step of hiring, at his personal expense,

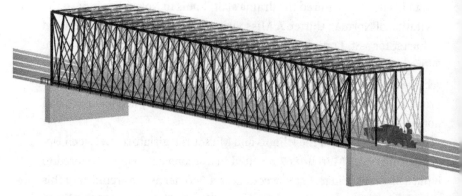

One of the two main channel spans proposed by Lucius Boomer and designed by Simeon Post for St. Louis. Along the centerline, a single right-of-way with four rails would have accommodated three track gauges then used by railroads serving the city. A Post truss flanked each side of the railroad. Outside those trusses, a two-lane roadway on each side would carry common roadway traffic. The design placed another Post truss outside each roadway, with a sidewalk outside the truss. As indicated by the locomotive shown here, Post scaled his patented design to massive proportions to carry the projected loads across the unprecedented 368-foot channel spans. Drawing by Richard Anderson.

two eminent engineers to survey river conditions and building sites in St. Louis.[32] Soon thereafter, he engaged Simeon Post to design the bridge. A founder of the American Society of Civil Engineers, Post had a long resume, including a stint as chief engineer of the Ohio and Mississippi line that connected St. Louis to the East. In 1863 he patented an iron truss suited for railway bridges.[33]

Unveiled in August 1867, the Boomer-Post plans for St. Louis largely reflected the new design orthodoxy for iron bridges that Keystone had originated at Steubenville, an approach adopted by many of Keystone's competitors. The design combined eight spans of varied lengths to cross the river. Two channel spans near the St. Louis levee, each 368 feet long, provided the clearance for river traffic mandated by the Omnibus Bridge Act. This plan put five piers into the river's flow and two on the levees. For each span, four parallel trusses would support a single deck to carry road and rail traffic. Although the four trusses in each span were unique, in its other aspects the superstructure design demonstrated the emerging paradigm of pin-connected railway bridges in iron.[34]

THE GREAT KANSAS AND MISSOURI BRIDGE.

Lucius Boomer's plans for St. Louis were echoed in the structure his company built across the Missouri River at Leavenworth, Kansas. Begun in July 1869 and completed in March 1872, it used the truss design patented by Simeon Post. Iron pneumatic pilings driven down to bedrock supported a superstructure fifty feet above the river's surface. Cut stone laid between the pilings and the stone ice-breakers improved the foundation's strength to withstand the wild Missouri River's rampaging spring currents, winter ice floes, and year-round burden of driftwood—which included whole trees. The structure proved remarkably durable, carrying auto traffic into the 1950s. Engraving from "Bridge Over the Missouri River," *Scientific American*, Mar. 15, 1873, 167.

To build investor interest for his venture, Boomer took the unusual step of hosting a convention of eminent civil engineers in St. Louis in August 1867. These distinguished men gave Boomer just what he sought, which isn't surprising since he was paying consultancy fees to six of the eighteen conventioneers.[35] He covered expenses for everyone, took stenographic notes of their deliberations over three days, then published a 100-page report signed by twenty-eight luminaries in civil engineering.[36]

The report offered facts worthy of emphasis. Post's trusses were already well proven in railway service. The loads and stresses in each truss were readily calculated, allowing a confident margin of safety. To build the essential piers in the river, Boomer proposed to use pneumatic piles. Pioneered in an 1861 bridge over the Harlem River in New York City, the method used a hollow cylinder of rolled iron to form a pier. Its descent to bedrock was aided by the use of compressed air. While Boomer's convention met, Octave Chanute was driving pneumatic piers into the tumultuous Missouri River at Kansas City, the first stage for the first bridge over that waterway. Boomer would employ the pneumatic pile method and Post trusses a few years later in the Leavenworth Bridge. In short, Boomer's report used the eminent authorities of American engineering to state that his design choices rested on proven technologies that were appropriate to the site and entirely feasible. Surely wealthy capitalists who considered his proposal would perceive these virtues too.

Eads at the Edge

The plans that James Eads unveiled in July 1867 broke in every respect from Linville's Steubenville crossing and Boomer's proposal for St. Louis. Before detailing the Captain's plans, we need to sketch how and when he entered this new field. In April 1866—before the competition among bridge promoters heated up—Eads had led the committee of the St. Louis Chamber of Commerce that offered guidance to the US Congress for a bridge at St. Louis. The committee's requested span length, six hundred feet, appears inconceivable on its face. But throughout his career, Captain Eads demonstrated blithe nonchalance, even contempt, for engineering precedent. Another recommendation from the Chamber indicates that he had begun even then to contemplate taking a direct role in building a St. Louis bridge. That is the best explanation for a crucial clause specifying that any bridge needed to give fifty feet of vertical clearance for steamer traffic, as measured *at the center* of the span. That curious stipulation suggests that Eads was already contemplating an arch bridge for the city.[37] Otherwise, the caveat made no sense, served no purpose. No other promoter would offer an arched design.

Boomer secured the Illinois charter for his new firm on February 21, 1867, immediately undercutting much of the actual and potential value of Cutter's original company. Eads picked that moment to act, encountering no opposition when he led new men into a stockholders' meeting of the Cutter company. On March 23, the stockholders elected Charles Dickson as the new president and Eads as the new chief engineer.[38] Cutter's company had become Eads's. With his old friend Dickson, Eads moved very quickly to make it a going concern. By May 1, Eads had named Henry Flad, an émigré with engineering training in Germany and a long resume in America, as first assistant engineer. And the firm had a new board of directors to represent its new stockholders. The most important addition was Tom Scott.[39] What Scott wanted from this connection will await a later chapter. What he brought was immediately valuable: connections to New York financiers, to the Keystone Bridge Company, and to the Pennsylvania Railroad. After Eads took over the Cutter company, Keystone's chief engineer, Jacob Linville, signed on as a consultant.[40] There was no better bridge man in the country.

Into June 1867, Eads and Flad worked up design drawings. The plans of that summer included all the major elements that would endure in the bridge as built.[41] Eads proposed a bridge of three arches, built largely of steel. The central arch was to be 515 feet in length. The others would have clear spans of 497 feet.[42] Those structural elements would carry a wide roadway above a dual-track railway deck. There was nothing like it in the world. The span lengths alone far exceeded any truss or arch bridge to that time in North America. While the conceptual aspects demonstrated inspired creativity, every detail in the superstructure required much further definition and clarity. For that reason, in early June 1867 Eads was pleased to send his preliminary tracings to Linville, asking for "any suggestions to aid me."[43]

Linville's fourteen-page reply amounted to a tutorial by a proud and accomplished designer. He outlined all the reasons why Eads's arches were a bad idea. He did praise "one desirable feature," the unobstructed views from the upper roadway. Every other element was problematic in Linville's view. The steel chords in each arch were far too light. If made to the specified dimensions, Linville believed, they would prove nearly as flexible as suspension cables. By stiffening them with struts and braces, the plans created inflexible arches that would suffer unacceptable rise or fall at the crown of each arch caused by daily and seasonal temperature variations. The planned use of steel was shot through with problems in availability, cost, and quality. Linville summarized his design review with this blast: "I cannot consent to imperil my reputation by appearing to encourage or approve of its adoption."

The letter went on to outline Linville's own ideas for a bridge at St. Louis. He claimed to have studied the problem since February 1866, a pointed contrast to the novice he was addressing. Linville closed with stirring notions. A bridge for St. Louis would "require all the ability, experience, and research we can command." Such a bridge "would stand for ages, an enduring monument to the public spirit of its projectors, and an honor to the profession and the country."[44]

The backstory here is mostly inference yet seems clear-cut. Eads valued Scott's participation for all the connections he brought. Scott likely insisted that Linville come aboard to bolster the engineering. Eads probably welcomed Linville's expertise. After all, his Steubenville Bridge held the record as the American rail bridge with the longest span. Scott and Linville had every reason to hope that Eads, once shown the shortcomings in his own approach, would turn the superstructure design over to Keystone. That would be the smart thing to do. And as a leading stockholder in the St. Louis Bridge Company, Eads could retain much of the promoter's role, renown, and reward. Everyone would make good money. Matters did not, however, take that path. Before the month was out, the board of directors of the St. Louis Bridge Company fired Jacob Linville from his post as consulting engineer. Within the week they accepted Eads's preliminary design. If he lacked relevant experience, Captain Eads clearly inspired confidence. Still, his associates may have harbored private doubts.

Ample evidence supports Linville's verdict that the plans Eads offered in the summer of 1867 were deeply flawed, woefully incomplete, and perilously venturesome. Or in Linville's phrasing, "entirely unsafe and impractical."[45] His critique focused on the steel superstructure, but Eads's plans for the river piers were also problematic. He specified just two piers, in contrast to Boomer's five. As Eads knew from his salvage career, those massive stone piers were an ideal choice to resist flood conditions and withstand ice gorges. But in the summer of 1867 the Captain scarcely had feasible notions of how to build them down to bedrock. In his first report as chief engineer (June 1868), he offered a makeshift plan for building the piers. A year later he would abandon that approach.[46]

His preference for just two piers dictated the 500-foot arches. Foreseeing the loads and stresses they would bear, Eads specified steel for the load-bearing chords in those arches. He had gained insight into steel during the war while designing his patented turrets. Its dual strength in tension and in compression captivated him. But his enthusiasm exceeded his knowledge. Eads claimed that the superstructure as then designed (June 1868) would safely carry both the roadway deck when densely packed with standing crowds of people across its length and the railway deck with its dual tracks "covered from end to end with locomotives."[47]

There is no point in evaluating the accuracy of these claims, dubious as they are. Any experienced civil engineer would flatly reject all this as ridiculous. The bridge would never carry such burdens, so why go to the trouble and, crucially, the expense of designing and constructing such an overbuilt structure? To practitioners, a bridge was a tool, not a demonstration of creativity. In this utilitarian view, any crossing should bear its expected daily loads while offering a calculated margin of safety.[48] For Eads, the St. Louis Bridge would be a monument for the city and proof of his genius.

When Linville offered his skeptical views of steel, the Bessemer process—the pioneering method of making steel in industrial quantities and quality—was well advanced in Great Britain. By contrast, the first American Bessemer plant had just opened, firing its initial, imperfect batches in May 1867.[49] A man who watched developments in steel very closely, the PRR's Edgar Thomson believed that Bessemer steel from any source would lack the requisite strength and assured quality needed for the bridge. Thomson thought Eads would need crucible steel, then available in the required quantities from only two sources worldwide: John Brown's Atlas Works, in Sheffield, England, and the Krupps plant in Prussia.[50] Despite Eads's confidence, structural steel had not yet been used to construct a bridge, building, or ship anywhere in the world.

In advocating for his novel steel arches, Eads was rejecting more than just iron trusses. By 1867, American Bridge, Keystone Bridge, and other companies were all joint creators of a distinctive American style for iron bridges.[51] Their new approach aggregated new knowledge in these new firms.[52] They standardized their products, design methods, tooling, and production processes, thus amortizing costs and boosting volume.

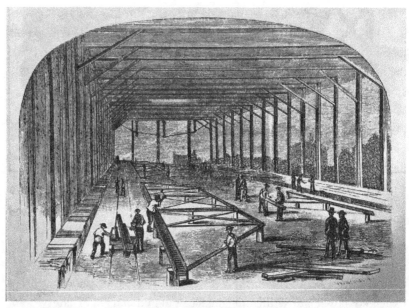

In January 1873, *Lippincott's Magazine* published "Iron Bridges, and Their Con-struction." The article described the high-tech revolution then remaking rail-roads, cities, and travel in America. This engraving shows men test-fitting com-ponents of a new truss bridge in the erecting shed of the Phoenix Bridge Company in Pennsylvania. Image from Howland, "Iron," 46.

By test-fitting their pin-connected trusses at the factory, they could assemble their iron truss bridges at the customer's site in just days or hours.[53] With that distinctive feature, their products became known as pin-connected bridges. Thanks to these standardizing methods (mirrored at American and elsewhere), Keystone would quote prices for its bridges by the lineal foot, as if it sold planed wood or muslin fabric.[54]

Given all his novel design elements, Eads would have to create new and demanding foundations in the knowledge required to make a steel arch bridge. Lucius Boomer or Jacob Linville would enhance and elaborate on established designs and processes even as Eads assembled his own team and developed its new expertise in design, fabrication, and assembly from a standing start.

In so many ways, then, the right solution for St. Louis in the fall of 1867 was Lucius Boomer's. Unlike Eads, he had a long resume in bridg-

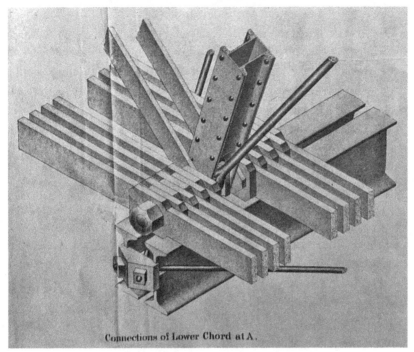

Connections of Lower Chord at A.

The components and connections in a Post truss. The beams and chords attained the requisite strength by using multiple identical components. Designers like Linville and Post preferred this approach, rather than making these parts to much larger cross sections. Heavier, bigger parts were difficult to roll or forge and more likely to have hidden flaws inside the metal. A single pin (with its hexagonal nut) joins twenty components here. Detail of a drawing in the author's collection.

ing. The Post design had proven its worth and value. Its strengths were easily computed. Its ironwork required simple elaborations of established techniques. Eads, not Boomer, was the late-arriving schemer. Furthermore, Boomer's venture held the exclusive right for a bridge into the city, thanks to the Illinois charter that had created the company. The overflowing confidence and charisma that Eads projected scarcely counterbalanced that asset.

Boomer versus Eads

After firing his consulting engineer before the end of July 1867, James Eads turned immediately to the Boomer challenge. In a savvy play for public support, he secured permission from the St. Louis City Council to start construction of the west abutment. Given its location on the levee, this was the easiest starting point. Bedrock lay close to the surface; the work proceeded on dry land. Just as important, every St. Louisan could see the project beginning, there on the doorstep of the city. As matters played out, those excavations stretched out for six interminable months before Eads's men could lay the cornerstone. An obstacle both perverse and ironic blocked progress. The digging uncovered the remnants of three burnt and wrecked steamboats, lost in the great fire of 1849.[55] In the aftermath of that catastrophe, the city had covered the wrecks with a levee expansion. Now Eads unexpectedly found himself back in the salvage business. Despite that frustrating delay, the hole showed everyone in town that the St. Louis Bridge Company was a going concern.

While the Captain's men dug away, Boomer's advisory board convened in St. Louis on August 21, 1867. The group included most of the country's leading civil engineers. Boomer had clear goals for the convention: to secure the imprimatur of the engineering establishment, give his St. Louis venture an air of inevitability, attract investors, and show the Eads project to be an implausible also-ran. His experts divided into groups to discuss the specialized issues entailed in this demanding project. Three committees delivered predictable reports: Post's superstructure was admirable; the bridge would not overly obstruct steamboat navigation; and the traffic crossing the new structure would provide ample toll revenues, sufficient to pay a profit to investors and service the bonds required to build a bridge estimated to cost $6.5 million ($136.4 million in 2023).

A fourth committee took up more difficult topics and reached ambiguous conclusions. This Committee on the Regimen of the River and the Character of its Bottom assembled useful information on the great floods that recurred typically in late June, caused by snowmelt 2,300 miles away in the Rockies. It discussed the "immense rafts of drift wood" carried in flood conditions and the threat from winter ice jams, which could cause an overnight rise of eight feet in the height or stage of the river at St. Louis. Drawing on the work of city engineer Truman Homer, the committee reported the results of four test borings, conducted over the previous winter, to determine the qualities of the riverbed and the depth through any sand or other "deposit" to reach bedrock. Just off the city wharf, the river bottom *was* bedrock, 23 feet below the low water mark. Moving 500 feet east into the river channel, the next test found a layer of deposit 50 feet thick that overlay bedrock there. Another 500 feet toward the Illinois shore, the deposit layer was 80 feet thick. The fourth bore, roughly 200 feet from Illinois, had to be abandoned before the drill hit rock. Working out on the frozen river, the survey crew fled for their lives as the ice of late winter began to break up.[56] So this committee and all other bridge advocates had an incomplete understanding of the river bottom and its substrata.

The convention adjourned after six days, having ably served Boomer's purposes. These eminent engineers aimed an indirect blow at the Eads design, arguing that Post's two main channel spans, each 368 feet long, together would cost half as much per lineal foot as a 520-foot span.[57] Half the cost and far less risky than Eads's untried approach. In short, Boomer's men offered reasonable conclusions on the merits. While Eads's name never appeared in the convention's report, none could doubt the target of this passage nor its attempt to discourage his backers: "In a Convention like that which has been assembled, the eccentricities of even the greatest minds would have been brought down to the consideration of the subject in its most practical form . . . restrain[ing] all tendency towards erratic but brilliant ideas. This consideration will have its due weight with capitalists who are also to perform their important part in the construction of the bridge."[58]

The Battle Joined . . . and Won

While many American engineers dismissed Eads's bridge, that verdict was not universal. In September 1868, the brilliant editor of the London journal *Engineering* published his views on Boomer's convention of engineers and on Eads's design. Zerah Colburn advocated for creative design based on rational principles. Who cared that the convention found no precedent for 500-foot spans? "If engineers waited for precedent, no great works would ever be carried out." Colburn praised Eads's "steel arched viaduct" as destined to become "one of the noblest monuments of the engineering skill of the nineteenth century."[59] The editor quietly overlooked the sobering fact that in his own adopted country, the British Board of Trade had explicitly banned the use of steel in bridges. Even in the birthplace of

Bessemer steel, the material was too novel, its qualities and durability largely unknown and unknowable. The ban remained in force until 1877.[60]

Chief Engineer Eads and his colleagues in the St. Louis Bridge Company took imaginative steps to counter Boomer's many advantages. In August 1867, a second assistant engineer joined the payroll. Charles Pfeifer was young, only 24, another well-born émigré from Bavaria. While Flad brought decades of experience to the project, Pfeifer's special talent lay in higher mathematics, then known as "the calculus." Before emigrating, he had served on the design staff for the Coblentz crossing that Eads took as a model. Using his advanced training and developing his calculations from equations worked out for Coblentz, Pfeifer's work in St. Louis would center on calculating the stresses that the bridge would have to bear.[61]

The Captain's plans went on public display in late September at the Merchants Exchange. The *St. Louis Republican* praised the drawings: "The steamboat men seem much pleased with its great arches, and the fact that but two piers will be placed in the channel."[62] Perhaps that was true. The *Republican*'s editor had quietly received a gift of five thousand dollars in bridge company stock, ensuring a stream of favorable publicity.[63] Other local editors enjoyed similar generosity.[64]

In an aggressive play for public support, Eads and his allies used newspapers to wage a campaign of accusation against his rival. Whatever his motives in St. Louis, Boomer could not deny that he was from Chicago. Eads's friend Henry Blow spun that fact into a dark conspiracy. Speaking to the St. Louis Board of Trade, Blow chastised the people of St. Louis who would "see themselves robbed of all the fruit that is to come from this enterprise [the bridge] by an overgrown and voracious rival city; a city that sends a man here for the purpose of prostrating this very enterprise; a city, sir, that is willing . . . to pay a quarter of million of dollars . . . if by that means they could postpone the building of the Bridge for five or six years!"[65] There is no record that any payment or bribe had actually been bruited about, or if one had, by whom, to whom, or for what end. The accusation itself provoked and agitated the boosters of St. Louis, which is to say much of the town.

Both companies bolstered their boards for battle. Boomer enlisted a new president for his venture. Daniel Garrison, the head of the Missouri Pacific, was nobody's patsy. His road was a serious rival to Eads's favored line, the North Missouri Railroad.[66] A phalanx of industrialists, railroad men, and bankers filled out Eads's board. Tom Scott remained a director even after Linville's discharge.[67] After an October meeting with his principals in Philadelphia, Andrew Carnegie wrote to Eads that Thomson and Scott were convinced that "you and your associates were *the* men of St. Louis to whom the Penna RR should adhere."[68]

That assurance mattered, as Linville's scathing critique surely had gotten the attention of Thomson and Scott. Still and all, they had reasons to remain aligned with Eads. In the first place, he was Scott's friend. Furthermore, Boomer had the backing of rivals to the PRR, sufficient cause to stay with Eads. Carnegie's letter tried gently to move the Captain away from his bet on steel. And he conveyed that "Mr. Thomson beleives [*sic*] you will find it necessary to modify the present plans as you proceed." In any case, such design questions would matter only after the Eads group bested Boomer. If Eads did prevail, Thomson, Scott, and Carnegie could foresee time and opportunity to persuade him to adopt a conventional design. That was the sensible thing to do. With the PRR against him, Boomer stood little chance of winning key railroad allies needed to build his bridge.

Although the fight between the rival companies had moved into a courtroom by January 1868, Boomer was losing interest. A month earlier his Cleveland friends, Amasa Stone and others, had lost out to the PRR in crafting the alliances needed to assemble their through line from Lake Erie to St. Louis.[69] Furthermore, Boomer had a big factory to run back in Chicago, and new jobs beckoned across the country. In September he had contracted to build a bridge nearly as prestigious as the St. Louis crossing. The Union Pacific had inked a deal with American Bridge to connect Omaha to Council Bluffs. This Missouri River crossing would become the culminating link in the transcontinental line.[70] There would always be more bridges for Boomer, while Eads had staked everything on a big win in his hometown. While Boomer moved on, Garrison would bear a grudge against Eads and his bridge for years to come. He could not countenance Eads's determination to push the North Missouri line into MoPac territory.[71]

As 1868 opened, a hard winter freeze struck the region. On January 11, the *Missouri Democrat* bore the headline "River Crossing Impeded . . . Give Us A Bridge At Once." Only one ferryboat could push across the ice-choked river that day, leaving hundreds of arriving rail passengers stranded in East St. Louis. For two days the mails did not move across the river in either direction. The article quoted a stranded traveler as saying, "Why, in God's name, don't the people build a bridge and cure this thing, instead of quarreling like a pack of damned fools?" Travelers arrived, got caught, and "go away cursing a city that don't know enough to take care of their own interests."[72]

The impasse was finally broken in February 1868. James Harrison, a prominent new director of Boomer's firm and a St. Louis industrialist, had opened talks with Eads in Washington, DC, a month earlier. Those conversations ended in St. Louis with a straightforward agreement to merge the two firms. On its face, the compact foresaw a merger of equals, with the consolidated company free to adopt any plans it desired. In reality, Eads and his associates carried the day. In exchange for a cash payment of $150,000 ($3.3 million in 2023), Boomer withdrew from the combined firm. The money recompensed his expenses and trouble. It also reflected the value of Boomer's Illinois charter, with its exclusive right to operate a bridge from St. Clair County for twenty-five years.[73] With the consolidation, Eads gained the freedom to alter the design and location of his bridge. By then, however, the Captain and his associates had committed themselves to his singular vision. "Drive on" remained his motto.

Why did the neophyte win in this contest with the country's leading bridge builder? Finance mattered crucially. Eads had close ties to New York money; Boomer did not. A leading New York financier, Edwin D. Morgan (concurrently a US senator and twice head of the Republican National Committee), wrote to congratulate Eads in March 1868 for vanquishing Boomer.[74] Ties to Tom Scott and the PRR provided other trump cards. Winning meant less than it might appear, however. In vanquishing his rival, James Eads had simply earned the right to try. His venture faced towering challenges at the site and in the river. Also unresolved were the problematic aspects of his design (highlighted by Linville and Thomson) and the murky unknowns that Boomer's convention had detailed. Failure in any single element would crash the whole effort.

The people of St. Louis desperately needed a bridge, but they were hardly sanguine about Eads's chances of success. After the consolidation, the St. Louis Bridge Company tried to raise new capital by offering stock

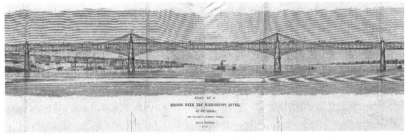

John Roebling developed a series of designs for St. Louis, using his suspension form to carry a double-deck bridge with stiffening trusses that were supported by shallow arches. He also devised towers that required far less mass and height than those planned for his Brooklyn crossing. Unlike Truman Homer, whose effort was rather slapdash, Roebling completed all the design work, calculations of loads and stresses, weight totals, drawings, and cost estimates. He hoped fervently to build a St. Louis bridge. His son published his data and drawings after his untimely death in Brooklyn in July 1869. Most of that booklet's illustrations were formal engineering plans. This view added an element of salesmanship, encouraging St. Louisans to envision this lovely amenity in their city. Detail from Roebling, *Long*, 70, Special Collections, Lehigh University Libraries, Bethlehem, PA.

directly to the public. For all its ties to Wall Street, the venture needed millions in new funds. For a month, "great efforts were made to obtain subscriptions large and small" across the city. No one signed up.[75]

Utter self-assurance had carried Eads through everything he had achieved since his salvage days. Now it guided him back to Congress. In July 1868 the company secured passage of a statute requiring that any St. Louis bridge have a clear span of five hundred feet. Here Eads used federal law to block competitive challengers. According to its congressional sponsor, "This bill is simply for the purpose of aiding this company in the money market."[76] With the statute, Captain Eads had transformed the Mississippi into his Rubicon.

Another leading civil engineer believed that Eads might fail to get across the river. In the summer of 1868, John Roebling presented complete plans for a new kind of suspension bridge for St. Louis. This "parabolic truss" used his familiar suspension cables to carry a deck stiffened by shallow trusses. A supremely confident man, Roebling was far better versed in mathematics than nearly all American-born engineers. He projected a 500-foot center span with a truss-suspension-arch combination tailored for railway loadings. Perhaps it met the requirements of the Omnibus Bridge Act, despite its prohibition of suspension designs.[77]

By the time Roebling offered these plans, Eads's St. Louis Bridge Company had charters, plans, stockholders, managers, engineers, and the foundations of the west abutment in sandstone and granite. Why Roebling came forward at this late stage requires an explanation. Like all successful engineers of his day, he knew that it was far easier to design a bridge than to amass the huge sums needed to build it. Design was only the first step.[78] Best to be ready if Eads could not entice investors.

Furthermore, Roebling knew what Eads did not know, although Linville had tried to tell him. The Captain's design was flawed. If it could be built at all, it might collapse into the river, overloaded by a heavy freight train. That catastrophe would grievously wound St. Louis. But John Roebling never had a chance to advance his St. Louis design. While preparing for his great work at Brooklyn, he suffered a freak accident that would prove fatal.[79] By then, Captain Eads was fully immersed in selling his bridge project to bankers and investors—and anyone else with deep pockets and a willingness to put money at risk.

CHAPTER 4

The Art of a Promoter

FIVE MONTHS AFTER vanquishing Lucius Boomer, James Eads published his first report as chief engineer. He portrayed the bridge project from every angle, placing its overall cost at $4,878,000 ($106.5 million in 2023).[1] Perhaps it was a bargain compared with Boomer's proposed structure (estimated to cost $6.5 million), but raising such a sum would prove difficult. St. Louis needed a bridge, but it lacked the money to build one. The men atop St. Louis Bridge knew they would have to attract capital from eastern American cities, London, or Europe. Heightening that challenge, the Captain's design was bedecked with enough red flags to scare off most prudent investors: the stone piers to an unknown depth, the novel arches in record lengths, to be made in experimental steel—all proposed by a novice.

From June 1868 onward, Eads and his associates in the bridge company devised various strategies to raise the funds they needed. Their methods, failures, and successes shed new light on the activities of promoters, investors, and bankers at this pivotal moment in US history. Eads and his allies in St. Louis Bridge grappled with timeless issues in finance, such as how to enlist investors while retaining overall control of the business and how to earn a premium for themselves from their financing plans. In developing their pitch to potential investors, Chief Engineer Eads and President William McPherson focused particularly on ways to quantify the risks in the venture. With the right pitch, they could present appealing returns, sufficient to outweigh pitfalls and hazards known and unknown. This was the art and achievement of an accomplished promoter.

By 1868 Captain Eads had ample experience in spinning entrancing visions of finance, risk, and gain. In 1857 he had transformed the partnership of Eads & Nelson into the Western River Improvement and Wrecking Company, a joint-stock corporation. During the war, he had used interest-bearing loans from friends to secure working capital to build his ironclads, a perilous investment, to say the least.[2] And before launching his bridge promotion, he bought a bank. In an assessment written after Eads's death, his old friend from steamboating days, Emerson Gould, credited him with innovative genius, perseverance, charisma, and some sharp dealing. Atop all his qualities, Gould highlighted Eads's "ability as a *financier*."[3]

Some Considerations of Money

By April 1866, Captain Eads was already considering the elements needed in a rail bridge for St. Louis. Six months later, he took up a different challenge in which he would demonstrate the promotional talents that could eventually help to finance his crossing. He joined a consortium of seven men to buy the State Bank of Missouri. The state legislature had chartered the bank in 1857, providing $1 million of its initial capital. In 1866, the legislators resolved to sell the state's interest, nearly eleven thousand shares with a market value then totaling $700,000. There were only two bidders. The Eads pool did not have the cash, but the Captain had a sophisticated understanding of finance and warm ties to New York bankers. In an echo of modern takeovers financed by debt, "the pool borrowed State bonds from various parties" to swap for the state's equity in the bank, thus conveying ownership to the Eads group. The largest block of borrowed bonds came from the vaults of New York's Bank of Commerce.

Once in control of the State Bank of Missouri, the Eads pool then rebranded their acquisition the National Bank of the State of Missouri. Concurrently, the consortium secured a loan of $1 million from its friends at the Bank of Commerce, which it paid to the state of Missouri in lieu of the bonds. (The well-traveled Missouri bonds then returned to the vaults of the Bank of Commerce in New York.) This loan provided long-term financing for the takeover. A top institution in New York finance, the Bank of Commerce demanded no collateral from the Eads pool for this million-dollar loan, and it charged an interest rate of just 3 percent. The Eads consortium would pay the same interest rate for borrowed money that the Bank of England received for its bonds, the most reliable security in the world. Eads, his bank, and his bridge would prove riskier. We cannot know why Robert Lenox Kennedy, president of the Bank of Commerce, led his own board to approve of this arrangement, which the directors did unanimously.[4] Nevertheless, the incident reveals much about finance in the period. Capital was available to men who had the right connections, made the right promises. Boldness secured especially worthy prizes. Oversight was minimal.

All types of postwar investments in America were shaped in various ways by the monetary policies the Union had adopted during the war. Federal disbursements in gold were suspended, except for obligations legally denominated in gold. Many banks emulated that policy. In 1862, Republicans introduced a form of paper money not redeemable in gold or silver, known popularly as greenbacks. These measures aimed to pay for the war. The conflict also touched off inflation in wages and the necessities of daily life. All these developments in turn made gold, and gold-denominated securities, highly attractive. Seemingly guaranteed to retain their value, such assets also served as a hedge against inflation.

From 1865 to 1873, the economy at large boomed, with a variety of securities serving as money in large deals. Much of the country was still awash in greenbacks and other types of financial instruments.* But investment funds and working capital were comparatively scarce in the West and South. So promoters and politicians in those areas wanted

---

* Circulating media included gold and silver coin; US treasury notes (backed by gold); paper money, or greenbacks; various kinds of federal bonds; currency issued by the new (from 1863) national banks; and notes issued by old and new state banks. States had their own bonds. Private-sector corporations also issued bonds. If the corporation was "sound" and its future auspicious, its bonds were negotiable. In good times, businesses (mostly partnerships) and individuals typically made payments with "drafts," akin to commercial paper today. A simple promise to pay, these drafts were often endorsed by a third party who had good credit. Payment on drafts was commonly due ninety days after issue; "sight drafts" required payment upon presentation. Either way, a draft was effectively a form of money. Bank loans amounted to another kind of money. The defeated states did not share fully in this financial cornucopia. Federal currency reverted to the gold standard on January 2, 1879, even as a substantial quantity of greenbacks remained in circulation, now redeemable at par in gold.

particularly to maintain greenbacks and other inflationary innovations. Meanwhile, the highest-quality capital, taking the form of gold (specie) and gold-denominated securities, was concentrated in the vaults of established bankers in East Coast cities or London.[5]

The means and ends of monetary policy became a burning political question across the country after 1865.[6] Behind the controversy, the advocates for hard money—a currency backed by gold—largely triumphed. In a 180-degree reversal, the federal government enacted deflationary policies, aiming to remove greenbacks from circulation, even as the economy at large boomed. The result was a monetary environment that focused on gold and proved highly favorable to lenders and investors in those high-quality bonds that specified repayments in specie.[7] Concurrently, eastern and British financiers grew fascinated with bets on the fast-growing West, fertile territory in their eyes for making a quick fortune by capitalizing on the gold in their vaults.

From 1865 to 1873 it seemed as if the whole country fell into a fevered mania to amass wealth as quickly as possible. New industries and western expansion created new demands for capital. In this environment, James Eads, his friends and associates, and other dealmakers across the country created capital on their own. The no-collateral loan that the Eads pool received from the Bank of Commerce demonstrated this kind of financial legerdemain. In a growing economy rife with solid investing opportunities, these forms of contrived capital often produced attractive returns that paid off loans, grew businesses and markets, and employed workers. Or they could fuel speculative plays that aimed only to harvest quick returns.

Placing his own bet on the West, on December 18, 1867, Robert Lenox Kennedy joined the initial group of twenty stockholders in the Cutter/Eads company. More precisely, he subscribed for an allotment of shares bearing a face value of $20,000.[8] To subscribe was merely an expression of interest. It cost nothing, at least initially. This first round of stockholders signed up for a total of $400,000 in equities, a number that provided good copy for newspaper stories but not much more. As a practical matter, the figure was insignificant against the project's total estimated cost of $4.9 million.

The huge need for long-term capital to build the bridge took center stage after Eads and his associates drove Boomer from the field. The company could not begin to call for payments from stock subscribers (current or new) until it had solid plans to meet an array of challenges: to build the piers, secure the steel, hire a prime contractor to erect the superstructure, and tie key railroads to use the new crossing. A complete and viable financing plan was the essential precondition to making all this possible.

Under no conceivable scenario would the necessary millions come from selling more shares of stock to the general public. During the Gilded Age, most Americans, even eastern sophisticates, expected a fleecing in stocks. Then (as now) a stockholder literally owned a portion of a company. Insiders like Robert Lenox Kennedy or Tom Scott saw the stock of the bridge company as a kind of wager, one that could pay well if the company's plans were fully realized. That was the crux. Perhaps valuable someday, such stock was little more than a bet on the future.

By this time, some Americans were venturing cautiously to invest in railroad bonds. Unlike stock, a bond was a debt obligation with clear terms detailing its interest rate and its protections (its security) in case of default. In their hunger for capital, US railroads enticed private banking houses to market their bonds, chiefly on Wall Street. In their hunger for wealth, a narrow segment of investors, speculators, and market operators met them on the Street and bought their offerings. In all, the bonded debt of American railroads grew from $416 million in 1867 to $2.23 billion just seven years later.[9] Half of that total came out of American pockets; the rest flowed in from Europe, chiefly London. Still, railroads were one thing. A bond on a bridge was something else.

Public Guarantee or Private Opportunities

The Boomer and Eads companies merged on March 5, 1868. Five days later, James Eads presented a financing plan to the mayor of St. Louis. Local newspapers published the details soon thereafter. He had estimated the cost for the bridge, the tunnel, and the land (for the approaches) at $4,878,000.[10] Under this new plan, private investors would subscribe for $1 million in stock. Eads proposed that after that equity was committed, the city of St. Louis should hold a referendum by which its citizens would decide whether to guarantee a $4 million issue of the company's 6 percent gold bonds. With the city's good credit behind those securities, they would surely sell readily. With a total of $5 million in hand, the bridge would go up, and the trains would cross. Over the succeeding twenty years, the bondholders would be paid in full out of the bridge's toll revenues. To be sure, the guarantee put the city on the hook if the bridge proved less than lucrative, if it failed to pay the interest or principal on its bonds. Nonetheless, the state legislature and the governor agreed to the Captain's plan just fifteen days after it was published. Their speed reflected the clamoring demand for a bridge. Then, with his own financing plan nearly consummated, Eads let it all fall away.

His proposal simply faded into memory. Explaining that curious fact would prove a challenge for the first biographer of the bridge. Calvin Woodward approached the riddle imaginatively, writing that "mature deliberation . . . led the Bridge Company to decline the proposed guarantee." As Woodward noted, it would have given city officers oversight of the finances. That power might "in unscrupulous or unfriendly hands have greatly embarrassed the Company."[11] With that bland phrasing, Woodward deftly sidestepped another plausible scenario. Public oversight might have uncovered in the books the hands of bridge company insiders, watering its stock or self-dealing in its contracts.* The possibilities were endless and not purely hypothetical. By October 1868 one notable stockholder in the St. Louis Bridge Company, Tom Scott, had a secret interest with the contractor holding the bridge company's contract for stonework for the first river pier.[12] And Scott had only begun to multiply his returns. By abandoning the bond guarantee, the bridge company kept the city's auditors out of its business.

Another strong argument worked against the public guarantee. Financing could itself produce handsome returns, legitimate profits without resort to fraud. The many bankers who held stock in the Cutter/Eads company certainly hoped to bid on a future bond issue—buying a portion or the whole issue at a discount—to then sell those securities at a markup to investors. This was the core business of investment bankers. Among the shareholders were Amos Cotting (a partner in a New York private

---

* Some railroad promoters sold stock to then divert the proceeds to personal projects or accounts. Total capitalization could grow impressively even as the real value of individual shares fell. Traders in horses and cattle did much the same thing by making their animals drink water to boost the weight (apparent value) of the stock in the eyes of buyers.

bank), Robert Lenox Kennedy (Bank of Commerce), and E. D. Morgan & Company (a New York investment house).[13]

Captain Eads had bankers and investors in mind when he drafted his first engineer's report (June 1868), especially in these paragraphs on finance. In projecting revenues "after its completion in 1870 or 1871," he exulted that "it is barely possible for the mind to keep within cool, calculating bounds when contemplating the future of St. Louis." The city was even then extending its railroad links to the West: the North Missouri, the Missouri Pacific, and the Kansas Pacific. He confidently asserted that those connections would in turn give to St. Louis "the two great trunk lines which run to the Pacific Ocean."* In truth, nothing was running to the Pacific quite yet, but it was best to be ready. Returning to a cool and calculating mind-set, he predicted that the bridge would rake in gross receipts totaling $1,136,260 ($28.9 million in 2023) for the single year 1871. As the city grew, thanks to its new rail connections, so would annual revenues. Such munificent results would surely pay back investors in the bridge and the tunnel.[14] For all its analytic force and persuasive power, however, the chief engineer's report failed to achieve its most important purpose: motivating investors and their money.

Amid that frustration, the project took two unexpected turns during the summer of 1868. Eads invited William Milnor Roberts to review his plans as revised to that point. At 56, Roberts had done everything in civil engineering, including route selection and construction for a score of railroads, as well as serving as superintendent of improvements for navigating the Ohio River and construction chief for a major Brazilian railway.[15] Roberts knew that Eads's arched design "had been criticized and condemned by a large number of civil engineers, some of them gentlemen of great professional experience."[16] After a searching review, he concluded that the core elements of the design were eminently feasible.[17] Heartened by that verdict, Eads hired Roberts to join the engineering staff on July 9, 1868. He brought considerable stature to the whole enterprise.

Roberts was invited to join the project because James Eads had grown desperately ill in June 1868, stricken by a "severe and sudden pulmonary

---

* Eads was probably alluding to the Kansas Pacific and the Atlantic & Pacific. As matters played out, neither line got anywhere near the Pacific Ocean.

attack."[18] His doctor cautioned that the illness could prove fatal. Only a recuperative ocean voyage and complete rest might save his life. That news drove Eads to propose that Roberts replace him as chief engineer. Roberts declined that post, unwilling "to place myself in any position that would tend to deprive [Eads] of the credit and honor due to the originator and designer of this important structure."[19] Instead, Roberts received the title of associate chief engineer. Eads prepared to leave the complicated work in his charge, confident that Roberts would carry his plans to fruition if death carried him off. Otherwise, Eads aimed to recover, return, and drive on.

Directed by his doctor, Eads embarked on a voyage from New York to Liverpool on the Cunard steamship *Cuba* on July 22, 1868.[20] His wartime friend Gustavus Fox wrote with concern after learning "that you are going abroad again, having broken down by work. That everlasting brain of yours will wear out two or three bodies."[21] This recurring illness came at a bad time. Eads's lung problems had begun during his salvage career; overwork and exhaustion may have triggered this bout. He "found the ocean air most beneficial for his lungs."[22] Undoubtedly, on his travels to New York and Europe he sought far more than restorative rest. Those places had money, and Eads needed investors. He traveled for five months in Europe, returned to New York for five weeks in December 1868, then spent three more months in Europe before finally returning home to St. Louis in April 1869. Henry Flad wrote that during this extraordinary interregnum "all the life of the Company seemed to go out."[23]

The Pennsylvania Connection

Despite the false starts and frustrations in finance, the St. Louis Bridge Company held two aces. The Illinois charter that Boomer had secured (now the property of the consolidated company) promised the lucrative returns of monopoly if Captain Eads could get his bridge up and running. The other ace could make that happen. Tom Scott, vice president of the Pennsylvania Railroad, remained a stockholder. The PRR had exemplary standing in the credit markets of New York, Philadelphia, and London. Surely the best path ahead in St. Louis ran directly through the railroad's Philadelphia headquarters.

On the other hand, precisely what Scott wanted in St. Louis Bridge re-mained a bit of a mystery to the bridge men. With Eads traveling in Europe, William Taussig tried to find out. On September 17, 1868, Taussig, chairman of the Finance and Executive Committee of the board of directors of St. Louis Bridge, wrote to the PRR's president, J. Edgar Thomson. Taussig proposed that Thomson's road agree to buy $1 million in a new issue of the bridge company's 6 percent gold bonds.[24] Those funds from that source would have provided an ideal foundation for the project. The PRR enjoyed universal regard as the best-run, most influen-tial railroad in the country. An alliance with the carrier would make all things possible. A day later (before Thomson had received his letter), Taussig solicited information about Scott, writing confidentially to Mil-nor Roberts, the new associate chief engineer:

> I have good reason to believe that Col. Thom. Scott is interested in one of
> the large Iron & Bridge building establishments in Pittsburg and that that
> establishment (whose name & firm I do not know) at one time last year was
> very desirous to obtain the contract for the superstructure and that a hint was
> at that time thrown out, that, in case the contract was given to that party, the
> Bonds of the Bridge Co . . . would be endorsed by the Penn. RR. & the endorsed
> bonds taken as payment by the [unknown bridge construction] firm.[25]

Taussig appears to have been uninformed at best, naïve at worst, given that he knew nothing of Scott's tie to Keystone Bridge. Eads surely remem-bered that connection, still smarting from his clash with Keystone's Jacob Linville a year earlier. But Eads was off in Europe. Two weeks later, Taussig wrote again to Roberts, then in Pittsburgh, with instructions to visit Philadelphia, and "see Col. Scott himself and ask him, (he being a stockholder in our Co.) what he thought of the 'Keystone Co.'" As if Tauss-ig's intentions were not plain, he went on to describe his "main object . . . to find out, how, in what shape or manner, we can make Col. Thompson's [*sic*] or Scott's interest identical with ours, either by participating in the profits of the [superstructure] contract or by issuing to them certificates of full paid stock, etc. It is a matter that must be handled with tact."[26]

At this moment, Tom Scott was becoming the essential man in Ameri-can railroading. For Scott, the St. Louis Bridge was a trifling thing, a

Thomas Alexander Scott. This portrait accompanied a flattering profile in *Harper's Weekly* (July 12, 1873) that cast Scott as the selfless servant of national progress. It is doubtful that his friends recognized him in the florid prose. Wood engraving from a photograph by F. Gutekunst, author's collection.

bagatelle compared with his big plays across the country. In the West, he was pushing two potential transcontinentals: the Kansas Pacific and the Atlantic & Pacific. He had interests in the North Missouri Railroad and, to the north, in the Lake Superior & Mississippi Railroad. He was a prime mover in two fast freight lines, firms that carried long-haul freight in their own cars thanks to insider deals with the railroads. With others in the Philadelphia interests, he had a hand in Keystone Bridge, Columbia Oil, a bank, an insurance company, iron works, a locomotive builder, and more. He invested in California oil fields and Arizona silver mines.[27]

Simply keeping track of it all was a challenge. An 1871 profile described Scott as "a man almost shaggy in loose chop whiskers and throat hair . . . a tall, long, thin man with nearly the look of an invalid. . . . He talks promptly, answers questions without an instant's hesitation . . . and he is as cool as an automaton." In addition to the broad investment portfolio, there was his day job as the vice president of the largest corporation in the world. Working from his office in the Philadelphia headquarters of the PRR, Scott dictated to half a dozen secretaries at once while eating his lunch, "instantly, confidently, making up his mind and dispatching work."[28]

Such a profile in the press was rare, as many considerations drove Scott to prefer a powerful anonymity. As second-in-command at the PRR, he left the public laurels for President Thomson. By veiling their many investments and speculations, both men might avoid criticism that they were neglecting their official duties at the railroad. And with discretion they could sidestep troublesome charges of self-dealing. Scott's stealthy approach in all his transactions frequently led friends, adversaries, and competitors to overestimate his power. Consider this description from the *Cheyenne Daily Leader* in June 1871, saying that Scott "finds relaxation in running the Pennsylvania and Virginia legislatures, and finds entire repose in directing Congress, the president, and the supreme bench."[29] All towering hyperbole, yet the appearance of power mattered as much as the reality. To achieve his far-flung purposes, Scott could act as a lone agent, a leader of the Philadelphia interests, or on behalf of the Pennsylvania Railroad.[30]

Taussig's overtures did not produce an immediate response from Scott or Thomson, driving the bridge men to consider other ways to raise the necessary capital. Perhaps they could ink an alliance with the B&O, another trunk line from the Eastern Seaboard to East St. Louis.[31] While he toured Europe, Eads likely met with financiers and talked up his bridge. In September 1868, Zerah Colburn's lavish praise for the project in the pages of *Engineering* coincided with Eads's trip to London. The good press could boost the venture among potential investors there. But no deals beckoned.

So Eads took a truly significant step. He boarded a transatlantic steamer in December to return to New York. The timing suggests that he traveled aboard the *City of Paris*. In just her third season, this luxurious and fast ship had an iron hull and the new screw propeller instead of paddle wheels. Her typical voyage required nine days. In the end, the crossing took twelve, delayed by the all-too-predictable gales of the North Atlantic in winter.[32] Only pressing need drove passengers to embark on the Northern Ocean in December. The invalid Eads, chronically seasick even during his summer voyages, was a driven man that month.

Money drove Eads to New York; most likely a transatlantic telegram had called him back. The PRR men had put an offer on the table. On December 19, the board of St. Louis Bridge authorized him to make binding contracts with the PRR or with "other persons" to facilitate construction. By early January, the parties had hammered out a tentative deal.[33] The terms obligated St. Louis Bridge to issue bonds, but the PRR would not guarantee that security. The "other persons," Thomson and Scott, had agreed to commit the carrier to use the bridge. In return, their road would get a preferential rate for its traffic. Their arrangements probably included a side understanding to place the superstructure contract with the Keystone Bridge Company. Under this deal, St. Louis Bridge would raise the money, while Keystone, Thomson, and Scott made tidy sums. They were tough negotiators who knew their leverage. The president of St. Louis Bridge, William McPherson, saw little to like in this proposal: "If we take all the risk [in the bond issue, which might not interest buyers] there is no reason that they should have all the cream." An experienced investor and booster of western railroads, McPherson also held a significant equity position in the PRR. As Scott's friend, he hoped to eventually reach an equitable deal with "the Pennsylvania parties." For now, he thought they had tried "to cut it a *little fat*."[34]

This deal died almost immediately, rejected by the bridge company board and by the directors of the North Missouri. Eads, McPherson, and Taussig wanted to balance a tricky equation. They needed a lot of capital but were loath to lose control of St. Louis Bridge or its profit potentials. Rather than taking this one-sided deal with the PRR, McPherson urged Eads to raise new commitments among New York investors, while he would redouble his efforts in St. Louis. His letter included a worrisome prediction: "We must get the Bridge going this year or I fear St. Louis will dry up with the [new] Bridges north of us spanning the river every few miles."[35] The significance of that prophecy was that freight, especially the long-haul loads, would bypass St. Louis, forgone revenues for what would surely be the most expensive bridge on the river.

Plenty to Go Around

With his daughter Josephine, Eads returned to England in late January 1869 after their stay in New York. Eads likely spent considerable time in London, with extensive conversations at 22 Old Broad Street. That address in the City housed the offices of J. S. Morgan & Company. There is little doubt that he had met with Junius Morgan's son Pierpont during his December visit to New York. The Morgans played a prominent part in a new financing plan that had been thrashed out in that city before Eads's return to Europe. This bold blueprint, crafted to attract top money men in the United States and Great Britain, aimed to provide all the funds required to build the bridge. The work of many hands, the new plan reflected William McPherson's input and Eads's qualities above all.[36]

In this audacious prospectus, Eads and McPherson were engineering the financial foundation required to build the bridge. To date, the Captain had taken estimates of its cost as his starting point in planning for financing. Now he aimed to raise a much greater sum, basing the new target on the bridge's ultimate income potential. Written in Eads's hand, the first line of the new prospectus read, "It is assumed and believed that the tolls on the Bridge will safely pay interest on ten millions [$254 million in 2023]."[37] From that starting point, the plan proposed that the company would issue $4 million in stock, $4 million in first mortgage bonds, and $2 million in second mortgage bonds, as follows:

$4,000,000: stock bearing a par value of $100 a share. A narrow group of insider investors and bankers would subscribe to $3 million in these equities. The company would retain the balance of $1 million in stock.

$4,000,000: first mortgage bonds. Par value $1,000, paying 6 percent interest annually in gold, to be sold on the bond markets of New York or London.

$2,000,000: second mortgage bonds. Par value $1,000, paying 7 percent in US currency, to be given at no cost to the stockholding insiders.

Eads and McPherson laid out these plans in a prospectus dated February 1869, although groundwork had been laid in January. The figures at left were par values, not to be mistaken for actual worth or real cash raised. Typically, bonds appear at the top of a capital stack, as they have the highest claim on the company's assets (they are the best protected), with subordinated debt below and common stock beneath that. In this plan, the equity investors were the mainspring driving the whole capital structure.

If events unfolded as Eads planned, half of the $10 million raised by those securities would cover the cost of building the bridge. The remaining $5 million could compensate investors personally. That munificent return reflected the considerable perils and unknowns in Eads's venture. For financiers and capitalists willing to gamble, this deal might prove compelling. Matching risk with sufficient enticement was the art of a good promoter.

The plan encompassed a range of possibilities for the company and its investors. Eads and McPherson pegged the cost to construct the bridge at $4 million, with "ample allowance for all contingencies." Even so, they tacked on another 20 percent to cover delays or problems. In that case, construction would require upwards of $4.8 million. The funds would come from selling—and awarding—a mix of stocks and bonds.

The plan's first step, a new stock offering, would raise the funds needed to start construction. A selected group of thirty investors would receive the opportunity to subscribe for $3 million in that issue. The new stockholders would buy with the assurance that the company planned to call on them to fund only 40 percent of par value. Thus, the stock issue would produce $1.2 million in cash for St. Louis Bridge. That block of money would build the piers and abutments, tangible success in besting the river, visible progress to reassure all involved.

Step 2 was the sale of first mortgage bonds after the piers and abutments were safely on bedrock. The prospectus assumed that the company would place the full $4 million issue with an investment banking house, receiving 90 percent of par value.[38] Therefore, the bond issue would return $3.6 million to St. Louis Bridge, exactly the sum required to build the steel superstructure and complete the project. If construction proceeded smoothly, step 3 anticipated the outright gift to the stockholders of the unissued $1 million in stock. Once the bridge entered service, those shares could command a good price if sold or attractive dividends if held. Step 4 called for the stockholders to receive at no charge their proportional allotments of the second mortgage bonds, an issue of $2 million (par value). Again, they could sell or hold as they saw fit.

In all, the prospectus offered a potentially lucrative package. An investor who subscribed for 1,000 shares of stock would pay in $40,000. With that ante, he could come away, if all went well, with three classes of securities bearing a total par value of $200,000.* Better yet, if he put $100,000 in, he could receive $500,000 back. Here was a fine blueprint to enlist capital. Eads anticipated needing two years to complete the bridge. Even if it took three, this was a heady return on investment.

We do not know if this kind of financing was unusual because historians have not explored this kind of venture. Two inferences suggest that the prospectus was no freak or one-off. Eads pitched this plan to sophisticated men with experience in finance. Their enthusiasm indicates that they were comfortable with its elements, perhaps having done similar deals in the past. And even if Eads *had* pioneered this capital structure, those money men surely would have replicated his approach when possible. It worked.

Given all its novel aspects, the St. Louis Bridge was a challenging project to finance, especially because the assets could prove very difficult to sell (and impossible to liquidate) if matters took a dire turn during construction. But city gas companies, street railways, and urban waterworks must have relied upon similar strategies to raise capital, as few were publicly owned—or even regulated in the modern sense. By holding chartered monopolies, they offered the rich plum of guaranteed revenue with no competitors.[39] Furthermore, those companies lacked financing options developed in the twentieth century, such as lines of credit or long-term loans from commercial banks.

Pieces of a Good Thing

The board of directors of St. Louis Bridge gave the plan its enthusiastic approval at its January 25, 1869, meeting.[40] By the end of February, fifty-five men had signed the subscriber rolls.[41] Clearly James Eads and William

---

* An investor who paid in $40,000 for his equity stake would net the original shares of stock (par value $100,000), the gift of unissued stock ($33,000), and the gift of second mortgage bonds ($67,000).

McPherson were compelling promoters. With funds flowing in, the Captain returned to Europe at the end of January. Signaling his confidence in the prospectus and his commitment to the bridge, Eads signed up for $300,000 in stock. His son-in-law John A. Ubsdell committed for $200,000. The president of St. Louis Bridge (and the coauthor of the plan), William McPherson, took $150,000. Eleven investors subscribed for $100,000. Forty-one other men took the remaining shares, valued at $1.33 million.

Those signing for blocks of $100,000 ($2.3 million in 2023) had demonstrated prowess as investors and financiers:

Gerard B. Allen (St. Louis), general investor, railroad director, ironworks proprietor

James Benedict (NYC), stockbroker

John Copelin (St. Louis), general investor, director of the North Missouri Railroad

Dabney, Morgan & Company (NYC), investment bankers (J. P. Morgan's firm)

Thomas Eakin (NYC), banker and broker

Edwin Hoyt (NYC), commission merchant, general investor

Jameson, Smith & Cotting (NYC), bankers and brokers

Robert Lenox Kennedy (NYC), president, Bank of Commerce

James Low (NYC), commission merchant

E. D. Morgan & Company (NYC), merchant bankers (Morgan was a US senator from New York)

John J. Roe (St. Louis), general investor, director of the North Missouri

Today, nearly all these men are obscure figures at best. In their own era they stood atop American finance. Their participation suggests that they saw enticing prospects in Eads's deal, for themselves and for the railroads and people of St. Louis.

Two overlapping groups became subscribers in the reorganized company. Much of the St. Louis contingent had interests in the North Missouri Railroad. Of that line's twelve-member board, ten subscribed for shares in the bridge company. A completed bridge could transform that carrier into a powerful interregional artery. The second group was made up of New York financiers, brokers, and bankers. This group included Adrian Iselin, Morris Jesup, and Morton, Bliss & Company, in addition to the bigger subscribers listed above. Another financier, London-based Junius Spencer Morgan, pledged $25,000 for shares.

The participation of so many New York private banking houses among the subscribers is noteworthy. Business historians portray these firms as capital intermediaries, working to ensure a flow of funds from investors to companies, chiefly by selling the bonds that built railroads. With St. Louis Bridge, however, they were trading on their own accounts, entering this new venture at the ground floor and for their own benefit. Histories of the Morgan banks largely ignore their private-account transactions.[42] The number of bankers in this deal suggests that it is a mistake to draw a sharp distinction between their public functions and their private investments.[43]

All the big-money investors in St. Louis Bridge were financial sophisticates, and their participation imparted momentum to the whole project. We don't know how they saw the venture at this moment, beyond the obvious appeal of its beckoning returns.[44] They likely cast themselves simply as investors in the United States and its bright future. Sure, they sought enrichment. That was the American genius. Their investments promised to lift St. Louis, improve commerce, create jobs, speed the country's business, and connect its people. They had the satisfaction of believing that their money was the mainspring of progress itself.

Many of the bridge company's new owners had large holdings in railroad stocks, the high-tech sector of their day. Their new investments in St. Louis Bridge could enhance those other ventures. For example, William McPherson envisioned that the bridge would bolster freights, profits, and dividends on the North Missouri, the Kansas Pacific, and the Pennsylvania Railroad, and he owned stock in all three. Thanks to the breadth of their portfolios and the depth of their experience, these investors knew to expect some losses alongside their winners.

Rewards for Risks

The capital structure and returns that Eads sketched may appear to have been overly generous to the investors. We simply do not know enough about finance in this era to make that judgment. Modern-day private-equity investors in start-up companies can expect similar results, relying on stock warrants to rake in early and generous gains if the business does well. Many don't. Eads's prospectus dangled fine returns for investors, but with a caveat. All involved knew that the document sketched a best-case scenario, with construction meeting deadlines and budgets. Then as now, the proffered returns reflected the hazards that would-be investors anticipated they would bear. The prospective rewards amounted to a statement in dollars of the incalculable perils of Eads's project, especially in its engineering and in the river.

The plan also reckoned with risks. After all, a prospectus presents a calculus into an unknowable future. In the document, Eads forecast that the bridge would cost $4 million. To that amount he added an allowance of $800,000 (20 percent) in case of unforeseen contingencies. The rosy 500 percent return had that contingency baked into the numbers. The document then considered a darker scenario, with expenditures rising another $600,000. If that happened, thanks to "an improbable mistake occurring in the estimates of costs," the proposed capital structure would still return 400 percent to investors.[45]

In considering potential pitfalls, these men had little basis on which to judge the difficulties of pioneering in steel. Finance, not metallurgy, was their métier. Even so, they could reckon with technical issues by using benchmarks of time and interest. For example, they surely considered the added interest burden that St. Louis Bridge would have to bear if technical problems added two full years to Eads's best-case construction schedule (which promised completion in 1871). In that case, a total debt of $5 million would require interest payments during a two-year delay totaling $600,000—just what Eads had sketched in his worse-case scenario.

The prospectus included another clause that offered some rainy-day protection for its participants. Until the bridge opened for business, the company would retain all the securities it had promised to investors (steps 3 and 4 on p. 102). That proviso encouraged each individual to stay with the group, not to sell stocks (their original subscriber shares) early. Their wealth, connections, and self-interest would sustain the enterprise until completion. And by holding those securities until traffic flowed over the river, St. Louis Bridge looked unlikely to ever need a new round of financing.

On the income side, the prospectus offered figures to please the most cautious banker. Anticipated revenues drove the whole plan. To pay annual dividends and interest on securities worth $10 million would require a clear profit of at least $700,000 each year. Drawing from publicly available data, Eads developed a forecast of $742,000 in annual revenues for St. Louis Bridge. He derived that total from the 1867 traffic to and from St. Louis of just three railroads, its top three connections to the East.[46] The income figures omitted the other eastern lines, all the western lines, and all the roadway traffic that would cross on the upper deck. Presented this way, Eads's data on revenues were compellingly conservative.* Once it opened, the bridge would draw new railroads to the city. The company's securities equated to an investment in the shining future of the region and the country. Sure, the Captain's bridge bore risks. It also had the potential to generate munificent returns to investors.

Many spoken understandings must have accompanied the document, given that it failed to address some notable questions. In his first engineer's report (June 1, 1868), Eads had pegged the bridge at $4.9 million. Why had his estimate fallen to $4 million seven months later? And who would pay for the tunnel under downtown? Without it, the bridge was useless. The prospectus offered only silence about the Wiggins Ferry Company. Presumably the finances had enough fat baked into the numbers to pay off or buy out the Wiggins owners if that became necessary. These omissions were highly relevant to the investors, sophisticated men who knew to ask the pointed question. Perhaps the prospectus ignored Wiggins because Eads had confidently asserted that the bridge would kill the ferry, taking all its business.[47]

The 1868 engineer's report and the prospectus also were silent on topics crucial to the people and businesses of St. Louis. After Wiggins

---

* The bridge company would have expenses besides its debt service costs, but there was no reason to clutter the prospectus with those comparatively modest numbers.

disappeared, the bridge company would have a comparatively free hand in setting its rates and tolls. Would it keep fees low, relying on economies of scale to grow revenues and profits? After all, its efficiencies (compared with the ferries) in moving railcars would surely prove its greatest advantage. Or like a classic monopolist would it raise rates? Thanks to the Illinois charter, no other bridge could be built into the city for twenty-five years. Without the ferries, the bridge company could have every expectation of monopoly pricing.

Then as now, economists described this as monopoly rent: extorting disproportionate returns thanks solely to market dominance, provided here by the state of Illinois. For average Americans of the Gilded Age, monopoly became *the* focal point of criticism of the new corporations growing to dominate their lives. Nobody liked the Wiggins company, with its high charges, miserable service, maddening inconvenience, and unchallengeable power to block any competitors. Time alone would reveal whether a bridge monopoly would take its place to leech off the commerce of city and region.

That issue apparently did not come up when William McPherson returned to St. Louis and announced the financing success in the press.[48] The news brought rejoicing all around. On February 21, 1869, St. Louis Bridge held a reception for all comers at the Wedge House Saloon in East St. Louis. After brief remarks, the crowd settled into a lunch provided by Heim Brothers. Mostly seasonal fare, the menu included oysters brought up from the Gulf of Mexico, washed down with Semon & Krug's lager beer. With vocal and instrumental music, it was quite a party.[49]

The House of Morgan and the Bridge

The subscriptions of Junius Morgan and of his son's New York banking house—Dabney, Morgan & Company—in this stock offering raise fascinating questions. St. Louis Bridge had scored a coup by drawing in these new owners, ranked among the top bankers in London and New York. We can only infer what the Morgans sought by their participation. Although superb records survive for the London house, that complete paper trail starts in 1870. Adding to the mystery, no historian of the Morgans or of St. Louis Bridge has even mentioned the bankers' connection to the bridge company's stock issue. One practical matter explains the oversight. This was a small transaction for the house of Morgan measured against the worldwide business it was already doing in 1869.

A larger conceptual issue also comes into play. Judging from his oft-repeated exhortations, Junius Morgan had compelling reasons to stay far away from this venture. In this era, the equities of American railroads often carried a speculative whiff. St. Louis Bridge wasn't even a railroad, and it wasn't even built. Now the Morgans, father and son, owned a piece of it. It was as if a pair of Baptist preachers had joined promoters over on the wrong side of town to buy into a new dance hall. They could make good money, but this was not what one would expect. And not the image they wished to project.

Over their lengthy careers, Junius Spencer Morgan (1813–1890) and his son John Pierpont Morgan (1837–1913) acquired towering reputations in international banking. Born in the United States, J. S. Morgan in 1854 became the junior partner at a London merchant bank. A decade later that firm had become J. S. Morgan & Company. He groomed his son to become his American affiliate. Fulfilling that plan, in 1864 Dabney, Morgan & Company opened its Wall Street offices with Pierpont as its junior partner. Both firms joined a banking elite then engaged primarily in financing international trade.

Reliable partners on both sides of the Atlantic were essential in this business. In letters spanning many years, Junius Morgan hammered away at Pierpont to fulfill his expectations for probity and reliability. As members of interlocking partnerships, the father in London was fully liable to pay any debts incurred by his New York affiliate, and vice versa. That was mostly a hypothetical concern, although the London house had a rule requiring "no speculations without the consent of all the parties."[50] A clear rule, although just what constituted a speculation could prove debatable.[51] And when Pierpont was starting out in the business, equities were a real concern for Junius. In 1858 the son reported that he had purchased shares in the Pacific Steamship Company. His father replied: "You must act upon your own judgment but I do not like your buying stock or having your mind turned in that direction. How many have been shipwrecked on that one thing—Speculation in stock."[52]

A decade after Junius wrote that paternal directive, Anglo-American finance had changed in fundamental ways, thanks to the enormous capital

Junius Morgan sat for this carte-de-visite in the Paris studio of Disderi & Company, probably in the 1860s. Photograph from The Morgan Library and Museum, New York, ARC 2295.

needs of American railroads. Seeking to grow with that shift, the Morgan firms added investment banking to their original focus on financing international trade, marketing new bond issues after 1867. Bonds posed far less downside to investors than stocks. Still, the debt instruments of these young American railroads were hardly guaranteed. Junius Morgan saw profitable opportunity for his firm in acting as the middleman between British investors and the right sort of American railroads. Judicious care in selecting those bond issues was essential. According to its chief biographer, the house of Morgan remained keenly aware that "its reputation—its single greatest asset—was tied to every security it endorsed."[53]

As it turned out, however, even Junius was not immune to temptation. In July 1866 he took a plunge in the London stock market, making a quick profit "by heavy sales of Atlantic Telegraph shares." Word got around. A partner at Brown Brothers (a competing merchant bank) wrote privately that Junius Morgan was a banker of "an undoubtedly speculative turn."[54] That assessment suggests that he was much more than a steady guardian of other people's money.

No historian has even attempted to explain why the Morgans bought equity stakes in St. Louis Bridge. As they made the rounds of New York bankers and brokers in January 1869, Eads and McPherson likely called on Dabney, Morgan & Company. Perhaps Pierpont, whose firm committed for shares bearing $100,000 in par value, persuaded his father to come into the deal with his $25,000 pledge. By their stock purchases in the bridge company, the Morgans gained inside knowledge of its finances and an inside track to market the securities envisioned under its financing plan. The house that underwrote those bonds could expect a healthy profit.

Another aspect of the financing plan is worth noting. Tom Scott no longer appeared as a stockholder in St. Louis Bridge. From December 1867 he had owned shares in the Cutter/Eads company, also serving on its board. With others in the Philadelphia interests, he was fully committed to the North Missouri and the Kansas Pacific. Under the new plan, he remained committed to St. Louis Bridge. While Scott held no equities in the reorganized firm in his own name, Andrew Carnegie had eight hundred shares.[55] Just 33, the smart and energetic Carnegie had already proven a loyal subordinate who advanced the investing goals of his mentors, Edgar Thomson and Tom Scott. St. Louis Bridge would soon provide new opportunities for all three men.

The February prospectus quickly attracted American equity investors, whose funds would buy progress on the river. As they signed up, Captain Eads returned to Europe in search of technical solutions needed to build his bridge. The knowledge and contacts he gained in his travels would result in progress on some knotty problems. In Britain and France, Eads

This view of the west abutment illustrates the essential role of money in engineering. Eads had begun excavations here during Boomer's convention in August 1867. Working on the levee, the men used a traditional cofferdam as they dug down to bedrock, twenty feet below the low-water mark of the river. But lack of funds then slowed further progress. By the time of this image (September 20, 1870), St. Louis Bridge had new capital, and the masons were back on the job. Even here at the starting point, much work remained, eased somewhat by the low stage of the river, typical of fall. Engraving from Twombly, "Illinois."

met with top men at leading iron and steel works, who endorsed his plans to use steel.* And in a significant stroke of good fortune, his visit to the Loire works of Petin, Gaudet & Company led to a radically new approach for putting in the piers.

---

* While they approved of his plans, those steelworks did not execute contracts with Eads (then or later) to supply the metal—a noteworthy if inexplicable fact.

The largest steelworks in France, Petin, Gaudet also made structural ironwork for bridges.[56] Eads was delighted when its chief, Hippolyte Petin invited a leading French bridge designer to review his plans. After a painstaking review in March 1869 in Paris, Félix Moreaux offered his enthusiastic approval. A prolific designer and builder, Moreaux was then supervising construction of a new bridge over the Allier River in Vichy, France. He invited the American to visit the site, where Eads could observe how the French engineers intended to place their stone piers on bedrock beneath the flowing river. He would see a technology, by then well developed in Europe, known as pneumatic caissons.[57] Within months, Eads would introduce this audacious suite of innovations to the United States. In its size, depth, and difficulty, the Captain's pioneering work would far exceed anything yet attempted or accomplished in Europe.

After nine months spent mostly in Britain and France, Eads returned to St Louis in May 1869, and the project shifted from preparations and incremental steps to full-bore execution. Financing made all the difference. Cut limestone from quarries at Grafton, Illinois, began arriving on the St. Louis levee. Granite blocks to face the two river piers were en route from the East. Eads's old salvage partner, William Nelson, started constructing the caisson for the east pier. The company bought tugs and barges, fitting them out with custom derricks and traveling cranes to place the stonework. By June 15, one thousand men labored at St. Louis and in the quarries.[58] After all the painstaking preparations and maddening delays, construction was about to begin in earnest.

This new phase posed far greater challenges than Eads, his engineers, and his men had encountered in building the west abutment on the levee. The two river piers required deep excavations to reach bedrock. The piers amounted to massive stone towers, and the masons had to work on—and in—the Mississippi. Success was far from assured.

CHAPTER 5

To Bedrock

THE CHIEF ENGINEER RETURNED to the river convinced that he had found the best path ahead for his bridge inside an iron box, or caisson. This "pneumatic plenum" method to lay stone foundations would have its North America debut at St. Louis. Raising the stakes, Eads was proposing the largest and deepest caissons yet attempted in the world. European precedents were reassuring, although conditions there scarcely forecast the ferocity of currents and the threatening ice on the Mississippi. As always, Eads projected confidence. Arguably, desperation drove this choice. His 1868 engineer's report had proposed complicated makeshifts to dig down to bedrock and build up his stone piers, using cofferdams and staging the work during periods of low water.[1] His willingness to abandon those untried plans almost immediately suggests his own doubts. But the caisson method also posed daunting problems. Assistant engineer Henry Flad doubted that it was even feasible.[2]

To this point, James Eads had confronted—and overcome—a diverse array of obstacles to advance his bridge: Linville's critique, the contest with Boomer, and his own bouts of bad health, as well as the difficulties of developing viable financing plans, promoting the project on two continents, and finding a method to build the river piers. As the work pivoted to constructing those piers, everyone had to grapple with new perils that were qualitatively different and much more threatening. Even as they began this new phase, all involved knew that catastrophic failure threatened at every turn.

Eads knew it better than most. In 1851, he had been working a salvage job on the lower Mississippi below Cairo, Illinois. Normally a half mile wide at that point, the river had been in flood at the time. Although much of its flow and force had spilled out across the adjacent plains, it rampaged down in the deep channel. After descending sixty-five feet in his iron diving bell into the cold, black torrent, Eads had found "the bed of the river, for at least three feet in depth, a moving mass, and so unstable that . . . my feet penetrated through it until I could feel . . . the sand rushing past my hands, driven by a current apparently as rapid as that at the surface."[3] This episode lived on in his mind, a stark demonstration of the river's power—as well as of his physical courage. Two decades later, Eads was convinced that the Mississippi's forceful currents at all depths and the shifting instability of the riverbed posed the fundamental design challenges for his bridge. These intimate experiences with the river drove his determination to land the piers of the St. Louis Bridge on bedrock. The years inside the diving bell had also prepared him to embrace caissons.

Engineering under Water

The pneumatic plenum resulted from a history of interlocking innovations and incremental improvements. The concept had three notable contributors, each brilliant. In 1779, a French military engineer and theoretical physicist, Charles-Augustin de Coulomb, had presented the essential ideas in a paper to the French Academy of Science. He had described a box strongly constructed of oak and braced with iron, like a giant shoebox but inverted and open on the bottom. Coulomb had proposed placing this airtight box in a river or harbor requiring bridge or dock construction. Temporary weights on its top face would force the box to the bottom. Using hand pumps and leather hoses, workers at ground level would pump air into the box, driving the water out under its edges. Entering through a primitive air lock, excavators could then dig away at the riverbed to reach stable bedrock. No evidence survives of widespread use, but Coulomb had laid out the conceptual core of the pneumatic caisson. In 1831, Lord Cochrane had taken the next important step, patenting the modern air lock. A daring officer in the Royal Navy, Cochrane had forecast its use for "building and working under water."[4]

The French engineer Fleur Saint-Denis brought the essential elements of pneumatic caissons into mature form in building the piers for this 1861 iron railroad bridge over the Rhine at Kehl. It had a difficult service life, reflecting the fraught conflicts on this border. The superstructure was dynamited by the French in 1870, rebuilt, then dynamited again in 1939. Retreating Germans thoroughly destroyed its east abutment in 1944. Despite the mayhem, the piers by Saint-Denis remained unmoved. Nineteenth-century postcard, author's collection.

In 1858 another brilliant Frenchman, Fleur Saint-Denis, had put these elements into mature form in constructing piers for the first railway bridge over the Rhine.[5] To place its foundations on bedrock beneath the flowing river, Saint-Denis had relied upon airtight caissons of iron plate. The largest at Kehl had a footprint of 77 by 23 feet. While it descended, masons laid cut stone atop the caisson, forcing it down while building up the pier. It landed on bedrock at sixty-six feet below the high-water level of the Rhine.[6] In all, Saint-Denis achieved a tour de force. To execute this kind of significant work, Coulomb's designs of 1779 had needed the iron plate and steam-powered air pumps of the industrial age. Located on the Franco-Prussian border, the crossing at Kehl carried a double-track railway line.

Eads's visit to Vichy in March 1869 offered a study in contrasts. He spoke no French, lacked any secondary education, and had graduated from a

80 VICHY. — *Le Pont sur l'Allier.* — LL.

The Pont Allier at Vichy, completed in 1870. Its cast-iron arches carried a common roadway across this lovely tranquil stream. Observing pneumatic caissons here, Eads came away determined to adapt the method to the demanding conditions of the Mississippi River at St. Louis. He took that path partly because every other option for building his foundations looked less feasible. Nineteenth-century postcard, author's collection.

school of hard knocks. By contrast, Félix Moreaux embodied the traditions, theories, and practices of what was arguably the best engineering college in the world. Founded in 1780, the École des Arts et Métiers was one of France's *grandes écoles*. That Eads could enlist Moreaux's input suggests that the American projected equal measures of engaging charisma and open-minded humility. His traveling companion, a St. Louis doctor named Charles Pope, translated his conversations with the French engineers. A graduate of the University of Pennsylvania, Pope had had two postgraduate years of medical studies in Paris, acquiring fluency in the language. The Americans spent several days observing the methods used at Vichy.

Eads came away convinced that this pneumatic plenum method was the right choice for placing his piers on bedrock. He did not lack for confidence. The Allier was a placid stream, less than 600 feet wide. The

Mississippi was a wild river, 1,500 feet across at St. Louis. The Vichy caissons descended through 23 feet of silt and muck. Eads needed to put the east pier down in the flowing river, then dig through 82 feet of sand, gravel, or stone to reach bedrock.[7] He would need greater pressures and larger caissons than any European engineer had attempted. Furthermore, Moreaux had been designing and building bridges of all sizes over a professional career spanning two decades.[8] Eads could not say the same.

Ice and Destruction

The Captain knew that weeks or months of river ice at St. Louis greatly exacerbated the challenges of fast current and deep scour in the riverbed. Each winter, ice formed for hundreds of miles upstream in the Mississippi and the Missouri. Because those rivers rose or fell every day, the ice along their banks broke up into large cakes that floated down to St. Louis, where the river narrowed. In a fortunate winter, all that ice flowed through the city with scant ill effects, perhaps damaging the steamers chained to the levee. Or the cakes and bergs could choke the Mississippi.

During prolonged hard freezes, the ice gorged. Colonel J. H. Simpson, of the Army Corps of Engineers, described the condition: "When the river is full, heavy masses [of ice] are cemented together under pressure by 're-gelation.' When moving, the mass is often several feet in thickness and capable of crushing vessels by pressure." At the city, "the mass is compacted together on the surface and solidifies very quickly under pressure; other ice following until a complete dam is formed and the ice is said to be gorged." More ice and water would then back up upstream of that dam, but the river would always find a route. For that reason, the currents and scouring action increased beneath the ice field, tearing away at the riverbed, even as the river level (or stage) rose behind the gorge. At some inevitable, unpredictable moment, the latent force of the backed-up river overwhelmed the gorge. As the ice and torrents broke free with a roar like Niagara, "no human device can avail to save whatever is exposed." Ice gorges measured upward of twenty-two feet in thickness, with rushing, scouring current beneath. When the river gorged, its height in St. Louis just upstream of that barrier could exceed by eight feet or more the river stage downstream of the city.[9]

MISSOURI.—THE GREAT ICE-GORGE ON THE MISSISSIPPI—STEAMERS AND COAL BARGES BLOCKADED IN THE ICE NEAR ST. LOUIS.—FROM A SKETCH BY A. L. HUFNAGEL.

This scene appears mythic, although river ice often proved this destructive. After three months of subzero weather, on February 25, 1872, ice across the frozen rivers of the Upper Plains states began to move after a sudden rise of the Platte River in Nebraska. Near St. Louis, the moving bergs tore away at steamers and coal barges long frozen in place. Then the ice gorged into a solid mass at the city. The steamers here with smoking chimneys had tried and failed to get away from the floes. Ice like this would assail the piers of any bridge. Engraving from a sketch by A. L. Hufnagel, from "The Mississippi Ice-Gorge," *Frank Leslie's Illustrated Newspaper*, Mar. 16, 1872, 12.

All this was challenging enough. By limiting his bridge to two river piers of massive stonework, Eads designed a structure to survive these winter onslaughts. The east pier presented a further complication. It had to go much deeper than its western counterpart, as bedrock fell away toward the Illinois shore. Measuring 197 feet in total height from bedrock, that pier would equal the height of a twenty-story building. Only there were no twenty-story buildings in the United States in 1869. New York City's Equitable Life Building (1868–70) held the record, with nine floors. At the river's normal stage, half the pier's height would lie beneath the surface. Pneumatic caissons really were the only possible way to land such a pier on bedrock.

Preparations

In using caissons, Eads should properly share credit as the US pathfinder. In 1868, a talented American civil engineer, Octave Chanute, published illustrated descriptions of bridge construction in Europe using this technology.* That three-part series in a leading technical journal detailed many projects, including the bridge at Kehl.[10] Eads knew about the series, since Chanute's second article appeared in the same issue of the *Journal of the Franklin Institute* (July 1868) that reviewed his own engineer's report. Soon after reading the article, Eads left for Europe, suggesting more than a mere coincidence of timing.[11]

Certainly the St. Louis Bridge was the first venture in North America to use pneumatic caissons. The Captain must have read Chanute's articles with a mix of excitement and concern. Chanute noted that at one deep foundation in Britain air pressures reached three times atmospheric pressure, resulting in serious injury to workers. On the basis of such experiences, European engineers believed that the "extreme limit" for this kind of work was twenty-five meters (82 feet).[12] Eads knew that his east pier would require men to work 100 feet below the surface, perhaps even 125 feet if the river was in flood.[13]

During the summer of 1869, pier construction began with two bold decisions. Eads resolved to start caisson work with the deeper east pier. The company directors matched that nervy choice by authorizing expenditures so that construction could begin concurrently on the west pier. The duplication of machinery would raise costs, but speed mattered more to the directors, all shareholders. Faster construction would lower the overall interest burden and hasten the day of completion, when income and profits would replace expenses and worries.

---

* At this time, Chanute was constructing a rail bridge at Kansas City, shown on p. 183 in ch. 7. There he used pneumatic pile foundations, the same technology that Boomer wanted in St. Louis and that he would employ for the Omaha Bridge. Whatever its pros and cons, that method to build piers was unsuited to Eads's arched superstructure.

Writing in late August, Eads was "confident of accomplishing the work before interruption from ice occurs."[14] Nonetheless, this delayed start raised the chances that ice could become a problem. The Captain's decision to press ahead despite that threat reflected the general pressure, felt by all, to produce visible progress on the river. His own headstrong confidence drove the decision. Time alone would reveal whether this choice was daring, lucky, foolish, or disastrous.

The summer had been devoted to preparations: buying work boats, designing and building the caissons, developing and making the sand pumps to excavate the sand beneath the caissons, contracting for masonry, and planning how to place the cut-stone blocks into a developing tower aboard a caisson anchored in the living river. According to Milnor Roberts, it all "demanded a vast amount of careful thought while the ordering and preparation of the different materials and appliances required great executive skill and energy."[15]

James Andrews, of Pittsburgh, was the prime contractor for the piers and abutments. Edgar Thomson had recommended this close professional friend, who had built the foundations for the Steubenville Bridge.[16] Thomson may have had a piece of Andrews's contract with St. Louis Bridge. Scott certainly did.[17] For the most part, the piers would be built out of magnesian limestone from nearby quarries in Grafton, Illinois, just forty miles upstream. By September, the stone began arriving at the levee on barges. For better protection against river ice, courses of granite (far more costly than limestone) would cover the piers and abutments.[18] The initial contract went to the Richmond Granite Company, but deliveries from Virginia lagged badly. As a result, Milnor Roberts traveled to New England to select durable and fault-free stone. Within months, new contracts were let for Maine granite.[19] Schooners loaded with stone sailed from Buck's Harbor and Blue Hill, Maine, to New Orleans. Steamers then hauled barges laden with rough-cut blocks upriver to St. Louis.[20] Red Missouri granite for the approach piers came out of quarries owned by Gratz Brown, then briefly a private citizen between terms as a US senator and Missouri governor. Friends in high places were always helpful.

By September 1, 1869, St. Louis Bridge had spent $851,000 ($19.6 million in 2023) on all its preparations.[21] The high start-up costs partly reflected

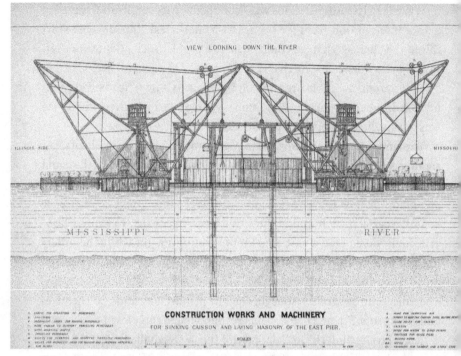

VIEW LOOKING DOWN THE RIVER

ILLINOIS SIDE

MISSOURI

MISSISSIPPI   RIVER

CONSTRUCTION WORKS AND MACHINERY
FOR SINKING CAISSON AND LAYING MASONRY OF THE EAST PIER.

SCALES

This complicated engraving depicts the key elements of Eads's innovative methods to build the two river piers. At its center, the sheet-iron caisson for the east
pier floated in the river, pinned into place by pilings driven into the sandy bottom.
On either side, a pontoon boat was tied to the pilings. Outboard of those boats,
a barge on each side brought out the first cut stones. Each pontoon boat had six
double-beaked cranes (because this is an elevation view, they show as a single
profile). A crane operator in his control shack aboard the pontoon boat at right
has lifted the first block of stone off the barge. From that perch, he will place that
milestone aboard the caisson. Detail of plate 9 from Woodward, *History.*

Eads's plans to mechanize the ancient art of masonry construction. To
speed the laying of stone, he designed four pontoon boats, each outfitted
with custom-designed machinery. The caisson for the east pier would be
served by the *Gerard B. Allen* and the *Ben Johnson,* while the *Alpha* and the
Omega tended the west caisson. Bearing a resemblance to Eads's old
Submarines, each pontoon boat sprouted six tall frameworks much like
double-beaked cranes. Fifty feet tall, the cranes each had a traveler suspended on wire rope from its jibs. Like the shuttling trolley moving back
and forth beneath the boom of a modern tower crane, the traveler was

This perspective view shows the two pontoon boats *Alpha* and *Omega* building the west pier. Stone laying was well advanced here, forcing the iron caisson down into the river. Photograph by Robert Benecke, MHS, N 13937.

controlled by an operator in a cabin perched in the framework. From that lofty spot, he could pluck a seven-ton stone block off a barge, transport it over his own pontoon boat, and place it on the caisson. The boats had boilers and winding engines to power the cranes and travelers, hydraulic pumps and air compressors, storerooms and an office.

With all that, the travelers were their raison d'être. With six travelers on each pontoon boat and two boats serving each caisson, the men and machinery could grab, lift, move, and place twelve massive blocks continuously. Eads projected that they would lay pier stone at the rate of 2.5 vertical feet per day.[22] In all, the pontoon boats were masterpieces of mechanical ingenuity. In addition to its speed, this method of placing stone improved safety and accuracy in positioning. While men would die in building the bridge, none were crushed to death by stone, an all-too-common fate on other big construction jobs. The boats would perform another essential task. While their jib cranes and travelers grew the stone tower above the caisson, their steam-driven compressors would supply the caisson with air under pressure.

If the pontoon boats appeared to be complicated makeshifts, Eads's other major innovation for constructing the piers was elegant simplicity. A device with no moving parts, the sand pump solved a problem that had vexed builders of caisson-dug bridges in Europe. Once that inverted box landed on the riverbed, it had to be entirely airtight to drive out the water and provide a safe working space for the men. The men would come and go via air locks, but how to get the spoil out? Eads proposed pumping a stream of water by iron piping connected to the sand pumps in each caisson. In turn, that flow would induce suction inside the pump to lift sand, gravel, even small rocks up through a discharge pipe. The laborers, or "submarines," in the caisson needed only to shovel the spoil to a suction line; the sand pump would then eject it to the surface. Tests during the summer gave Eads and his engineers confidence in this approach. As built, the east pier caisson had seven integral sand pumps; five would go into its somewhat smaller western counterpart.[23]

Another crucial step underscored the threat of ice and Eads's determination to defend against it. All involved knew that winter was an inauspicious time to undertake this work. Even if the ice didn't gorge, the cold presented difficult challenges. The Portland mortar used to cement the limestone and granite blocks would not set during a hard freeze, so masonry work would have to stop whenever the temperature dipped below twenty-eight degrees. Furthermore, ice floes coursing down the river would slow or imperil the frequent movements of tugs and barges delivering men and new stone to the pontoon boats. Beyond those predictable issues, the simple threat of a gorge called for an expensive defense. Just upstream of each pier site, Eads installed a multipart icebreaker of his own design.

In winter, floating ice at St. Louis often passed through the city as a carpet of bergs and shards, an interlocking grinding expanse stretching from bank to bank, upwards of a foot thick and moving at three miles an hour. During the initial stages of stone laying, the caissons were very vulnerable to that inexorable mass. If ice displaced a floating caisson by as

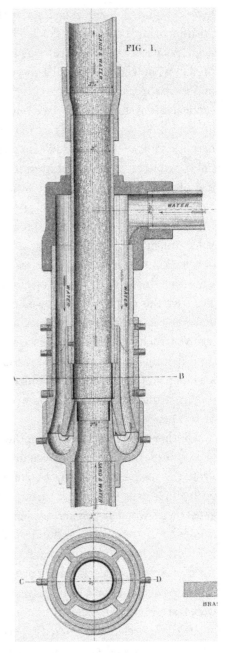

FIG. 1.

James Eads patented this three-foot-tall sand pump to remove spoil from the caissons. A water pump aboard a pontoon boat supplied the sand pump with river water under pressure (via the inlet pipe at right). The passages inside this device first directed that flow downward, then turned it upwards, creating a strong vacuum to draw the sandy spoil from the suction line at bottom. The pump ejected the mixture to a discharge tube outside the developing pier. Stones up to three inches in diameter did not choke it. Larger rocks were brought out via the air locks in the caissons. After their success here, sand pumps entered the standard toolkit of civil engineers. Engraving from Woodward, *History*, plate 12.

little as a yard, no tugboat on the river had the power to push it back into position against the winter flood. The bridge alignment would be sacrificed. A more ominous threat was perhaps just as likely. No one could predict what would happen if a gorge formed and then broke up, with its masses of ice crashing into a floating caisson laden with twenty or thirty vertical feet of stone pier. Still well short of the riverbed, that work in progress would present a broad face to the moving ice, while lacking any inherent stability. It was the stuff of nightmares. Arguably, Eads should have put the work off at least until spring of 1870, although the high water typical of that season would raise other problems. Characteristically, he chose to drive on by installing his novel icebreakers in the river.

The core challenge for the icebreakers lay in finessing a contradiction. While under construction, the caisson-pier combination needed a shield of some kind to ward off ice. The ice, however, would simply overwhelm a massive shield, or the scouring current could undermine and upset it. The trick was a defense that slowed the river, letting water through while deflecting ice away from the caisson and pier. In all, Eads created three defensive structures to counter the ice at each caisson site. About three hundred feet upstream of a pier's location, the floes would first strike an ice bumper of sheet iron, driven into the riverbed by a pile driver. Fifty feet downstream of that bumper, a clump of nine massive wooden piles would deflect heavy floes to either side.

Another fifty feet downstream lay the main defense. Viewed from above, this series of pilings, driven into the riverbed, outlined the letter A. From the apex, each side was two hundred feet long, made up of pilings with ample gaps to let water through, the pilings thoroughly bolted together with wooden stringers. An enormous "ice apron" hung from each side of the structure. Two hundred feet long and sixty feet wide, the aprons hung from the pilings that formed the A. The aprons' bottom edges were oriented upstream, resting on the riverbed. The design echoed the classic cowcatcher on a locomotive of the era. Made of oak, sheathed in iron, the aprons were held at the correct 45-degree angle by the current itself. All three elements worked to perfection. Ice nearly filled the river in late December and early January. But the icebreakers deflected the threat away from the piers and buffered the current. All those years in river salvage had given Eads the knowledge to manage the incalculable

Taken from the Illinois shore, this photograph gives a sense of the deflecting structures that Eads designed to ward off the ice threat. The initial ice bumpers are out of view to the right. Standing alone out in the river, upstream of each developing pier, were the second defenses: clumps of massive pilings wired together (both are visible here). Adjacent to the east caisson and pier, with its serving pontoon boats, was the main defense, the A-shaped structure of pilings. Driven into the riverbed and wired together, those posts supported the wooden aprons that deflected ice from the caisson—if all went well. Photograph by Robert Benecke, MHS, P0078.

forces at play. Luck too played a big role, as the river did not gorge during the winter of 1869–70.[24]

Building the East Pier

The caisson for the east pier was launched at the Carondelet yard of William Nelson on October 17, 1869. Looking down on it, the pier would have a hexagonal shape, its leading edge parting the currents. The caisson took the same shape, measuring eighty-two feet long by sixty feet wide. To hold it in place, six guide pilings had been pre-positioned in the river, driven deep into the sand. With two pilings for each side and two more at the upstream edge, the caisson was pushed into place by tugboats. Then the pile driver pinned it into position with two more massive posts at the caisson's downstream face. If there were no gorging, it would probably stay in place. No one could be sure.

It snowed heavily on October 20, followed by freezing temperatures on the twenty-fifth.[25] Aboard the *Gerard B. Allen*, a crowd of friends, stockholders, directors, and engineers braved the miserable weather that day to watch as a traveler operator gently placed the pier's cornerstone aboard the floating caisson. The great work had now entered a new phase. This milestone became national news, covered by the *New York Times*.[26]

All the preparations paid off, with Eads's novel cranes and travelers speeding the once laborious work of laying stone. After final dressing ashore, each block was marked with a number assigning its position in the pier. For three weeks, tugs shuttled barges laden with cut stone out to the *Ben Johnson* and the *Gerard B. Allen*. The traveler men swung the blocks onto the caisson, where masons built up the pier. On November 15, Eads wrote a friend, "I am just about landing my big pier on the sand 37 feet below the surface of the water. Yesterday the caisson had reached within 5 feet of it, the masons laying stone day and night."[27] Writing to friends and investors, Eads gave an impression of humdrum routine. In truth, great peril bore down at just this moment. The pier, a thirty-foot column of heavy stone, had yet to reach the riverbed, while the Mississippi current continuously buffeted its upstream face. Air under pressure inside the caisson supported this ponderous, inherently unstable mass of stone and iron. It amounted to a tower floating on a bubble.

Throughout the stone laying for the piers, the foremen and engineers had to manage three challenges with vigilance. First, the caisson received pressurized air continuously, supplied by compressors aboard the *Gerard B. Allen* and admitted by iron piping laid in the masonry. The quantity of air inside the caisson helped to regulate its descent. Therefore, the four air compressors needed constant adjustments. Second, the caisson had to remain perfectly level as it descended. Failure in that regard would stress the guide pilings, while air would escape from the caisson on the high side. Close attention by the traveler men and masons was essential to maintaining balance as they placed cut stone on the growing pier.

The third challenge arose from the rate of stone laying. The key was to track the growing weight of the pier against the desired height of its top, or working, face. To simplify the movement of stone, Eads, his engineers, and the traveler operators all would have liked the pier head to remain six to ten feet above the river. It was not to be. The massive weight in the developing tower pushed the masons' daily workplace below the river's surface within weeks. The engineers had foreseen this problem. As the caisson descended, as the stone pier grew, men progressively built up a kind of envelope, or skin, of plate iron to surround the masonry and keep the river out. It did not need to be perfectly watertight; pumps could deal with any leakage into the narrow air space between the pier and the encircling wall of plate iron. Nonetheless, the masons saw little to like when working increasingly below the river. Open air above but lapping water all around. By mid-November they had built the stone pier up to thirty feet of vertical height. At that time, the river's level was ten feet higher still.[28] The slow deliveries of Virginia granite contributed to this vexing problem, which constituted a worry if not a full-blown threat.

As the east caisson neared the sand of the riverbed, a typical workforce of nineteen laborers and twelve masons moved and laid stone on the pierhead. The work went on day and night, brilliantly lit by calcium lamps, the limelight of theaters. On the afternoon of November 14, the workplace routine, which was always tense, became instantly threatened. For unknown reasons, volumes of air began to escape from the caisson's bottom edges, bubbling to surround the pier in a froth. With that loss of air, the whole assemblage of caisson and pier descended twenty-nine inches in just forty-five minutes. The normal rate of descent was fifteen to twenty-four inches per day. Henry Flad immediately speeded up the compressors to regain the lost buoyancy. Working in miserable cold and rain, the men fully restored order the next day. Investigation revealed that the air loss and the worrisome lurch had resulted from concentrated current flows at the upstream face of the caisson as it neared the bottom. The solution, much appreciated by everyone, was to hasten the work to land the pier quickly on the riverbed. On November 25, the caisson reached that initial destination, thirty-seven feet down, exactly one month after the first stone came aboard.[29] Now it was the submarines' turn to work in otherworldly surroundings.

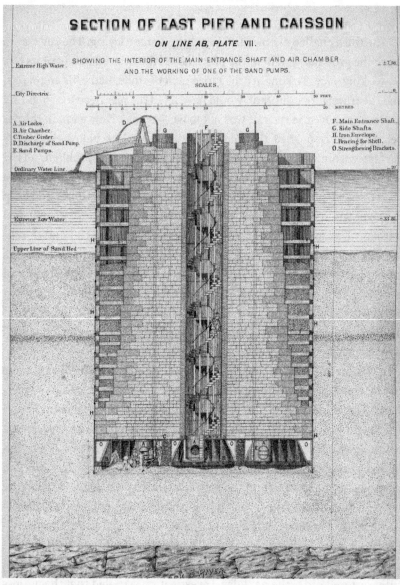

SECTION OF EAST PIER AND CAISSON

ON LINE AB, PLATE VII.

SHOWING THE INTERIOR OF THE MAIN ENTRANCE SHAFT AND AIR CHAMBER
AND THE WORKING OF ONE OF THE SAND PUMPS.

SCALES.

Extreme High Water.

City Directrix.

A. Air Locks.
B. Air Chamber.
C. Timber Girder.
D. Discharge of Sand Pump.
E. Sand Pumps.

Ordinary Water Line.

Extreme Low Water.

Upper Line of Sand Bed.

F. Main Entrance Shaft.
G. Side Shafts.
H. Iron Envelope.
I. Bracing for Shell.
O. Strengthening Brackets.

RIVER.

This sectional view shows the progress of the work by February 1870, but it appears at this point in the narrative (November 1869) to show the elements that made pneumatic caissons so audacious. A central iron staircase built into the masonry gave access to the caisson, which bore the growing weight of stone above. Limestone blocks made up most of the pier; larger granite blocks protected its exterior surfaces from ice and erosion. At the moment shown here, the work had passed two milestones: the caisson and pier were well rooted in the silt and sand, largely safe from ice, and the masons had built the pier up above the normal level of the Mississippi. Weeks earlier, the iron envelope surrounding the pier had protected the masons while they worked below the river's stage. From the moment shown here, the caisson had to descend another twenty feet or so to land on bedrock. Engraving from Woodward, *History*, plate 13.

Digging to China

There was no shortage of men seeking work in the caisson, digging away beneath the rushing river while a massive tower of stone grew over their heads. At least that was true before the mysterious pains began. After a boat ride out to the *Ben Johnson* or the *Gerard B. Allen*, the submarines climbed to the pierhead. To arrive at their unique workspace, they descended inside the masonry via a circular iron staircase built into its core. The pier also contained air and water piping as well as discharge lines for the sand pumps. In the gloomy half-light at the bottom of the staircase, the men entered an air lock and closed an iron door behind them. Standing in a dimly lit space capable of holding upwards of ten people, the workmen waited while the chamber operator opened a valve, admitting pressurized air from the caisson with a shrieking rush. Minutes later, the pressures equalized, and the entry door to the caisson swung open easily. The men emerged from the air lock to stand on yellow sand, their ears and sinuses strained by the sudden increase in air pressure. Lit by oil lamps that smoked badly in the compressed air, the caisson's interior scarcely seemed of this earth. It barely was. One visitor later recounted the sights: "Shrouded in a mantle of vapor, labored the workmen there loosening the sand; dim flickered the flames of the lamps, and the air had such a strange density and moisture that one wandered about almost as if he were in a dream."[30]

Inside the caisson, the engineers, foremen, and submarines endured dim, noxious conditions. In the compressed air, kerosene lamps gave "a very dull flame," and the air grew thick with particles of floating carbon. Twice, accidental fires started from the simplest cause, hand lamps igniting men's clothing. In this strange, pressurized demiworld, the flames proved very difficult to extinguish, and one man was severely burned. From that point, Superintendent William McComas banned oil lamps; candles provided the only light. As the depth and pressure grew, even candles proved tricky. At eighty feet, a man could blow out a candle, yet there was so much oxygen in the air that the flame would immediately reignite the wick if it had maintained a glowing ember. At one hundred feet, Eads repeated this conjuring thirteen times in half a minute.[31] A diverting trick to some, the phenomenon could worry a man in a sealed box

far underwater. Once started, any fire proved quite difficult to extinguish. Fortunately, the east pier caisson was mostly plate iron.

By February 1870, eighty submarines were working in the caisson's digging galleries each day, divided into gangs of eight to ten men. A foreman
hired and oversaw each gang, preferring younger men in good health,
muscular if somewhat short in height. About a third were native-born
whites, German emigrants accounted for another third, while Irishmen
made up most of the balance. Most were in their twenties; just six men
were older than forty. All could read and write. The doctor who surveyed
the workforce believed that many were "using strong drinks rather too
freely, and most of them chewing tobacco." He wished they would drink
less and eat more. Even so, Doctor Alphonse Jaminet believed them a
hardworking, healthy, and intelligent class of laboring men.[32]

At first the work was not difficult. The men dug away at a bed of yellow
sand, carting the spoil in wheelbarrows to the sand pumps. Normally,
three sand pumps—one for each digging gallery in the caisson—were in
operation simultaneously, each served by a gang of men.[33] Those patented
devices worked nearly flawlessly to eject the spoil into the river. At all
depths, the digging uncovered occasional pieces of brick or charred wood,
even bone fragments.

Those artifacts spoke to a fundamental question: was there a level,
short of bedrock, that the river currents never reached? This evidence of
life suggested otherwise, confirming Eads's conviction that the sandy riverbed had eroded or increased capriciously, even in the recent past. This
suggested in turn that the currents here were far too powerful to allow the
use of pile foundations, no matter how deeply driven into the sand. Pilings
commonly supported the piers of bridges upstream on the Mississippi. At
St. Louis, bedrock alone offered security.

Day after day, the gangs dug away, undercutting the sand at the caisson's perimeter sill and along the two massive timber girders that ran
across its center sections. That framing transferred most of the tower's
weight to ground. A visitor found it "a strange sensation to feel the massive pier sinking beneath one's feet, and the descent of an inch gave one the
impression of a much greater fall."[34] With the elevated air pressure,
the men sweated profusely even with the interior temperature at forty-
eight degrees Fahrenheit. In the early days, each gang worked three two-

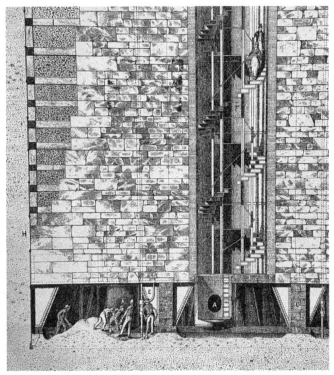

This detail illustrates the caisson's construction and the work of the submarines. It shows the main air lock (A) and two of the three digging galleries. The men dig away, carting sand in barrows to the sand pump (E), whose operator suctions the spoil. A fashionable lady and gentleman descend the circular staircase to observe these strange sights. Detail of an engraving from Woodward, *History*, plate 13.

hour shifts daily, with two-hour rest periods between the shifts.[35] The work went on around the clock. For three weeks the digging proceeded well. The masons too made good progress laying pier stone above the river stage. Then a prolonged cold wave settled over the area, halting their work for twenty days in December.

In those first weeks on the riverbed, curious and well-connected friends of the bridge company frequently descended to see the work unfold. Their accounts provide a window onto this subaqueous, underground world. For many, the passage through the air chamber caused no distress. Others suffered "pain in the ears, bleeding at the nose, or a feeling of suffocation." Once they were inside the caisson, the sights were interesting enough for about fifteen minutes. It did not take long "before the wish to escape again from this strange situation gained the upper hand." Still, hundreds made the trip, men and women alike, and all returned with some of "the most remarkable reminiscences of their whole life."[36] As the depth increased, the number of visitors fell off. For his part, Eads visited the caisson nearly every day, claiming and showing no ill effects.[37] At any depth, he hurried through the air lock, impatiently ordering the operator to equalize the pressures as quickly as possible. Eads knew the dangers that deep work under pressure posed for his men. His own stamina set an example that he hoped all would follow.

By December 22 the ice ran thick in the river, nearly blocking the passage of the *Little Giant*, the company tugboat bringing a shift change out to the pier. That day, Eads described the progress to an old friend:

> The cold weather stops my [masonry] work on the pier except that of undermining it which goes on steadily and securely at the rate of about 6 inches per day. This is not as fast as it will move when I get more sand pumps in operation, but as I don't have to go through to China with it, but only some 50 or 60 feet towards it, I have no fear of getting to the rock ere long. I must stop, however, soon or lay more masonry as the top is only 8 feet above water. The boats are provisioned at the pier for a siege of ice: today it was as much as the tug could do to get through the ice to it. The pier is some 13 or 14 feet deep in the sand, and with the ice breaker I have above it [upstream], is perfectly safe from all danger. It [the caisson] is 43 feet deep from the surface of the water. Everything has worked just as I expected but more slowly.[38]

Eads's confidence covered a somewhat tenuous reality. For two days the ice ran thick and hard; no boat dared the dangers to relieve the men. During this period, the gangs in the caisson worked their regular shifts, then ate and slept aboard a tug tied to the pier. These periodic ice blockades recurred in January, marooning the men for a day or two at a time. If a gorge had formed, the pier probably would have survived, given its growing weight and depth in the sand. The breakout of a gorge, however, would have carried off or crushed the pontoon boats, a serious setback. Instead, their luck held and the cold snap broke.[39] By then, another concern was growing.

A New Threat

By early February 1870, the caisson had reached a depth of sixty feet, and its internal air pressure had risen to 45 pounds per square inch (psi). As Eads had learned in France, and as Chanute had forecast two years earlier, at that depth and pressure many submarines began to feel a range of disturbing symptoms. Most of the men reported no problems. Among those afflicted, a temporary paralysis in the legs became common. A day or two later, most regained full motion. More serious cases paralyzed the arms, bowels, and sphincter. Some were stricken with acute pains in the joints or stomach.[40] The men were suffering from decompression sickness, known then as caisson's disease.

In the twentieth century, doctors and engineers came to understand its cause. In normal conditions, the bloodstream carries nitrogen gas. When deep-sea divers or caisson laborers undergo prolonged exposure to a high-pressure environment, that gas dissolves. If they return too quickly to the surface and normal atmospheric pressure, the nitrogen creates bubbles in the bloodstream. In turn, the bubbles can block normal blood circulation, starving muscle tissue of oxygen and causing acute pains and paralysis. In time, the bubbles can dissipate, along with the pains and paralysis. Or the afflicted can die.

Eads knew to expect men to be stricken at this depth, but understanding the causes of the disease or its treatment was difficult. Essential humanity gave motive enough to seek answers. Eads, his engineers, and the foremen were in the caisson every day alongside the laborers. No one could ignore their groans or their paralysis at the pierhead, the strong young men stricken just in the short ride to shore. Brutal practicality motivated the search for a cause. Eads knew that the pier had to go much deeper.

During the first weeks in February, the submarines themselves made light of their problems. The common nickname for the disease, *the bends*, amounted to modest ridicule of the afflicted, often stooped by the pain. Some men placed great faith in massaging paralyzed muscles with Abolition Oil. At its own expense, the company issued "galvanic armor" to all the submarines. Made of alternating scales of zinc and silver, these armor bands were worn at the wrists, arms, ankles, and feet. A newspaper ad described this as "a scientific and rational method of curing diseases

Men stricken by caisson's disease often walked doubled over in pain. Friends made a joke of their condition, describing it as "the Grecian bend." That term was first directed at women of the day, their posture altered by their fashionable hooped skirts. The publisher of this sheet music applied the joke to a different target. Image from author's collection.

originating in a disturbed condition of the electrical or vitalizing forces of the body. The only sure prevention of pier cramps and paralysis."[41] For his part, Captain Eads believed the armor was a "valuable" preventive measure.[42]

On February 15 the caisson reached a depth of seventy-six feet, and the problems grew worse. From then on, "severe cases of cramps and paralysis were frequent, and several cases were sent to the hospital." The submarines finally reached bedrock, at ninety-three feet, on February 28,

1870. The pressure inside reached 44 pounds above atmospheric, three times the normal condition for the human body at sea level. News of that welcome arrival at bedrock traveled immediately by the telegraph line that connected the caisson to the superintendent's office aboard the *Ben Johnson*. From there, a temporary line ran to the bridge company's office downtown. When word broke on the street, cannons boomed and steamboat whistles raised their shrieks. Down in the caisson, Superintendent McComas led his submarines in celebration, but he noted that "seven of them . . . are not enjoying it, on account of their suffering with the cramps."[43]

Their ordeals had barely begun. In early March the men's work turned to concreting. After cleaning the bedrock of all sand, they laid down layers of concrete made of water and crushed limestone, each layer less than a foot thick to encourage even hardening. Successive layers would fill the air chamber to its roof, and more concrete would seal the air lock and forever plug the staircase. Well before that work was completed, the submarines began to die. James Riley was the first, on March 19. After a two-hour shift, he returned to the surface apparently no worse for wear. Fifteen minutes later, he gasped, fell over, and died nearly instantaneously. Later that same day, a second man died at the hospital. Four more fell before the end of the month.[44]

We cannot know how these casualties affected their fellows, with death now crowding around them. They did strike for higher wages. At the time, they were receiving a comparatively generous $4 a day for working three shifts of two hours each. Common street laborers in St. Louis that year received an average of $1.65 for a ten-hour day.[45] Offsetting their meager pay, those ditch diggers and pipe layers did not fear death on the job. Within four days, Superintendent McComas broke the strike.[46] Replacement workers took their chances, weighing talk of pain or worse against the certainty of a wage that would feed families and pay rents.

On March 25, the *Missouri Democrat* carried the headline "Death from Compressed Air, The Coroner's Examination Yesterday, Startling Developments." This inquest into the death of a submarine, Theodore Baum, heard contradictory evidence at best. McComas said that he wanted the men to "take as much as fifteen or twenty minutes" in the air lock after their shifts, "but they almost always hurry." A doctor testified that "the sudden transition from the air lock to outside air has nothing to do with it."[47] If the inquest revealed little about why a man had died, the headline shouted a warning to anyone considering a job in the caisson.

At the end of March, the bridge company hired James Eads's family physician to take overall charge of the workforce. With the strike and the bad press, it was a wise move. Dr. Alphonse Jaminet immediately laid down new regulations for all the submarines. Shifts were shortened. A boat equipped as a floating hospital tied up to the pier. Men on their required rest periods stayed aboard; there were no more midday trips home or to a tavern. Beyond these recuperative measures, the doctor undertook scientific study of the men and conditions in the caisson. While the death rate slowed, nearly every day brought new cases of cramps and paralysis. One day, after visiting the submarines at full depth, the doctor himself became paralyzed, unable to move or speak. With his own experience and observations, Jaminet came to understand that the transition out of the caisson brought the most danger. He ordered the men to slow the release of pressure in the air lock; he even had its piping diameters changed to enforce slower decompression.[48] Perhaps it made a difference.

Finishing Below

After the submarines reached bedrock, the work of concreting slowed overall progress to an excruciating pace, the inevitable result of Jaminet's short shifts to counter the bends. During this difficult time (March 1870), the river rose, requiring even higher pressure inside the shrinking air chamber to keep the water out. When it reached 50 psi, Eads cut the working shifts to forty-five minutes.

The men were still concreting on April 13, when disaster struck. With the typical flood conditions of spring, the Mississippi swelled with rains and snowmelt runoff, rising well above the pierhead. The powerful current tore away much of the iron envelope surrounding the pier. The masons were again waiting for granite, so the only portions of the pier that rose above the high waters were the brick linings of the airshafts and the central stairway. The river quickly seeped through the brickwork, brown water cascading down the shafts and staircase. The submarines escaped in time, but it was a near thing. The caisson nearly flooded, and it took days to put everything to rights. The concrete filling of the east caisson was not completed until May 27. The masonry of the east pier finally rose to a safe level above the river in mid-August. Eads, the engineers, Superintendent McComas, and the submarines had prevailed, at a considerable cost in lives and delays to the overall schedule.[49] For the accomplished Milnor Roberts, the work was "the most arduous and the most interesting duty of my professional life."[50]

All the hard-won knowledge gained from work on the east pier paid off in building the west pier and the east abutment. Launched on January 3, 1870, the west pier caisson received its first stone on January 15, and it landed on the sand bottom on the twenty-fourth. Smaller in size, destined for shallower water, the west pier caisson reached bedrock, seventy-eight feet below the river's surface, on April 2. On May 8, the men finished the concreting inside its caisson and shafts, three weeks before work reached the same stage at the east pier. Crucially, the death toll on the west pier was dramatically lower, just one man, a young Dane named James Andrews. By contrast, the work on the east pier had killed thirteen. Shallower depth, improved knowledge, better screening of the workers, and Dr. Jaminet's rigorous procedures made the difference.[51]

Eads had originally planned to land the east abutment on wooden pilings driven deep into the sand of the East St. Louis shoreline. Pilings supporting heavy stone walls, piers, and abutments dated back to medieval cathedrals and Roman temples. River currents could have little effect at the abutment, which was built into the shoreline, the main reason that pilings would suffice. After his successes with the piers, however, Eads decided to use a caisson to bedrock here too. The change would raise costs by upwards of $175,000 ($4.2 million in 2023). It would increase the chances of death or injury, as bedrock was estimated at 136 feet down during high water—the deepest yet. That grim fact had steered Eads to pile foundations here in the first place. On the other hand, Jaminet seemed to be gaining the upper hand on the health and safety issues.

With another winter looming, the east abutment caisson—much larger than its pier counterparts—was launched in Carondelet on November 3, 1870. With this third foray into deep foundations, Eads and his team nearly achieved perfection in methods and results. The foremen had their pick of men, all experienced submarines. Eads developed a new way to light the interior brightly. His new glass lamps had integral exhaust pipes to carry off smoke and carbon. With its whitewashed walls and roof, the caisson interior glowed with light. In addition to the customary circular staircase, the east abutment had an elevator down to a pair of air locks. Instead of a wearisome trudge (it would total 190 steps eventually), men leaving their digging shifts rode back to the surface. Although the Otis patent safety elevator dated from the 1850s, twenty years later few buildings were tall enough to warrant this modern convenience. All these improvements made the east abutment caisson popular with visitors, who arrived in a continuous stream. One day so many Missouri legislators dropped in that the submarines had to knock off work. The legislature was debating a bill important to the bridge company, ample justification for their trip.[52]

As the work went deeper, McComas and Jaminet again cut the working shifts, down to forty-five minutes at 100 feet.[53] Furthermore, they strictly enforced new rules to slow compression and decompression in the air locks. When the men entered to start their shifts, the operator raised the pressure in the air lock at a rate of 3 psi per minute. When they left, the decompression rate was 6 psi per minute. These intervals were not nearly long enough, and, perversely, they had the problem and its solution exactly backwards. Jaminet needed to slow the *decompression* to ease or prevent the bends. Under his rules, men leaving the caisson at a depth of 100 feet (or 45 psi) spent five minutes in the air lock as its pressure bled off.[54] Modern practice for divers in the US Navy calls for at least 100 minutes to acclimatize the body to such a pressure differential.[55] Not surprisingly, Jaminet's new routines did not save the men from pains and paralysis. Still, only one man, John Hoots, died in building the east abutment.[56]

The improved mortality rate suggests that Dr. Jaminet's research and rules made a real difference. The modern author of *The Bends* credits him with significant advances alongside his notable errors. Of the 352 submarines, a countless number experienced the bends, suffering more or less gravely, then recovering.[57] Thirty were seriously stricken, some paralyzed for life. One poor invalid spent the rest of his days on a stool at the bridge, selling picture postcards to tourists. In all, 14 men died.[58] Despite those grim totals, Jaminet's investigations and procedures undoubtedly saved lives. He suffered the bends himself, and provided the foundational diagnosis of its cause. For the most part, he resisted the moralizing that often distorted medicine in the era, resisting as well the smug superiority common in middle-class views of working-class people. True, he rejected applicants who smoke or drank, preferring "single men of good habits." Modern medicine would confirm that preference. Unlike many of his contemporaries, he focused on physical causes of illnesses. His strict limits on time spent under pressure saved lives.

In 1871, Dr. Jaminet published a thorough report titled *Physical Effects of Compressed Air*. This significant work in the new field of industrial medicine proved its value immediately. Within a year, caisson disease began to strike the men working under New York's East River, building the piers for the Brooklyn Bridge. Like Eads, Washington Roebling put his men under the care of a doctor. Alfred H. Smith found the Jaminet study "exceedingly valuable."[59] Nonetheless, many men would die from caisson disease in New York. That toll seems especially tragic given that Smith too isolated rapid decompression as the likely culprit.

Although the two doctors largely diagnosed the cause correctly, three facts hindered an effective solution. Rather than seeking to advance medical science, they had a far more limited and practical goal: to strike the compromises that would improve safety while getting the caissons down quickly. The workers themselves amounted to a second problem. Their varied responses—from asymptomatic nonchalance to temporary pains, paralysis, or death—hindered understanding. Perhaps the biggest obstacle to solving the riddle of caisson disease lay in the culture at large. The submarines had ample cause to believe that injury and early death were inescapable and capricious realities of their times.

Two Blows

Notwithstanding all that the engineers had learned in building the river piers, Eads's decision to use a caisson for the east abutment tempted fate yet again. In his view, doubt plagued lesser men; he would always press ahead. This time his habitual drive led to a perilous gamble with other men's lives as he sent them to work at depths as great as 136 feet. And his decision to undertake caisson work as a second winter approached was a daring wager in the face of known perils, a bet he nearly lost. With the digging barely begun, a cold snap hit in mid-December 1870. Masonry work stopped for eleven days, then the river gorged. On January 1, that ice dam began to break up, wounding the icebreaker. After moving as far as fifteen feet, the gorge miraculously stopped and refroze. By dumping tons of stone riprap just upstream of the work, the men prevented the ice from shifting the caisson and the masonry.[60] A near thing. When the temperatures eased, the masons again laid brick and stone.

On March 8, 1871, another punch came without warning and landed hard. By then the masons had built up more than one hundred vertical feet of stonework. With the abutment eight feet above the river and the caisson well buried in sand and just ten feet from bedrock, a cataclysmic tornado tore through the city without warning. The East St. Louis docks and the rail yards along a half-mile stretch of riverfront received a battering. The pontoon boat barely remained afloat, all its crane timbers and cables broken, tangled, or lost. As Eads later described, "The men in the cabooses up in the frames of the 'Allen,' forty feet above the deck, went down with their little cabins." All work halted, the air pumps stopped, and the caisson filled with water. Amazingly, only one man died. Repairing the extensive damage took three weeks and cost more than fifty thousand dollars ($1.3 million in 2023).[61] With the repairs completed, the caisson reached bedrock on March 28.

Again the city rejoiced. Julius Walsh wrote to his wife, Josephine, the daughter of bridge company president, Charles Dickson, with the details: "The event was celebrated by the blowing of steam whistles, ringing of bells, firing of canon and other demonstrations of joy. . . . The bridge is now '*un fait accompli*,' all the difficulties have been manfully encountered and overcome and with their exchequer in a replete condition we shall soon enter the City of St. Louis over this grand work of art, the fabric of Capt. Eads's genius. The bridge will be completed in May 1872."[62] The reality was a bit more complicated. Eads's own report of May did predict a finished bridge just a year out, on May 15, 1872. The same report described the masonry work as nearly completed, although two thousand cubic yards of granite remained to lay. True, he noted some delays at the steel supplier and at the firm fabricating the superstructure, but Eads was sanguine, as always.[63]

Spurred by the March 8 tornado, Eads wisely chose to review his plans. Cyclones were a recurring threat across the Midwest each year, resulting from cold Canadian air clashing with weather flowing up from the warm

Gulf of Mexico as winter gave way to spring. Although they were common enough, the recent twister had visited unparalleled destruction. It had uprooted venerable sycamores three feet in diameter, hurled empty freight cars for hundreds of yards, destroyed every frame house in its path, and carried off a 25-ton locomotive.

Eads found a lesson in all this. With steel deliveries delayed, he set his engineers to redesigning the upper deck. Beneath its length and breadth, the roadway would now have a supporting system of riveted iron girders instead of the wood beams originally planned. That new substructure would double as a wind truss.[64] The young profession of civil engineering had just begun to design for wind loadings on bridges. This new feature would incur modest costs, but those were trifling concerns in the case of a structure built to defy time.

Yet time was starting to add up for St. Louis Bridge by May 1871. Everyone had performed brilliantly in their pioneering work on the foundations. Still, the timetable had slipped by a full year. And no one could really predict what challenges would come with the steel. It wouldn't be easy. The March 1871 cyclone revealed that luck and timing were shaping the bridge at St. Louis. If the Eads project had been "on time" on March 8, 1871, the tornado would surely have ravaged and wrecked its half-built steel arches. That grievous blow would have bankrupted the company. Eads and his team reaped a tangible benefit from delay. The new wind truss was a prudent choice in the face of an unknowable future.

CHAPTER 6

London and Real Money

THE FRAUGHT CHALLENGES of constructing piers and abutments preoccupied the men of St. Louis Bridge from the fall of 1869 to the spring of
1871. But that demanding work was only a foundation in two senses. The
steel superstructure would rise from the stone piers and abutments. And
as the prospectus of February 1869 had forecast, the bond issue to pay for
the superstructure would become feasible only after the company had
proven itself with masonry on bedrock.

Even as the piers went down, James Eads, William McPherson, William Taussig, the directors, and many stockholders turned to focus on
the bonds. They had to navigate a maze of interlocking issues. They
planned to secure $4 million from selling those securities, an exponen-
tially greater challenge than raising the $1.2 million paid in by stock sub-
scribers (see ch. 4). The new, larger sum fixed their sights on the bond
markets of New York or London. Any investment banker would need ample reassurance that the Captain's superstructure design was sound,
steel feasible, the schedule viable, and the revenue projections accurate. A
bond issue was essential for attracting and compensating the contractors
needed to complete the bridge: a steelworks, a structural ironmaker, and
a bridge builder such as the Keystone company.

Placing the bonds with a banker amounted to another design puzzle involving many players and elements, one that could only be completed if
all parts fell into place concurrently. The Eads-McPherson prospectus of
February 1869 would guide those players up to a point. But the work was

already a year behind schedule, altering the calculations of all involved. Timing itself would become another challenge.

Making the Cat Jump

Back in the fall of 1867, James Eads had received the welcome news that Edgar Thomson and Tom Scott backed his bridge venture in the contest with Boomer. From September to December 1868, Eads and William McPherson, the president of St. Louis Bridge, had negotiated with Thomson and Scott on a comprehensive deal to advance their complementary interests. Yet those negotiations had died in January 1869, with McPherson accusing his friend Scott of avarice. The bridge company's February 1869 financing plan brought new stockholders, funds to build the foundations, and the blueprint for issuing bonds to finance the superstructure. Despite that fresh start, St. Louis Bridge and the Philadelphia parties failed to hammer out terms during 1869.

Finally, McPherson had had enough. On December 9, 1869, he wrote to Scott's right hand, Andrew Carnegie, accusing the Philadelphia interests of bad faith. McPherson had learned that Scott and his associates were preventing two carriers allied to the PRR from signing contracts to use the bridge. Such guaranteed revenues would be essential for attracting any banker to market the bonds. McPherson wrote that "if the object is to hold on to contracts until you have something better than my word for the superstructure contract, why not say so at once for I prefer square talk to any such finesse." He then laid his cards on the negotiating table: "Supposing this was the way the cat was jumping, I got a meeting of the board yesterday, and have authority to deal fairly with you, and to give you a margin on [the] Superstructure of $250,000 . . . this is whether Bonds are negotiated or not and must be satisfactory. But on this you can rely, so no more of your delays in the matter."[1] His letter finally restarted the negotiations.

By February 5, 1870, the St. Louis Bridge Company had hammered out the agreement that McPherson sought. The other signatory was not the Pennsylvania Railroad nor the Keystone Bridge Company. Andrew Carnegie & Associates did the deal. The PRR president, J. Edgar Thomson, headed that triumvirate, which included the vice president, Tom Scott, and Carnegie. The deal would prove transformational for St. Louis Bridge, which was one reason why its terms had proven difficult to thrash out. The agreement also illustrates noteworthy elements that reflect on the larger economy. The PRR was the largest corporation in the world, yet its top two officers chose to act as free agents here, admitting no conflicts between their own personal interests and the corporate interest. In the agreement, Thomson, Scott, and Carnegie obligated themselves to put up a steel arch bridge. It was not a task they would undertake on their own. But St. Louis Bridge contracted with these men because Carnegie & Associates had the power to influence railroads, manufacturing corporations, and bankers. They exercised their power from behind a curtain, which explains why the principals chose to obscure their own names behind Carnegie's.

The power of these men partly reflected the undeveloped state of capital markets circa 1870. To be sure, railroad bonds traded on exchanges in London, New York, and other cities. But a new venture like St. Louis Bridge had no hope of tapping into those markets without clear commitments from powerful backers. Assured profits would prove essential for enlisting any banker to even consider underwriting the bridge company's bonds. As matters played out, only Junius Morgan proved willing to bear the risks of marketing those securities.[2]

The arrangements that the associates crafted with St. Louis Bridge and with the Morgan bank in turn provide a fresh look at Andrew Carnegie. In his account, written a half century later, Carnegie describes this as "my first large financial transaction."[3] Sources newly uncovered show that it was not his deal in any meaningful sense. It *was* large, although his responsibility in negotiations covered only a portion of the $4 million that he claimed in his *Autobiography*. A fair reckoning would place Carnegie as the agent for Thomson and Scott. His role made him more than a factotum but less than their equal. Andrew Carnegie was a man on the move in these years, always dreaming up more deals, pushing for more profit. Between 1867 and 1874 he transformed himself from the dutiful protégé of Scott and Thomson into a player in his own right.

Eads and McPherson would enlist Carnegie & Associates in plans for two other infrastructures essential to the overall project: the tunnel and the union depot. The bridge would not work without the tunnel. Washington Avenue could serve either roadway traffic or railway trains; it could not do both. For Eads and his colleagues, the tunnel and the depot were linked physically and organizationally. For the city and region at large, those improvements would eventually prove transformational.

Building a Deal with the Associates

McPherson's curt letter guaranteeing $250,000 ($5.7 million in 2023) in profits for Andrew Carnegie & Associates broke the negotiating deadlock. Shortly after receiving it, Carnegie wrote up a "St. Louis Bridge Memoranda" to lay out the profit potentials that beckoned to the trio. The memo proposed that Thomson, Scott, and Carnegie take the contract to make and erect the superstructure for St. Louis Bridge, then subcontract that work to the Keystone Bridge Company. He calculated that this alone would clear $380,055 for the three men.[4] The three associates owned roughly one-quarter of the shares in the closely held Keystone Bridge Company.[5] It is difficult to imagine a clearer case of self-dealing, but no one was in a position to object. Thomson and Scott had such benefits in mind when they founded Keystone Bridge in 1862.

As McPherson had intended, once the associates seized on the $250,000 guarantee, they became motivated to take several more steps essential to St. Louis Bridge. The superstructure contract was only the first. The second step entailed placing bridge company bonds with a banker or underwriter somewhere. In turn, the proceeds from issuing those bonds would provide the funds to pay the associates for the superstructure. In order to interest any banker or investor in those securities, St. Louis Bridge (and now the associates) would need a guaranteed revenue stream. Thomson's Pennsylvania Railroad would serve that purpose ideally. These component parts all advanced the blueprint that Eads and McPherson had drafted in their prospectus of February 1869.

Along with his memo for Thomson and Scott, Carnegie wrote up a second document under the letterhead of the Pennsylvania Railroad. It appears to be a summary of talks in Philadelphia among the associates in preparation for Carnegie's December 18, 1869, negotiations in New York City with the bridge company's Executive Committee.[6] That meeting would focus on the bonds; the associates wanted an option to market those securities.

Two merchant banking houses had expressed interest in the bridge company's bonds: a German firm and J. Henry Schröder & Company of

London. The British house knew risk, having marketed a complex $3 million bond issue for the Confederate States of America in 1863.[7] As Carnegie's summary document described, Schröder had offered to broker $1 million in bridge company bonds at a price of 65 in gold. At one-third off the bonds' par value, the bid suggests that Schröder saw plenty of pitfalls in St. Louis Bridge. The bank would be left holding a bagful of securities if London investors did not cotton to an American bridge bond from far-off St. Louis. Schröder appeared genuinely interested, but closing the deal would require sending an emissary to London to explain how the bridge revenues and bond payments would be secured by exclusive-use contracts with railroads.

The New York meeting on December 18 resulted in real progress. The bridge company gave Andrew Carnegie & Associates an option to market a tranche of first mortgage bonds with a par value of $2.5 million. The option had a narrow time window, and it specified that the bonds had to yield for the bridge company 90 percent of par value in US currency. On its own the option (later extended) could prove lucrative for the trio, although its stipulated yield of 90 percent appeared to foreclose a deal with Schröder. The real value for all parties, however, would follow a successful bond placement, which would generate the funds to build the superstructure. That work could prove lucrative for many.

At a subsequent meeting in New York on February 3, 1870, Carnegie and Eads came to terms on many topics. Their complicated, interlocking agreement deserves a detailed review for the light it sheds on three larger issues. The specific terms demonstrate the power wielded by the associates. Furthermore, the clauses created a web of self-interest to bind all involved. As a whole, the deal reveals how the parties intended to make a very real bridge out of an aggregation of promises. Here are the main points, taken from a memo in Eads's hand (recollect that these negotiations took place before much of the masonry work described in the previous chapter):

"Andrew Carnegie & Associates agree as follows:"

1. To "construct and erect the superstructure of the bridge at St. Louis in conformity with the plans and specifications of the Engineer [and] at the prices . . . named in his printed report [of]

May 1868." That report had estimated that the superstructure would cost $1,460,418.30.[8]

2. St. Louis Bridge would supply the requisite steel. The associates agreed to pay for steel at the rate of twelve cents per pound, with payments made directly to St. Louis Bridge.

3. "Carnegie & Co agree to complete the erection of the Bridge within 12 months from 1st December 1870."

4. Completion and erection by that date "shall entitle C. & Co to a bonus of $250,000."

5. "As a matter precedent" to this contract, Andrew Carnegie & Associates "will cause to be made" contracts with two lines allied to the PRR serving East St. Louis, obligating those carriers "for the perpetual use of the bridge."*

6. In another precedent matter, Carnegie & Associates agreed to place with an investment banker the bridge company's first mortgage bonds, with the obligation to net for St. Louis Bridge a price of 90 (in gold) for that $4 million issue. (To clarify, the associates agreed here that St. Louis Bridge would receive gold worth at least 90 percent of the face value of its bonds.) This was amended to $2.5 million to accord with the option that Carnegie had secured earlier.

7. For its part, St. Louis Bridge agreed that Carnegie & Associates would be credited with $800,000 in bridge company stock. As with all its stockholders, St. Louis Bridge expected the associates to pay only 40 percent of the par value of those shares. Under this deal, however, Carnegie & Associates could pay for their shares out of "credits for bridge work."[9]

There is much to be gleaned from each element of the agreement. One aspect of its first provision—the associates' commitment to use Eads's 1868 estimates to set their own price for the superstructure—is surprising, even stunning. Presumably, engineers at Keystone Bridge had reviewed those figures, but Keystone's chief designer, Jacob Linville, had grave doubts

---

* The St. Louis & Alton and the Vandalia (then under construction and completed in the summer of 1870).

about nearly every element in the Captain's drawings. Besides, Eads and his team had been revising the superstructure plans almost continuously in the twenty-one months since developing those estimates.[10] In accepting those dated figures, the associates revealed their own eagerness to close a deal.

The second provision, regarding steel prices and payments, could allow Eads and other insiders at St. Louis Bridge to add their own markup to the price of steel. The promise of the third provision, a completed bridge by December 1, 1871, would gladden many hearts. To any sober planner, however, that deadline was pure fantasy. Even as the parties were negotiating this deal (in February 1870), neither river pier had reached bedrock, and construction had not begun on the east abutment. As matters played out, masonry work on the two abutments and the two piers extended well into the summer of 1871. The fourth clause codified McPherson's bonus offer of a few months earlier.

The perpetual-use contracts covered in the fifth provision would prove to be the mainspring for the agreement and the entire project. Those obligations indicated that the bridge would be profitable and that the PRR stood with this venture, without spending a dime of its own money. Reflecting the strength imparted by the fifth clause, the sixth provision suggests that the associates had already received indications that a banker somewhere could be enticed to take the bonds at 90 percent of their par value. Without such an indication, both the sixth provision and the entire deal would have been stillborn.

The final clause produced more value out of thin air. Paid with shares in the bridge company, the associates could sell those stocks off as progress in construction created demand for them. Or Thomson, Scott, and Carnegie could hold those equities until the bridge was in business and paying dividends. Just what or how the associates would, in turn, pay Keystone Bridge for its work on the superstructure introduced whole new fields for creative capitalism.

The Eads memo survives among the Carnegie Papers, an archive that also contains a document in Carnegie's hand titled "Memorandum of an Agreement," referring to the deal between the associates and St. Louis Bridge.[11] The Eads draft has more detail, and the Carnegie version largely aligns with it, but the few substantive differences are important. In the

Carnegie memo, the associates committed to completion before December 1, 1871, provided that they received working drawings "without unreasonable delay." Furthermore, that completion deadline depended upon the bridge company's finishing the masonry foundations by July 1, 1870. A third source yields another crucial aspect of the final deal: the associates would bear all the risks in erecting the superstructure.[12] That liability was potentially huge if utterly opaque given the novel design.

The final, signed copy of this essential agreement has never surfaced. Even so, enough is known to reveal the deep hunger for profit motivating all the participants. Probably too deep for their own good. The terms suggest that the signatories felt no fiduciary duty toward their own corporations and stockholders. In their desire to close the deal, each side skipped over difficult issues in scheduling, testing, arbitration, and resolution of any matters in dispute. The parties must have envisioned the need for subsequent agreements, which came soon enough.[13]

It seems that Eads was negotiating in good faith; only this deal could get his bridge built. On the other hand, the associates had ample reasons to prefer a Keystone truss superstructure for St. Louis. That alternative would result in an adequate bridge, faster, at a lower cost, and with less difficulty than the Captain's untried ideas. Prudence and profit lay that way. Furthermore, Keystone's chief engineer had excoriated Eads's original design. And Jacob Linville had already laid conceptual and detailed groundwork for his own alternative for St. Louis. Linville believed that *his* structure "would stand for ages." Soon after he penned those words in June 1867, Scott and Carnegie had tried and failed to ease Eads away from his steel arches. This comprehensive deal of February 1870 put the associates at the center of the Captain's project. That position gave them immediate gains, while increasing their future leverage. With time, the men in St. Louis surely would come to accept what the men from Philadelphia knew to be true. Steel was an unknown commodity. Eads's arch design "was entirely unsafe and impractical."[14] And Keystone had already perfected suitable designs in proven wrought iron.

By modern legal and ethical standards, all the parties here—and their agreement—do not stand close scrutiny from any angle. But that verdict removes them from their own era. The historian Richard White has studied similar wheeler-dealers in railroad finance of the Gilded Age and offers insights into their attitudes and actions: "Like business, morality reduced to a bottom line that reflected an increase in wealth. Bad men were bears. Good men were bulls. Bad men lied and manipulated information to drive down values. They hurt investors to help themselves. Good men lied and manipulated information to maintain or increase values. In helping themselves, they helped investors. Men of character considered themselves the final judges of their own rectitude."[15] This agreement between the bridge men and the associates committed the signatories to executing specific, demanding objectives. While it took the form of a contractual agreement, many crucial agents fell outside the contract. It did not bind them.[16] And all the while, Eads and Carnegie aspired to different outcomes. Lawyers then and today would scoff at the nature of the agreement, while an investment banker would immediately perceive the value added by each provision.

J. S. Morgan & Company and the London Bond Market

In the unfolding sweep of business history in the decades after the Civil War, Junius Spencer Morgan receives credit as the man who made American capitalism safe for British investors. Because amoral scoundrels had often corrupted the finances of American railroads, Morgan fulfilled an essential function for his wealthy customers. He thoroughly vetted the companies that issued the bonds he underwrote, essential assurance to the merchant bankers, industrialists, landed gentry, and nobles in Britain's narrow upper class who bought those securities.

Those investors also knew that Morgan would stand by the issuing company if it fell on hard times or—God forbid—defaulted on its obligations. Two powerful motivations impelled that allegiance. His bank depended upon a continuous flow of new investing capital through the offices at 22 Old Broad Street. In short, the business depended above all on maintaining an unimpeachable reputation for himself, his partners and allied banks, and the companies they underwrote. Morgan stood by those firms in hard times for a second, practical reason. Believing them sound for the long haul, he could make good money from their short-term problems. Morgan extended interest-bearing cash loans to troubled companies, while his firm reorganized their debt, also taking fees for that valuable service.

While protecting his own reputation, Morgan stalked risks continuously. Hazard or uncertainty could bring reward, more business at higher rates of return.* As equity holders in St. Louis Bridge since February 1869, the Morgans had privileged insights into the company. They also had an inside track to underwrite the bond issues forecast in the Eads-McPherson prospectus. A deal still had to be negotiated, however.

In December 1869, Andrew Carnegie & Associates had secured an option on the bridge company's bonds. In the February 1870 agreement, that trio had driven themselves into the heart of the bridge project. Placing the bonds with an investment banking house was the next challenge. This proved more difficult than the associates had anticipated. Tom Scott had predicted that they would place the bonds with a Philadelphia banker.[17] That hope quickly evaporated. As Carnegie wrote to Eads on February 25, those bankers "seem afraid of anything in the shape of a Bridge—[to them] Railway Bonds with not a tithe of assured revenue are preferred."[18] He began to pack his bags for a trip to London, the world center of finance.

Carnegie's bond-selling mission to London features in every biography of his life, in every history of St. Louis Bridge, and in most accounts of the Morgans. It makes an appealing story: the Scots immigrant, up from humble origins and just 34, meeting in a well-appointed London office for talks with one of the world's leading financiers about a multimillion-dollar loan. Although widely repeated, the portrait in Carnegie's *Autobiography* is inaccurate in nearly every detail and misleading in its larger impressions.† He claimed that the bond deal had originated when the St. Louis promoters knocked at Carnegie's door, that he had opened the negotiations

*   On October 24, 1870, Junius Morgan demonstrated his tolerance for a gamble by underwriting a loan of 10 million pounds sterling to the French government. Paris was then besieged by the armies of Prussia, and the emperor of France was a prisoner in Germany. France's long-standing bankers in London had refused to consider a loan, but Morgan did the deal—which was essential to France and profitable to the bank.

†   The aggrandizing claims of the *Autobiography* were not the faded memories of an old man. They reflected Carnegie's fundamental character, described in private recollections by his friend Mark Twain: "Longheaded . . . in many and many a wise small way—the way of the trimmer, the way of the smart calculator . . . keeping a permanent place on the top of the wave of advantage while other men as intelligent as he, but more addicted to principle . . . get stranded on the reefs and

ANDREW CARNEGIE.

This engraving accompanied Andrew Carnegie's debut on the national stage in a front-page profile in the *Railroad Gazette*. He made a striking appearance: short (barely five feet tall), youthful, with a shock of white hair. The profile claimed that he was worth a million dollars and held the presidency of the Keystone Bridge Company. The source of those falsehoods goes uncredited. Engraving from J. D. Reid, "Andrew Carnegie," *Railroad Gazette*, Nov. 19, 1870, 169.

to place the issue, and that he had acted alone in taking the deal to London.[19] All false.

When Andrew Carnegie took passage for Great Britain on March 9, 1870, a well-established New York banker, James D. Smith, accompanied him aboard Cunard's SS *Russia*. Smith represented the bridge company's stockholders, while Carnegie stood for the associates. Smith was a partner in Jameson, Smith & Cotting, a private New York bank that, like the Morgans, had taken an equity position in St. Louis Bridge a year earlier. Both men carried letters of introduction clarifying their goals. Pierpont Morgan penned the document Smith carried; Thomson, the one Carnegie brought.

------

bars." Twain, *Eruption*, 40. Twain wrote this biting character sketch in his own autobiography, unpublished until after his death.

Smith and Carnegie were stepping into the next act of a drama largely plotted in advance. Junius in London and Pierpont in New York operated as a single unit, being in close touch by letter and telegram. Indeed, the letter by Pierpont that Smith carried noted that "some of the facts connected with this bridge are now known" by the head of the London house—his father. Even so, Smith and Carnegie needed to present themselves in London as emissaries for the deal. Junius wanted to look a man in the eye before staking his reputation, and the money of his investors, on the business acumen and integrity of any new debtor.

Those foundational steps mattered particularly in this deal. From Morgan's perspective, St. Louis Bridge would seem to have been a perilously improbable bet. Smith and Carnegie were asking the house of Morgan to fund the novel designs of a neophyte civil engineer committed to an untried metal. Construction would unfold in a wild river nearly 4,200 miles from London, far from Morgan's purview. If St. Louis Bridge fell short of the mark in any way, financial failure would be hard to avoid. Morgan had plenty of experience investing in the expansive future of the United States. This time, the challenges appeared nearly incalculable.

Yet Junius Morgan had clearly signaled his interest in this project. Smith and Carnegie did not cross the Atlantic on a vague hope. The men would negotiate over the price that Morgan would pay for the bonds. Their face value, $4 million if he took the whole issue, represented the debt obligation for St. Louis Bridge, the amount it would have to pay out, plus interest, as the bonds matured. Morgan would pay less than face value since he would then shoulder the burdens involved in selling the bonds to investors. Indeed, he might not take the whole issue.

The letter Smith carried conveyed Pierpont Morgan's enthusiasm for St. Louis Bridge. After listing its most prominent financier-stockholders, Morgan wrote that with "a judicious management" at the company, the bonds were "unquestionable." He then offered this assurance: "Whatever possibility of accident at one time existed has now been entirely removed, for the east pier—where the danger lay—reached the rock [bedrock] last week. The permanency of the structure is therefore secured." The ice gorge and the cyclone lay in the unknowable future. Pierpont boosted "the value of the property," as "several of the leading railroads" from the East had contracted to use the bridge in perpetuity. He closed by "asking your favorable consideration of the propositions which Mr. Smith may lay before you."[20] Eads and McPherson could not have asked for more. Smith represented the stockholders in the London negotiations. On their behalf, he hoped to place upwards of $1.5 million in bonds.

Edgar Thomson opened his letter by introducing "my young friend Andrew Carnegie Esq. . . . who now visits Europe for the purpose of disposing of Two and a half Millions of the St. Louis and Illinois Seven per cent Bridge Bonds." He highlighted the progress of the east pier, the strength of the steel superstructure (although no steel had yet been ordered, supplied, or tested), the contracts with eastern railways, and the growing commerce of St. Louis with its regional population of three hundred thousand. Requesting a hearing for Carnegie, Thomson closed with this: "Of the soundness and value of the security that he has to offer, there can be no doubt."[21]

Smith and Carnegie arrived in London in mid-March. Negotiations began within days, proceeding at a leisurely pace. The first man to die from caisson disease succumbed on March 19; others soon followed.[22] It is doubtful that such sour tidings reached London. Sometime before March 25, Morgan made an offer to Carnegie for a $1 million tranche of bonds. The price is unknown, but Thomson thought it too low when consulted by telegram. On March 25, Carnegie happily responded that Morgan had raised his bid, offering 70 in gold—which translated to 85 in US currency at the New York exchange rate.[23] With that offer, Carnegie fired off cables seeking guidance or approval from McPherson at the bridge company and Scott in Philadelphia.[24]

In considering how to approach the deal, Junius Morgan weighed many issues. The profit prospects of St. Louis Bridge appeared straightforward and strong. As he wrote to Pierpont on March 24, "There can, we think, be but little doubt that it will be a financial success." Even so, it nagged at Junius that the bridge was "liable to great risks which cannot be provided for." His conversations with "one or two friends" offered assurances that other bankers might form a syndicate to share the burden of underwriting the bonds. But "it is extremely difficult to form an opinion as to how the public would view a mortgage bond upon a *Bridge*."[25] That unknown shaped two key questions: how many bonds would Morgan take, and how much would he offer?

On March 26, Junius wrote Pierpont that he would "decline to go further" if the Americans rejected his offer on the table, which was to pay 70 in gold for bonds worth a million in par value (i.e., he offered $700,000 in gold).[26] Two days later, a telegram from McPherson to Carnegie at Morgan's London office brought welcome news. The west pier had reached bedrock that day, and the St. Louis investors in the bridge had acquired control of the Missouri Pacific Railroad.[27] Carnegie and Smith knew how to cast these developments in the most favorable light, and the talks continued. On March 31, the parties agreed to terms: J. S. Morgan & Company would pay 70 in gold for the $1 million tranche, and the banker took an option for a second tranche of $1.5 million in bonds at 72.[28] Carnegie sought approval from his principals, who quickly signed off by the Atlantic cable.

What are we to make of the associates' deal? If Carnegie *was* a principal, the bond salesman negotiating his first big deal, perhaps Morgan had gotten the best of him.[29] In their December option, the associates had guaranteed that they would return a price for the bonds of 90 net in currency to the bridge company. Yet the price Morgan paid amounted to 85 in US currency at the current exchange rate in New York. Carnegie may have done the best he could. Nonetheless, the associates came out of this bond deal under water. That fact would heighten their desire to profit in other ways from St. Louis Bridge. In a letter to Thomson on April 2, Carnegie highlighted those lucrative stratagems. Even though they started a bit behind on the first mortgage bonds, Carnegie predicted that the three men could clear a profit of $900,000 ($22.9 million in 2023) from their wide-ranging roles in the bridge project—with "not a cent invested" of their own money.[30]

On April 2, 1870, the London house reported to Drexel, Morgan & Company in New York that the first tranche of bonds was selling briskly, so well that J. S. Morgan & Company found it impossible to reserve a portion of the issue for sale by its allied house in New York. The letter noted that "the lists were opened today and the subscriptions are quite as large as we expected."[31] As usual, success had many fathers. Carnegie was one. He wrote to Thomson: "I have been engaged daily expounding this enterprise to various members of the Press—writing prospectuses, etc. It grew better daily."[32] He detailed his colorful manipulations of British financial journalists: Because Congress had declared the bridge to be integral to a national postal route (untrue), the company was under the "direct jurisdiction" of the US Supreme Court (untrue). Therefore, neither St. Louis Bridge nor its investors need fear that corrupt state judges had any power over the company (also untrue). Carnegie's gleeful embroidery played upon all-too-real fears among London investors. They knew that Jay Gould had wielded compliant New York judges to raid the assets of the Erie Railroad.[33] Carnegie's were the classic lies of a good man. He wanted only to increase values.

Soon Morgan and a syndicate of London bankers were selling the second tranche to investors, the remaining $1.5 million of the associates' option, for which Morgan had paid 72 in gold. That price amounted to $87.88 in US currency in New York, nearly matching the trio's obligation to the bridge company. With sales still strong, Morgan then took a third tranche, $1.5 million, from James Smith at 90 percent of par (in US currency).[34] Junius Morgan was now all in for St. Louis Bridge. In May, he wrote to Jameson, Smith & Cotting seeking to enlarge his personal stock position in the company. The New York bank replied that a diligent search had failed to uncover any stockholders willing to sell, even at a 25 percent premium over the shares' par value.[35]

J. S. Morgan & Company did well on the bond deal. Word filtered back to William Taussig that Morgan, having paid 72, then sold the associates' tranches within ten days to a broad array of bankers and investors at 90 in gold.[36] The bank's gross profit of $450,000 ($10.7 million in 2023) demonstrated why bond underwriting could prove an attractive line of work, at least for an elite with an astute grasp of risk and markets. The security enjoyed strong demand on the London exchange, trading at 87 a year later, with the price rising to 99 in April 1872.[37]

This bond issue ultimately grew out of a massive, consequential shift in the decades after 1850 as wealthy Britons shifted capital out of landholdings and into financial assets, investing around the globe. Profits earned in shipping and by manufacturers added to that flow overseas. Domestic realities pushed some funds abroad. As land values fell in England (driven down primarily by cheap imported American wheat), the gentry redeployed capital. Attracting more money abroad, alluring foreign projects encouraged UK investors to diversify their portfolios. As beneficiaries of

a worldwide empire—governed by British law, operating on the gold standard, and protected by the Royal Navy—Britons could and did invest
nearly anywhere. But American assets and companies topped their shop-
ping lists.[38] This context explains why British buyers saw opportunity in
far-off St. Louis. The two hundred–odd bondholders in St. Louis Bridge
included a covey of landed gentry, a rear admiral and a major general, the
Rothschild bank, and Queen Victoria's physician in ordinary, Sir William
Gull.[39]

Those investors looked forward to receiving interest payments at the
rate of 7 percent of their bonds' face value, in contrast to the 3 percent return on the bonds of the Bank of England. Attractive to venturesome investors, the high rate was a numeric representation of the many potential
pitfalls embodied in St. Louis Bridge.[40] Interest payments, due in gold, began to accrue on April 1, 1870. This was a problem for a business without
customers or income. Because the company could earn nothing until it
completed the bridge, the board authorized a 20 percent call on its stockholders. That pot of funds would service the bonds. For the first time,
those men were being asked to pay something for their ownership stake
in St. Louis Bridge.[41] To this point, the Eads-McPherson financing plan of
February 1869 was unfolding exactly as crafted.

A Man on the Make

The London bond deal represented a turning point in the career of Andrew Carnegie. For many reasons, however, it is difficult to understand
exactly who or what Carnegie was in March of 1870. To April 1865, he had
been a rising star in management at the Pennsylvania Railroad. From 1874,
he became the hard-driving prime mover of an innovative steel company.
During the interval, the man was a bridge salesman and an opportun-
ist, speculating in stocks, searching for deals, and profiting from "crony
capitalism at its most basic."[42]

Repeatedly in these years, Carnegie took the role of front man for
Thomson and Scott, another factor obscuring his actual position and
power. He became famous later in life, but in 1870 both portions of the
moniker "Andrew Carnegie & Associates" were opaque ciphers. Thomson
and Scott clearly valued his quick mind and sound financial judgment. In
letters and memos, Carnegie assiduously deferred to his principals. They

made him a partner in deals like those arising from St. Louis Bridge because they had a railroad to run and he was so brilliant in dreaming up new ways to make money. He anticipated that the bridge would result in eleven distinctive sources of profits for the three associates.[43] As noted by his best biographer, all of "his business adventures," resulted from "the umbilical cord that connected him to Thomson, Scott, and the Pennsylvania Railroad."[44] That connection explains why the ventures often made good money, and why they needed a front man in the first place.

Some evidence, cloudy but intriguing, suggests that Carnegie struggled in these years to reconcile his lucrative work for Thomson and Scott with his own sense of self-worth. In December 1868, he wrote out a list of assets. Ever since, biographers have made much of the document, although it is unclear why he totted up this reckoning.[45] He listed interests worth $400,000 in nineteen named firms that generated an annual income to him of $56,000 ($1.2 million in 2023). The historian John Kouwenhoven believed that Carnegie compiled the list to gauge "exactly how dependent he was for his handsome income upon continued subservience to Tom Scott."* This is only inference; the memo has no heading or explanation. The speculation seems justified, and it may suggest that Andrew Carnegie had moments when he found it difficult to know what to make of himself.

For the most part, he radiated prosperity and optimism. In January 1870, he wrote to the general manager of the PRR's subsidiary line to Chicago requesting an annual pass to travel at no charge. "As for paying [the] fare, that is altogether too bad a habit for one of my standing to be guilty of."[46] Two weeks later, Carnegie sent to Scott his no-cost block of shares in St. Louis Bridge, bearing a par value of $150,000, one element in

---

* Kouwenhoven offered further inference to support his impression that Carnegie chafed to become a player in his own right, not simply the agent of bigger men. Carnegie wrote another memo to himself in December 1868: "Thirty three and an income of 50,000$ per annum. By this time two years I can so arrange all my business as to secure at least 50,000 per annum." Wall, *Carnegie*, 224–25. From this, Kouwenhoven posed an obvious question: if Carnegie was already making $50,000 a year in 1868, why would he need two more years to earn the same income? Kouwenhoven's answer: "If one accents the word 'my' in Carnegie's phrase about arranging 'all my business,' the question disappears." JAK, notes titled "Carnegie's 1868 Turning Point," K8/12.

the comprehensive deal with the associates. With the stock he included a prediction: "within five years it will be worth double."[47]

Components of a System

From the beginning, James Eads's innovative bridge has commanded the attention of St. Louisans, journalists, visitors, and historians. Although nearly invisible, its connecting tunnel made the Captain's design possible by taking trains off the bridge and routing them under the city. In turn, the route created by the bridge and tunnel, as well as the high cost to build them, argued for a single station to serve all the carriers converging on the city. With its monopoly grant from the state of Illinois, the bridge would gather up the business of the eastern lines and should fill the pockets of its stockholders. Eads, McPherson, Taussig, Carnegie, and other bridge insiders planned for a consortium of railroads to fund the tunnel and depot. Again, those men would take insider seats and profits in promoting this new venture. Again, the customers—freight or passenger—would bear the ultimate costs, passed along by the railroads' pricing. Eads's unified system could wrest advantages for himself and his city from St. Louis's delayed connection to the national railway map.

In May 1870, the St. Louis City Council unintentionally illustrated the advantages of direct track connections linking different carriers. The Missouri Pacific, newly under the control of bridge insiders, requested permission to lay rails down Poplar Street to the river, then up the levee. This half mile of new track would have connected the MoPac and the North Missouri, both controlled by Eads, McPherson, and their associates. The councilors said no, perhaps reasonably enough. Trains on the levee would disrupt freight handling for steamboats at the very least. The veto created a problem for the North Missouri, however, which had leased some sleeping cars to the Atlantic & Pacific. Instead of running those cars across a short connection in town, the sleepers had to travel from St. Louis to Kansas City on the North Missouri, then return to St. Louis on the Missouri Pacific (for delivery to the A&P). As the *Railroad Gazette* reported, "The actual connection between the North Missouri and [the Missouri] Pacific railroad depots *in St. Louis* is 554 miles."[48]

At this moment, James Eads laid groundwork for yet another grandly innovative idea. In his May 1868 report, Eads had sketched plans for the

Architect's rendering, "Proposed Grand Union Passenger Depot in Saint Louis."
Portrayed without the crowded streets of downtown, this westward-facing image reflected Eads's approach to design. Unshackled by precedent, he offered
up brilliant novelty on a largely blank slate. The station would have fronted on
Fifth Street. The long side elevation on Washington Avenue included an arched
train shed. The four-story structures would have housed a trading floor for the
Merchants Exchange, a hotel for salesmen and drummers, and offices for local
businesses. This grandiose, impractical structure never advanced beyond these
plans. Description and Plans of the Grand Union Passenger Depot in St. Louis, frontispiece,
MHS, B38/1.

tunnel under downtown and for a union depot in the Mill Creek Valley.
That single station would welcome the passenger trains of every St. Louis
carrier (see illustration on p. 5). The Missouri Pacific already had its
westward-reaching tracks and its own station there, so the plan had sound
logic—which Eads simply laid aside in 1870. Instead, he led the bridge
company directors to propose a huge new union station on Washington
Avenue just five blocks west of the river.

Top architects in St. Louis, Barnett & Piquenard, executed the plans for
the new depot. Its general conception and precedent-shattering details
certainly originated with Eads. The drawings detailed a station with
eleven tracks, laid well underground at twenty feet below street level. Its
capacity would have exceeded the accommodation for trains at the new

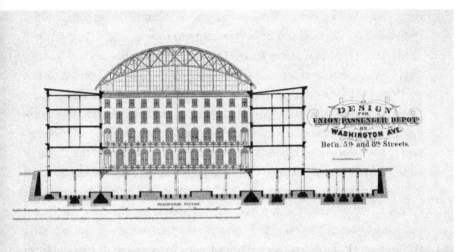

This sectional view shows how Eads's grand ideas came together. In this eastward-looking perspective, three tracks for through freight trains run beneath Washington Avenue (*at right*). The arched roof creates a vast atrium, flooding natural light all the way down to the eleven tracks for passenger service (ten were to be stub, or dead-end, tracks). Their raised platforms were rare in this era. The four stories of windows shown here would share the atrium's light with the interiors of the proposed Exchange room and hotel, to rise just east of the train shed. Offices would have filled the floors on either side. *Description and Plans of the Grand Union Passenger Depot in St. Louis, following p. 13, MHS, B38/1.*

Grand Central Depot then under construction in New York City (it would open a year later).[49] Passenger trains would use the tunnel to access the station. Given its subterranean tracks, the proposal envisioned "smoke consuming" locomotives. A marketing pamphlet claimed that "by deadening the floors and operating the trains with signals [rather than whistles, to control train movements], no noise will be created, and the occupants and guests above will suffer no inconvenience from that cause." The upper floors would accommodate a large hotel, the trading floor of the Merchants Exchange, and 330 offices for rent. In its conception and details, it broke with every other railway station in the world.

These bold plans were utterly impractical in the age of steam locomotives. Although this design never left the drawing board, it deserves notice here given that this grand station received serious consideration for a time. Architects drew enticing plans, and William Taussig collaborated with Andrew Carnegie to draw investors to the project.[50] The venture sheds further light on James Eads. He envisioned another masterstroke of engineering, one that would boost the commerce of St. Louis and the wealth of those insider-promoters who would make all this happen. As matters played out, the proposal barely left a shadow. Twenty-five years later, William Taussig saw many pitfalls in this first draft for a union depot: "passengers to be landed in a dark hole at tunnel level," the station's stub tracks causing miserable problems in switching cars and locomotives for both arrivals and departures.[51] Although it died quickly—local landowners stood resolutely opposed—the proposal for this "Grand Union Passenger Depot" speaks to larger issues that recurred periodically throughout the bridge project. Eads's creativity occasionally veered from brilliant to febrile, and his scorn for precedent could lead him into trouble.

Six months later, the backers of St. Louis Bridge made a fresh start on a depot. In March 1871 the Missouri legislature enacted an authorizing statute to guide the formation of union stations within the state. That the concept needed this enabling legislation reflected its novelty. A union depot company for St. Louis was organized within the month.[52] Once again, this was capitalism shaped by the state. An initial planning meeting for this new station, held in Carnegie's office at 57 Broadway in lower Manhattan, underscored the strength of the Pennsylvania connection. At the April 1871 meeting, plans and estimates for a new depot and the tunnel were reviewed. The PRR's chief assistant engineer had designed the station; Thomson and Scott approved those plans. Practical railroad men were now calling the shots. In June 1871, William Taussig hosted in St. Louis the first organizing meeting for the new company. Ten railroads subscribed for shares; Tom Scott signed for two of those carriers.[53] A PRR ally, the Vandalia, joined the effort. William Taussig was elected president, and Tom Scott became vice president. In all, the subscribers pledged for shares worth $1.5 million. It was a fine start, though not one actual dollar changed hands.[54] The ten carriers that committed to the depot and tunnel company had also signed contracts to put their traffic over the bridge upon its completion. Eads, McPherson, and Taussig, along with Thomson,

Scott, and Carnegie, had much cause for celebration. Carnegie began counting more profits to flow from financing the new depot and tunnel company. He figured their total cost at $2.25 million—plenty of stocks and bonds to hold or sell.[55]

The station project reveals a profound difference between Carnegie and Eads. Both were consummate promoters, highly capable architects of the deals behind this era's big infrastructure projects, including the bridge and many railroads. Unlike Carnegie, however, Eads had in addition to his promotional skills unique talents in designing and building these ventures. For the Captain, means and ends were one. The bridge—its fitness, its beauty—was as important as the deal.

Problems

Along with this good progress, complications were developing. In negotiations for the new station and tunnel, the Missouri Pacific stood aloof. Its bridge-friendly directors had sold their interest in the line by March 1871; the new owner-managers saw no reason to support the union depot effort.[56] Their precise objections are unknown, although geography suggests a fundamental schism at the heart of the Captain's planning. Powerful arguments could drive the eastern carriers into cooperating to run their trains directly into downtown St. Louis. Cooperation would get them out of the mud in East St. Louis, Illinois; cooperation would share the considerable costs to execute Eads's grand vision. A western carrier like the MoPac had far less to gain. The Missouri Pacific already had its station in the Mill Creek Valley. Furthermore, it had no interest in spending money on a tunnel that reached beyond its own lines.

The Wiggins Ferry Company was another growing problem in 1871. By that year, Wiggins had begun a full car-transfer service over the river. In the past, the railroads terminating in East St. Louis had needed to "break bulk." An independent freight forwarder, the St. Louis Transfer Company, unloaded rail cars in the yards there; carted the freight in bales, barrels, and boxes to the river; loaded its wagons aboard Wiggins ferries; then hauled the shipments to consignees in St. Louis. For all this, the transfer company placed its own charges atop Wiggins's fees and the railroads' freight rates. As the bridge evolved from chimera to masonry, two freight-forwarding services began to move loaded rail cars over the river on specialized ferries. By the early 1870s, thirty thousand rail cars were crossing the Mississippi this way each year.[57]

We don't know how the bridge men interpreted a letter they had received in September 1870 from Andrew Carnegie. The associates had been happy to receive a block of stock carrying a face value of five hundred thousand dollars, their compensation for executing the bond deal in London.[58] But Carnegie had ignored the bridge company's April 1870 call on all outstanding shares, levied to raise cash to service the interest on those bonds. In his September letter, Carnegie had bridled at the presumption that the associates should pay anything for their stock, and he threatened to sell those shares rather than pay the call. He wrote to the bridge company treasurer: "In my opinion . . . until our enterprises were earning successfully, it was well to keep AC & Associates firmly interested beyond contingency."[59] The erstwhile ally of St. Louis Bridge was determined to have his due and his way.

CHAPTER 7

Troubles with Steel

THE CAPTAIN'S DECISION to rely upon steel for the main structural members of the bridge resulted in a tangled web of problems. It could not have been otherwise: a new metal for a novel design, reliant on suppliers lacking useful precedents to guide their work. The prime contractor, Keystone Bridge, had never built a metal arch bridge, never worked with steel. The situation with a materials contractor was worse. Over three years of searching, Eads could not find a steelworks in the United States, Great Britain, France, or Prussia willing to bid on his requirements.

Finally, St. Louis Bridge and Keystone turned in desperation to a small Philadelphia company, the William Butcher Steel Works. As Carnegie wrote in October 1870, "After inviting every steel manuf'r in the world to bid upon the Captain's specifications, only one party bids + he the most sanguine of men, one who has broken up every concern with which he has been identified."[1] With no alternative in sight, Keystone signed a contract with the sanguine William Butcher two weeks later. A pleased but dubious Carnegie wrote to Eads, "At last we have succeeded in getting that steel matter closed."[2]

As Carnegie knew all too well, the contract closed little. It really signaled the start of new challenges. Over the next two years, troubles with steel would bedevil St. Louis Bridge, Keystone Bridge, and the Butcher works. The delays originating there metastasized into further problems throughout the complicated project. While difficulties were inevitable, how the three companies would tackle them became the key issue. Committed to producing components well beyond his abilities, William

Butcher simply thrashed about without plan or method. Caught in the middle, Keystone's men groused endlessly and complied grudgingly with their contractual duties. It fell largely to Eads and his engineers to develop the goals and methods required to succeed in steel. They became path-breaking innovators in yet another realm.* Caissons and piers to bedrock were their first challenge met; steel became the second. Triumph there, if they could achieve it, was only the precondition to a third novel challenge: putting up the arches.

"Pioneering doesn't pay" became a guiding principle for Andrew Carnegie after 1874, when he entered the business of making steel rails. St. Louis Bridge taught that lesson. Captain Eads sought fame and wealth in his innovations, while Carnegie mostly saw costs. Although no engineer, Carnegie had an accurate view. The manufacture of steel would involve a steep and expensive learning curve, assuming that it was even possible to make it with the qualities that Eads demanded and in the quantities he needed. Testing its strengths and machining it to finished dimensions would require more laborious innovation, time, and expense.

Steel and the William Butcher Works

After Keystone contracted for the steel, the William Butcher Steel Works needed eight months simply to produce the first components. Those anchor rods were massive steel bolts as long as thirty-six feet. Passing through each stone pier, they would fasten the wrought-iron skewbacks that would connect the arches to the piers and abutments.

After the months of delay, Andrew Carnegie conveyed grim news to William Taussig in St. Louis. Butcher was a mess. Its first steel anchor rods had broken during testing, unable to bear even half the strains safely carried by quality wrought iron. Those failures meant that "Keystone is kept waiting, at great loss, for the material (Steel)—the tools, Special force [of trained machinists we hired], etc. all idle." Carnegie complained that

---

* In all matters relating to steel, Eads and his engineering team defined the specifications, designed the components, and oversaw testing—dictating these choices to Butcher and Keystone. When Butcher fell short, St. Louis Bridge typically presented those failures as costs that Keystone had to bear. While legally justified (Keystone had contracted for steel directly with Butcher), this buck-passing would infuriate Keystone's engineers and managers. Eads, "Report," 1871, 590–94.

This detail shows how the steel chords landed at the piers. The first tube of each chord screwed into a skewback, a 7,000-pound fabrication in wrought iron. Anchor rods fastened the skewbacks to the pier. Those massive bolts, appearing here in dotted lines, ran through the stone pier to fasten the skewbacks on each side. Suitably massive nuts held the skewbacks in place. The figure on p. 3 also shows the braced ribs, skewbacks, and anchor-rod nuts. Detail from Woodruff, *History*, plate 20.

St. Louis Bridge, "with its eyes wide open, & against the best judgement of men skilled in Bridge building, elected to undertake a gigantic experiment" with this steel bridge. His letter asked "whether the best *practical* Bridge building talent should now be called to your aid + a new departure taken." The time seemed ripe for "abandoning present experimental plans + returning to the beaten path of a good permanent structure of ordinary construction." Carnegie asked Taussig to raise this matter with the leadership of St. Louis Bridge.[3] For the second time, the top men at Keystone Bridge wanted to push aside James Eads and forget all about his offbeat ideas. With him gone, they would put up their own proven design for an iron truss bridge. This open revolt by the prime contractor foretold the grim future all too accurately.

Carnegie's attempts to replace the chief engineer or his design failed—and so did the next batch of Butcher's steel anchor bolts.[4] That did not surprise Carnegie. As he wrote to Eads, "I don't see how Mr. Butcher can stand the treatment received at the hands of his partners" given that they "have no more regard for his opinions or his promises than if he were totally unknown to them." Butcher had learned the art of making crucible steel in Sheffield, England. And because that art was a rare attainment, his partners tolerated him, albeit rudely.[5]

Known in antiquity, crucible steel had been perfected by artisans in medieval times, melting pig iron in a closed clay vessel to refine the iron's carbon content. The crucible controlled the amount of oxygen reaching the molten metal. From those origins, steel production depended upon hard-won craft knowledge grounded in local ores and fuels and refined over centuries. As a result, regional centers such as Damascus, Syria, and Sheffield, England, grew famous for their products. Until the industrial age, steel was widely known though rarely used, chiefly in knives, swords, and cutting tools. Reflecting those limited applications for this expensive product, "steelmaking was a cottage industry, the work of veritable alchemists."[6]

Mechanical engineers of 1867 dealt with steel all the time, but many had trouble defining it. Even today, the common definition lacks a measure of clarity: steel is iron with less than 1 percent carbon in its composition. A definition by chemical composition, however, was beyond the capacity of nearly all steelmen of the 1860s, who lacked the knowledge or the equipment to undertake chemical analysis. As late as 1875, the *Engineering and Mining Journal* published an article titled "What is Steel?"[7] Until chemical analysis became widespread in the 1890s, creating good steels depended on equal measures of trial and error, craft, and mystery.

During the mid-nineteenth century, entrepreneurs and engineers in Britain and America were developing new processes to transform steel production into a business of massive industrial scale. But the makers of Bessemer, or open-hearth, steel found it challenging enough to produce steel rails in industrial quantities. They saw little benefit in pioneering for James Eads. In 1867, Edgar Thomson organized one of the first American Bessemer steelworks, focused on producing rails and located directly on the PRR main line.[8] He could have thrown the contract for bridge steel to his own firm, but the Pennsylvania Steel Company wanted nothing to do with Eads and his demanding requirements.

What Eads Wanted

To build his bridge, Eads required four things unknown in steelmaking to that time. He wanted a great deal of product; in 1868 he reckoned the total at 2,544 tons.[9] He needed very large components, such as his thirty-six-foot anchor rods. Yet he insisted on uniform strength and hardness throughout his orders. And he required extensive testing to ensure those qualities. In all, he demanded novel industrial standards for this ancient craft product. No wonder most steelmakers declined to wager their shirts and their reputations.

His initial design choice for long and shallow arches had driven Eads to select steel in the first place. Contemporary English tests by William Fairbairn had shown that the best British steels had twice the breaking strength of quality wrought iron. Steel was especially strong in compression, the primary loading in an arch.[10] Better still, steel offered higher strength at lower weight than wrought iron. Given the record-breaking spans that Eads planned to build, strength without excess weight was a worthy prospect. To confirm Fairbairn's data, the Captain approached an old friend from the war. Using the facilities of the Washington Navy Yard, Chief Engineer William Shock tested samples of rolled cast steel for Eads. The data were promising: the samples withstood a compressive force of 75,000 pounds per square inch before deforming.

For a skeptic, the conclusions drawn from the eight sample bars, one inch in diameter and a foot long, proved nothing relevant to building a bridge.[11]

The sections on steel in Eads's first report as chief engineer (June 1868) mixed confidence, caution, and deception. He conceded that there was much to learn about steel, while reassuring stockholders and potential investors that "leading steel makers" of the United States and Europe had offered to furnish steel and "to guarantee its strength fully up to the standard required."[12] In fact, those offers did not exist, and those standards remained unstated by the bridge company, as they were largely unknown at that time. No one in the churning turmoil that was the American steel industry of 1868 had any real basis for offering such assurances. Eads knew the truth, even as his prose papered over this critical deficiency in his effort to lure investors.

Because potential suppliers knew comparatively little about their own product, before closing the contract with Butcher, Eads established three specifications for the steel destined for the bridge. A mandate for "elastic limit" measured how much strain any component could bear before it became permanently deformed. Such deformation was called "permanent set." Eads required his steel to bear an elastic limit of 60,000 pounds per square inch under compression and 40,000 pounds under tension. He also required an "ultimate tensile strength" of 100,000 pounds, the load at which the steel simply broke apart. Finally, he stipulated the "modulus of elasticity." This ratio between stress and deformation was a revealing measure of the steel's resilience under load.[13]

Eads and his college-trained émigré subordinates did not originate the concept of an elastic modulus. It appears in Herman Haupt's *General Theory of Bridge Construction*, a leading American text on the theory and practice of bridge design, published in 1851.[14] Until Eads, however, few designers or builders of metal bridges had paid much attention to the core issue embodied in the modulus. They had focused on the ultimate strength of iron, measured by its breaking strain, and were little concerned that structural members needed resilience as varied loads passed across a bridge over time.

These specifications equipped the engineers to begin their work designing the components and configurations for the steel arches and their iron bracing. That iterative process began in earnest in 1868 and continued for five years. The complicated issues inherent in this novel metal arch bridge required that burdensome effort. In the first place, the original design was simply too weak and flexible, as Jacob Linville had warned. In May 1868, the engineers projected that 4,322 tons of steel and iron would be needed for the three arches. As built, they required 5,550 tons.[15] As they increased strength and weight, the designers had to reckon with the stresses that heavier chords (the load-bearing steel tubes in each arch) and braces imposed across the structure. The live loadings anticipated for moving trains required more calculations. Day-to-day and seasonal variations in temperature would impose further stresses, because the metal arches would shrink and flatten in the cold of winter and expand and rise with the summer heat.[16] The designers also grappled with the effects of wind loadings. Assistant engineer Charles Pfeifer's skill with the calculus was essential in integrating these forces. Verifying the calculations and results required more time.[17]

Their pioneering specifications for steel guided the engineers in sizing components and ensuring quality. It was one thing to mandate strengths, another to verify them. Before inking the deal with Butcher, St. Louis Bridge bought a testing machine designed by Henry Flad. Installed in its St. Louis office, the machine was more than fifteen feet long. It could exert upwards of one hundred tons of force, while its graduated scales showed how much a sample shrank when compressed or lengthened in tension. Readings extended out to a hundred-thousandth of an inch (0.00001).[18] Under its contract with St. Louis Bridge, Keystone acquired its own testing machine, which it set up at the Butcher works in Philadelphia. Its operators were busy. St. Louis Bridge required compression tests for every piece of steel destined for the arches that were the bridge's structural backbone. The bridge company sent its own inspector to Butcher to ensure that each steel part was tested properly, with its modulus of elasticity stamped into the metal.[19]

Working with his engineering staff, James Eads established with these standards and methods a thorough regime to guide Butcher and Keystone and ensure that they met his exacting requirements. This level of rigor matched his creativity and broke entirely with the prevailing customs in structural ironwork and bridge fabrication. These demanding protocols gave Carnegie another reason for his short-tempered letter to William Taussig. But Eads had no choice. No one, not even an alchemist like William Butcher, knew enough chemistry to reliably create steels with the desired properties. Eads's only recourse was to mandate physical standards and tests to ensure that Butcher met the mark. Too often he did not.

Troubles with Chemistry and Contractors

From the start, the design team at St. Louis Bridge planned to use steel only for components requiring extra strength, relying on wrought iron for much of the structure. Initial plans called for steel anchor bolts, steel tubes in the arches, and steel couplings to fasten those tubes together, making the eight chords of each span. Steel production proved so troublesome, however, that in the end Eads got less than he wanted. Anchor bolts bearing the most strain, fifty-nine in all, would be of steel. The rest (fifty-three) were of wrought iron.[20] The tubes making up each arch *were* steel through and through. Each tube was 12 to 13 feet long and 18 inches in diameter. Eads had wanted steel for the "sleeve couplings" that would join the tubes, but two years of experiments by Keystone, Butcher, and other subcontractors to make cast-steel couplings produced only failure. Finally, Eads grudgingly agreed to accept wrought iron for the 1,012 couplings required to connect the tubes and create the arches.[21] As built, the super-structure had approximately 4.8 million pounds of steel, 6.3 million pounds of wrought iron (mostly in the bracing that connected the steel chords), and 1.6 million pounds of wood in the decks and sidewalks.[22] A steel bridge more or less.

The many steps required to produce the steel components resulted in more pitfalls and failure points. An anchor rod started as a cast steel bloom, 4 feet long and 15 inches square. To transform that bloom into a round bolt—25 feet long and 6.5 inches in diameter—a team of skilled craftsmen heated it to a white glow, swung it on a crane to the anvil of a steam forging hammer, then shaped it with forming dies.[23] Innumerable heats and bouts under the hammer were required to achieve the desired dimensions. That forging then traveled by rail to the Keystone factory in Pittsburgh, where skilled machinists turned it on a lathe and cut screw threads on both ends. If the hammer or lathe work uncovered flaws in the metal, Butcher went unpaid. Failure was expensive for all involved, especially given the heavy labor costs for each component, on top of materials and transport expenses.

William Butcher's first attempts to make steel anchor rods in June 1871 failed miserably when tested. He adjusted his "recipes" (as steelmen described their mixtures of ores and additives) and failed again. Failure could prove explosive. In one test at St. Louis, a twenty-foot, 1,000-pound bolt broke while under load in the testing machine. A portion "shot out of the machine like an arrow," falling fifty feet away. The other part wrecked the test apparatus. Since it was Keystone that had contracted with Butcher for steel, Keystone bore the costs of testing and the expense of failures, which continued in dismal procession.[24] Carnegie had counted on skimming a significant markup when Keystone took the contract for steel, even as he doubted Butcher's abilities. Pioneering wasn't paying.

In desperation, Eads turned to the Chrome Steel Company of Brooklyn, which had enjoyed some marketing success with a patented chromium-alloy steel. Like Butcher, this company knew little about the actual chemical composition of its steel. Nonetheless, it accepted a royalty payment of fifteen thousand dollars from St. Louis Bridge in exchange for sending its superintendent down to Philadelphia to take over the mixing from William Butcher.[25] Results improved immediately. In September 1871 upwards of one hundred steel-tube components "were all beautifully and perfectly rolled," and all stood the strength testing.[26]

By the fall of 1871 William Butcher had worn out the patience of his partners, who forced him out of the firm. A creditor and Butcher partner, William Sellers, reorganized the works as the Midvale Steel Company.[27] Sellers ran one of the best machine-tool works in the country. In the 1860s, he had promulgated the first national standards for nuts and bolts, the prosaic connections of the machine age.[28] They were widely adopted and are still sold in every hardware store in America. After taking full charge at Midvale, Sellers hired a chemist who had trained abroad. Charles Brinley's "first step was to label clearly and to analyze exactly every pile of raw material in the yard."[29] It was a turning point in American steelmaking. The new management at Midvale finally provided adequate components for St. Louis Bridge, although problems recurred into 1873. Chemical analysis undertaken many decades later found the chords in the arches to be of an "unusually high grade of chrome-alloy steel."[30]

Given the rudimentary state of the art in steel production and fabrication, the engineering team at St. Louis Bridge needed imaginative adaptations to create its record-breaking spans. Although each arch demonstrates a graceful curve, they were built out of straight tubes. Sleeve couplings joined the tubes; slight angles in those connections resulted in arches that appear to have a continuous curve (the illustration on p. 169 shows tubes joined by the couplings). The tube lengths varied, depending on their placement in the arches. But their lengths were hardly variable. Those in the lower chords of the center span measured 142.622 inches long. The segments making the upper chords of the side spans measured 158.387 inches.[31] Eads and his assistant engineer, Henry Flad, had scant tolerance for approximation.

Fabricating the tubes required more innovative adaptations. No steelworks in the world could produce solid drawn tubes in the dimensions that Eads required: twelve to thirteen feet long, eighteen inches in diameter, with wall thicknesses from 1.375 to 2.875 inches. But Eads had committed to steel tubes back in 1867, so his engineers developed methods to deliver on that promise. Six steel staves, akin to the staves of a barrel, would make up each tube. Resolving how to roll staves with the requisite qualities took ten months of effort at Butcher/Midvale, from May 1871 to March 1872.[32] To create a tube, the staves were then pressed into a round wrapper of quarter-inch sheet steel. Further complicating matters, the staves used to create each component needed a comparable elastic modulus to ensure that a tube bore its loadings nearly as if it were a homogenous component. Once formed by its staves and wrapper, every tube required precision turning on a lathe specially built for the purpose.

Eads had begun with a first principle: he would build a steel bridge, although the steel industry was in its infancy. His engineers then developed designs and methods to achieve that vision. The bridge had eight steel chords in each span, making twenty-four chords across the three arches. That translated into 6,216 staves.[33] And it fell on the contractors, Butcher/Midvale and Keystone, to execute Eads's pioneering vision. The work drove Butcher bankrupt. It maddened the leadership at Keystone, men who knew full well how to fabricate a perfectly adequate bridge for St. Louis—in iron.

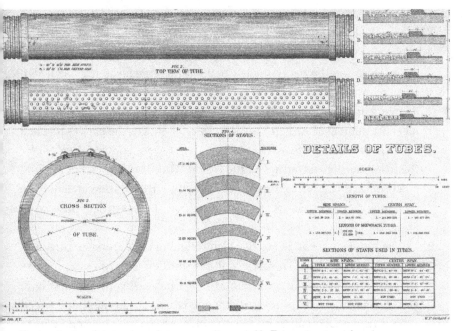

Each tube is a fabrication of six staves (*lower left*). Tubes destined for placement near the skewbacks bear heavier loads, so their cross sections are larger; those at midspan are thinner and lighter (*lower center*). To make up a tube, Keystone's men used a hydraulic press to force the assemblage of six staves into a steel wrapper, forming a solid tube. The wrapper had been rolled from sheet steel, its long edges riveted together with a cover plate to form a cylinder. Machinists then cut grooves into the tube ends to mate with the sleeve couplings. All this fabricating and machining required skilled work to exacting tolerances. Detail from Woodruff, *History*, plate 29.

Keystone's Carnegie in Revolt

St. Louis Bridge also applied its rigorous standards and testing to the wrought-iron components used in the structure. Each steel tube would be braced to a partner, another chord, twelve vertical feet away. That combination, what Eads referred to as an "arched rib," was braced to the adjacent rib, 16.5 horizontal feet away (see illustration on p. 3).[34] All those wrought-iron braces did essential work, ensuring that the steel tubes could not deflect under load.

Keystone subcontracted for the ironwork to a reliable supplier: the Union Iron Works of Pittsburgh. The legal name of that partnership—Carnegie,

Kloman & Company—reflected Andrew Carnegie's 39 percent owner-ship stake in the business.[35] Such close relations between an iron sup-plier and a bridge builder safeguarded quality and ensured timely deliv-eries. Here too, however, Eads's demanding standards imposed high costs and demolished one of Carnegie's plays to profit from St. Louis Bridge. Carnegie complained bitterly when the first brace bars failed to meet the mandated strength tests. In December 1871, the parties reached a mutually unsatisfactory compromise. Eads lowered his requirement for the iron's tensile strength, and the Union Works cut its price per pound and produced thicker bars to achieve the strength Eads demanded. Even so, it took six more months of work for the main braces to pass the tests in St. Louis.[36]

The heavier bars added to the dead weight of the superstructure, cre-ating a new problem. To carry that new burden in turn could require stronger steel chords in the arches, further increasing the dead weight. Every element in bridge design demanded a balance in cause and effect. Rather than exacerbating his steel challenges, Eads set his engineers to redesigning the wind truss, the iron fabrication beneath the roadway deck. With many tweaks, Henry Flad and Charles Pfeifer lightened that struc-ture, calculating that it would sustain the bridge against loads exceeding those in the violent tornado of March 1871. Eads hoped the bridge would endure against higher loadings, "which will perhaps not occur oftener than once in a century."[37] The remark again revealed how widely Eads di-verged from common practice for American engineers. Most saw no ben-efit, and much needless expense, in creating technologies to endure for generations.[38] The national genius expected and welcomed continuous in-novation in every facet of life. If better lay in the immediate future, then good enough was ideal for today.

The challenges with iron were maddening to Keystone's owners and managers, although those difficulties resolved more readily than did the troubles with steel. Far more was a stake here than Eads's demanding specifications and their evaporating profits. Keystone, a leader in the bridge-building industry, had committed much of its time, workforce, and shop space in 1871 to the St. Louis project. All the delays raised opportu-nity costs in Pittsburgh, with investment lying idle and other contracts forgone. Just as maddening, Eads's specifications gave him unprecedented power over Keystone's operations. In protest, Carnegie wrote confidentially to William Taussig that "the Captain . . . can only require from Keystone 'the custom of the trade.'"[39] That letter, written in late December 1870, spoke repeatedly of Keystone's frustrations, even at that early date.

Carnegie wrote to Taussig in confidence, already troubled by "all the difficulties . . . between our respective Engineers." He suggested to Taussig that the two men of business should "endeavor to moderate our respective belligerants." Keystone's engineers "consider themselves misled by Captain Eads." After reviewing a number of detailed engineering problems, Carnegie summarized as follows: "The Bridge Keystone undertook to build, + that now determined upon are so essentially different that our Foremen (Piper + Shiffler) report that the difference in time required [for fabrication and construction] is not less than 12 to 18 [more months]." All this was undoubtedly true. Carnegie showed insight into the ultimate cause of the difficulties: "Keystone is only experiencing the fact that of all men, your man of real decided genius is the most difficult to deal with—*practically*—but he will come out all right. . . . It is the very nature of genius to devise new, untried, methods even when old ways meet all practical requirements."[40]

Carnegie's equanimity evaporated as troubles with iron, steel, and Eads all mounted. Edgar Thomson and Tom Scott shared that growing frustration. On November 27, 1871, Carnegie wrote to Thomson that he had sold one thousand shares of stock in St. Louis Bridge for a miserable ten bucks per share (its par value was $100), explaining, "I do not believe Capt. Eads will get through without trouble—at all events I wished to get rid of so much liability." In his enticing prospectus of February 1869, Eads had predicted that the stockholders would pay only 40 percent on the par value of their stock (or $1.2 million). Now Carnegie foresaw that they would all be assessed for the full value of their shares, more than $3 million ($77 million in 2023), to complete the bridge. Conceivably, the costs and assessments could go even higher. If Thomson wanted out, Scott's Andy would execute the trades, "although it looks like a great sacrifice at present." Carnegie closed with these thoughts: "In short, I am disgusted with the affair + may have sold at panic prices. Still, this day [a] year [from now] we may have to buy an unfinished Bridge subject to the 1st Mtge. Bonds + let Linville + you, as Engineers, get a Bridge for us."[41]

Although he cast it as a fear, Carnegie likely would have welcomed that result. A foreclosure sale could force Eads out and give Keystone a free hand to build its preferred Linville truss bridge. Two weeks later, Carnegie wrote to Taussig that all three associates had unloaded a portion of their shares. He counseled Taussig that the stockholders should "insist on putting a practical bridge builder in charge."[42] To press that question, on December 20, 1871, Carnegie led a revolt at the New York meeting of the stockholders.

The difficulties with materials, delays in construction, and mounting costs prepared those owners to listen to Carnegie's concerns. After all, the March 1870 prospectus for the first mortgage bonds had forecast that the bridge would be earning lucrative toll revenues by December 1871. In fact, not a single component of the superstructure had been installed, while the project swallowed up funds insatiably. To secure more capital, the meeting authorized additional calls on the stockholders, just as Carnegie had feared. To rein in the Captain, Carnegie secured a motion to create a new committee of stockholders. It would engage a new consulting engineer "to ensure swift and cost-effective completion of the bridge." The historian Robert Jackson believes that Carnegie was angling for Keystone's Jacob Linville to step in again.[43] As matters played out, the committee hired James Laurie. An accomplished civil engineer, Laurie had laid out the Jersey Central Railroad, designed iron railroad bridges, and founded (with others) the American Society of Civil Engineers.[44]

After investigations in St. Louis, Laurie produced two reports that surely left Carnegie unhappy. His first offered modest suggestions for cost savings.[45] One idea was to omit the four stone towers of Eads's plans, masonry exclamation points to surmount the abutments. Another was to drop the elevators planned for each tower. Deriving their power from the river's flow (again Eads's creativity bubbled over), the elevators would have moved pedestrians between the levee and the roadway deck. A good idea, although costly. As built, the bridge would have no towers, no elevators. Eads preemptively offered other economies, taking much of the wind out of Laurie's report.[46]

In April, Laurie produced a detailed review of engineering and finances. In his view, the balance sheet suggested that St. Louis Bridge would have sufficient resources to finish the structure if the stockholders paid their

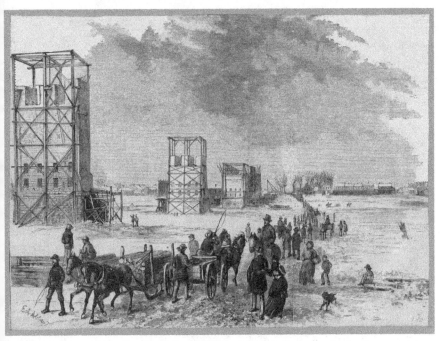

In December 1872, the Mississippi River at St. Louis froze hard, allowing people, teams, and wagons to cross on the ice. By this point, Eads and his men had been laboring for five years. The two piers and east abutment appeared solid but useless. Engraving by E. A. Abbey, "Ice Bridge," *Harper's Weekly*, Jan. 18, 1873, 52, MHS, N 45660.

full assessments. Estimating the total cost at $6.8 million, he figured that roughly 30 percent of the work (measured by expenses) lay ahead.[47] Eads greeted Laurie's input with bland acquiescence. The chief engineer now enjoyed a stronger position against his critics.[48] Carnegie had lost this bout, his third attempt to restrain or replace Eads. But Keystone, Thomson, Scott, and their young man remained essential to completing the St. Louis Bridge.

Challenges on the Map

The troubles with steel took twenty-eight months to resolve.[49] As a result, the people of St. Louis saw no progress on the arches until March 1873. The interminable delay gave everyone in town plenty of time to contemplate the two stone piers and two abutments, useless monuments mocking any deliverance from the ferries as the timeless Mississippi

This cartoon appeared in a new humor magazine published in St. Louis in 1872. James Eads pulls the strings while Henry Flad cranks the conveyor belt. Image from Joseph Keppler's Puck, *Illustrierte Wochenschrifte*, Nov. 1872, MHS, D O3667.

churned past or froze over. To many of the 240,000 residents of St. Louis, and to its visitors from the East, James Eads's bridge increasingly looked like a monument to folly or hubris. While progress stalled in St. Louis, engineers, promoters, and railroads across the Midwest built new facts on the ground. The railroad map evolved dramatically, with implications for Eads's project and his city.

Over on the western border of Missouri, Kansas City had grown by 1872 into a significant rail hub thanks to a bridge over the Missouri River into Kansas.[50] Octave Chanute was its chief engineer; Keystone Bridge fabricated and erected its iron drawspan (its fixed spans were largely wooden). Authorized under the same Omnibus Bridge Act (July 1866) as the St. Louis crossing, the Kansas City Bridge opened on July 3, 1869, to the great delight of this small town, whose population then numbered less than thirty thousand. This crossing was earning revenues and shaping the national railway network for years before St. Louis had its bridge. Chanute attempted far less than Eads. His single-track bridge lacked a separate

Octave Chanute's bridge at Kansas City suggests the radical nature of Eads's leap to a steel bridge. After requesting bids from many firms, in October 1867 Chanute chose a proposal from Keystone Bridge to supply composite wood and iron trusses for the fixed spans shown here. Keystone also supplied the pivoting drawspan; it alone was entirely in iron. The draw is shown opened for river traffic. Beneath it, fixed timber guide structures prevented steamboats from crashing against the open draw if the Missouri River's fierce currents overpowered those vessels. Photograph from Chanute, *Kansas City*, plate 11.

roadway deck, while its low fixed spans and the pivoting draw were economies granted by its federal authorization and forbidden for any St. Louis crossing.[51]

The Kansas City Bridge shaped the flows of midwestern commerce and travel in diverse ways. Linked to St. Louis by two railroads, the Missouri Pacific and the North Missouri, the new bridge benefitted Eads's hometown. It also boosted trade and travel to and from Chicago. Its promoter, James F. Joy, ran the Chicago, Burlington & Quincy.[52] With clear focus and sufficient capital from Boston investors, Joy built out a plan to link

Chicago to Kansas City and then penetrate Nebraska. With the bridge at Kansas City, the Burlington and its associated lines drew Texas cattle and Nebraska grain to processors in Chicago. To those ends, Joy had already promoted two Mississippi River bridges, at Burlington, Iowa, and Quincy, Illinois. Strictly utilitarian structures, both opened in 1868, just two years after their congressional authorization (see illustration on p. 67). Joy's strategic determination contrasted sharply with the railroad promotions of Eads and Scott. All too often, they were diverted by bright, shiny objects, turning from long-term goals when short-term bounties offered immediate personal gains. By 1871 Scott was the president of five railroads, vice president of twelve, and a director of thirty-three.[53]

While James Joy demonstrated vision, strategy, and discipline, the same cannot be said of the Captain and his major railroad venture. The North Missouri Railroad did rack up some accomplishments. After three years of construction, the road opened *its* bridge over the Missouri River in June 1871. The St. Charles Bridge had high, fixed spans with 80 feet of vertical clearance. Its seven main trusses, all iron, each exceeded 300 feet in length. It was an impressive structure. Costly too, placing a heavy burden on the independent company that built it.[54]

The St. Charles Bridge provided an essential link for the North Missouri, but to what end? The carrier had aimed originally for Council Bluffs and Omaha. Instead, it first opened a line connecting St. Louis to Kansas City in December 1868 (see again illustration on p. 67). That route competed directly with the Missouri Pacific, which was one reason for that carrier's opposition to Eads's St. Louis bridge. By contrast, Council Bluffs beckoned, with fine strategic prospects for the North Missouri. Instead of competing with the established MoPac, that link could have diverted transcontinental freights away from the granger roads, away from Chicago, to travel via St. Louis.

As matters played out, the North Missouri's tracks did not reach the Council Bluffs–Omaha gateway until 1879.[55] That painful delay reflected the North Missouri's miserable finances, which had become a scandal by August 1871, after the company defaulted on its second mortgage. The foreclosure had complicated roots, all dissected in the *Railroad Gazette*. People in St. Louis believed that "the failure of the road was owing to gross and corrupt mismanagement, and that individuals . . . had made fortunes by bankrupting the company."[56] It is impossible to assign culpability strictly to Eads or his New York partners, as blame attached to them all.

Reorganized early in 1872, the road again appeared to have bright prospects. Directors included T. B. Blackstone of the Alton, Tom Scott from the PRR, Solon Humphreys (a New York investment banker), William McPherson, and Eads. This line in turn provided the PRR with "a route in which it has a powerful, if not a controlling, interest from New York to Denver."[57] The reality never matched its potential, "thanks to the entire neglect of the parties involved."[58] With his far-flung interests, Scott consistently fell short in the focus and commitment that serious matters required.

While Eads struggled with steel in St. Louis, the national railway map was being shaped by Thomson and Scott, Carnegie and Keystone. Some of their efforts could aid the railroads of St. Louis; others hurt. In November 1871, William Taussig read in the newspapers that Thomson and Scott were promoting a union depot for Chicago. For Taussig, this was a figurative knife in the back. He wrote acidly to Carnegie, complaining about the Chicago station project: "Should that be brought about and ours [a union depot for St. Louis] fail, we can transfer our scheme to the Lake City. How does that strike you?"[59] The Chicago venture soon died. Elsewhere the Pennsylvania parties did well. In June 1872, Keystone completed a magnificent structure over the Ohio River. A hundred feet above the river, the Newport & Cincinnati Bridge had nine spans, all pin-connected trusses in Linville's "Keystone pattern." Its main channel span reached 415 feet.[60] With a free hand, Keystone could have achieved similar results for St. Louis by June 1872. Or earlier.

New-Fangled Notions and No Confidence

While James Eads devoted two years to achieving his requirements in steel, William Taussig worked tirelessly to advance the union depot and tunnel, key elements in this complicated project. After the false start of Eads's grand underground station, in January 1871 Thomson and Scott sent a PRR staff architect to St. Louis to develop plans for the new union depot (discussed in ch. 6). Joseph M. Wilson had just returned from a tour of Europe, where he had inspected the latest in railroad architecture.[61] By April, depot matters were advancing nicely, with Wilson's plans completed, a new corporate charter in hand, and options purchased for six blocks of land in the location Eads had originally planned on, near Chouteau's Pond in the Mill Creek Valley.[62] (The pond appears at left in the illustration on p. 5).

Then the churning mind of James Eads upset everything. During a visit to St. Louis in June 1871, Scott became captivated as Eads sketched another radical design for a depot. This time, he proposed a station perched above the St. Louis levee on iron stilts and made largely of glass. This crystal structure would shine over the bridge and the river. While Scott loved it, Taussig was horrified. He wrote confidentially to Carnegie: "Of the 'new fangled' notions, none is newer + more fangled that Col. Scott's of building a narrow long shed alongside the Bridge on high works 45 feet above ground, with positive legislative forbiddance, with danger of rousing the whole city against us + at Enormous cost! And all that the result of staying one night at the Captain's house and being carried away by his marvelous persuasive power!"[63] Carnegie knew how this dreamy vision had originated. He wrote to Scott that Eads "*must not superintend*" plans for the depot.[64] Again, Eads's originality and brilliance could lead to problems.

Before the summer was out, Carnegie and Taussig succeeded in returning station planning to Joseph Wilson's largely conventional design, to be located near the western mouth of the proposed tunnel. Then Scott himself became problematic. In September 1871, Taussig complained to Carnegie that "as Mr. Scott is the head and front of both Depot and [the North Missouri Railroad] I can do nothing until he says the word 'go.'"[65] In December, Taussig reported that major lines, including the Atlantic & Pacific to the west and the Wabash to the east, had committed to buy stock in the depot company. But the PRR's allied line, the Vandalia, was nonresponsive. Taussig wrote to Carnegie that "if you can put the Pennsylvania interest right, the day is won. I write to you and not to Col. Scott because I know he has no time to give to letters giving long Explanations. You must . . . get his promise to act."[66]

On the same day, Carnegie was writing a stern letter to Taussig. The associates had no confidence in Eads; they thought he would need two more years to complete the bridge. "Get a span of bridge up + then we can talk" about the depot, he wrote. In its early days, the aid of Thomson, Scott, and the PRR had bolstered the station project. Their withdrawal sent quite a different signal to the railroads, merchants, and investors of St. Louis. But Carnegie was undoubtedly correct. With the bridge much delayed and mired in problems, the officers of the eastern lines (those terminating in East St. Louis) had little reason to care about plans for an expensive station beyond that unbuilt bridge.[67]

The proposed tunnel created further obstacles that Taussig struggled to surmount. This elegant solution to run trains beneath downtown would cost a lot of money. Estimates started at $400,000 ($8.7 million in 2023).[68] In their original business model, Eads, Taussig, McPherson, Scott, and Carnegie all imagined that a single new company would finance the depot and tunnel.[69] The eastern lines needed the tunnel to put their passenger trains into the station and to move their freight into the western rail network. Time revealed that the western carriers had less cash and fewer incentives to support the tunnel.

And they came to suspect in 1871 that Eads—an investor, director, and partisan for the North Missouri Railroad—might betray his own plans, capture the bridge for the benefit of that road, and then dictate terms to other carriers. In May 1871, Taussig wrote to Solon Humphreys, a director of St. Louis Bridge: "I mean to leave no stone unturned to attain the object of having the Tunnel built in time for the opening of the Bridge, and to have the Depot and Tunnel matter determined in the interest of the *Bridge*, and not in the interest of any single line or combination of lines, to the great financial detriment of the bridge." Otherwise Taussig threatened to quit the whole thing. He feared that a "powerfull combination of the No. Mo. has been enlisted against the building of the Tunnel by the *combined* roads and against the location of the Depot at the Chouteau pond."[70] This May letter helps us to understand Eads's June proposal for a glass station at the bridge. Given the lay of tracks into St. Louis from the west, that site for a depot would advantage the North Missouri and burden its leading competitor, the MoPac.

Taussig succeeded in blocking these departures from the original plans. Knowing Eads's role in these upsets, he wrote to the chief engineer, "Do not forget that our Bridge will cost seven millions and will have to pay on ten millions"—so they could not have enemies or competitors.[71] Taussig wanted every carrier serving St. Louis to share the project's costs and benefits.

Concerned too about the slipping schedule, the St. Louis promoters de-cided to sever ownership of the depot from ownership of the tunnel, al-lowing excavations to begin for that essential artery. Conferences in New York in June 1872 with bankers (and bridge shareholders) Edwin Morgan and Pierpont Morgan laid the groundwork for a new tunnel corporation.[72] In August, Eads wrote an extraordinary letter to Junius Morgan in Lon-don. He wanted Morgan's help in selling a new security, $1.2 million ($30.5 million in 2023) in 8 percent bonds, to build the tunnel and the eastern approaches to the bridge. Eads claimed that "there is no doubt" that US investors would buy such an issue, but he wanted to offer Morgan the chance to market such a first-class security. The Captain conceded that the bridge was behind schedule and over budget. Eads suggested that such annoyances were nearly predictable with such an innovative proj-ect. Was Morgan interested in financing the tunnel?[73] Morgan agreed, laying out his terms in an October 1872 meeting with Eads in London. He had little choice, understanding that the tunnel would encourage all the lines, east and west, to send their cars over the bridge. Without a tunnel, everything could turn sour in a hurry. He had already staked his reputa-tion and his investors' funds on St. Louis Bridge to the tune of $4 million in first mortgage bonds. The pressing need for funds in St. Louis gave him leverage along with risk.

Convinced (as Carnegie was) that the Captain's bridge had to prove it-self, Morgan would not attempt to sell tunnel bonds to the investing pub-lic until that proof stood over the river. Unlike Carnegie, Morgan put his own funds at some peril to ensure that the tunnel was completed before the bridge opened. In the short run, he agreed to make cash loans at the stiff interest rate of 9 percent to a new company, the Tunnel Railroad of St. Louis. Morgan would hold its bonds as collateral for the loans. He stip-ulated that when the time was ripe to take the bond issue public, he would pay 70 percent of par for those collateral bonds.[74] The terms revealed that Eads was desperate, Morgan bargained hard, and St. Louis Bridge had a stalwart ally in London. The banker too had little choice but to drive on.

At this late stage, St. Louis Bridge confronted another challenge. Lucius Boomer had visited St. Louis in June 1871 "at the instance of parties pro-jecting a bridge across the Mississippi at Carondelet."[75] Located imme-diately south of St. Louis, this village had been annexed by the city in

1870. In February 1873 the US Congress took up authorizing legislation for this Carondelet crossing.* As senators and representatives considered that matter, the superstructure work in St. Louis remained stalled for lack of steel. The Missouri Pacific stood behind the new bridge. William Taussig stood resolutely opposed, appealing to Missouri senator Carl Schurz to fight the Carondelet authorization. Taussig complained that the bill "seeks to destroy, by obtaining special privileges, private enterprise."[76] Taussig was sailing with the political winds propelling the new Liberal Republican Party. Led by Schurz and Gratz Brown, the movement denounced all the legislative perks that mainstream Republican legislators had showered on the new corporations in railroading, banking, and bridge-building. Not surprisingly, Taussig's letter overlooked the monopoly privilege granted to St. Louis Bridge by its Illinois charter.

As this threat grew in Congress, Taussig wrote to Junius Morgan in far-off London. On one hand, Taussig professed that "we feel indifferent" about the Carondelet Bridge, even after its authorizing act became law. On the other hand, he needed eleven pages of closely reasoned text to convey his nonchalance. In brief, Eads, Taussig, and their allies in Congress had weighted the Carondelet bill down with burdensome terms to overwhelm the financial viability of the bridge that it authorized.[77] Again, here was a capitalism shaped by powerful men wielding the intrusive hand of the state.

By this point, April 1873, sporadic deliveries of acceptable steel had begun arriving in St. Louis. In this aspect, at least, it looked as if the Captain and his men would prevail. Even so, if Eads could not get his bridge up in a timely way, Congress would revisit the matter of bridges for St. Louis. Nor was it clear how much longer Junius Morgan would sustain St. Louis Bridge, with construction now in its sixth year. And now the project would take on difficult, uncharted challenges in building the arches.

---

* It is unclear whether this Carondelet bridge would have violated the exclusive right held by St. Louis Bridge for a crossing that originated in St. Clair County, Illinois.

CHAPTER 8

Arches over the River

ON MARCH 13, 1873, a staff engineer from Keystone led a team of men in the fateful first steps of erecting the superstructure of the St. Louis Bridge. Salesman Carnegie charmed, negotiated, and complained. Linville designed. Colonel John Piper drove the men and machines back in the Pittsburgh shops. Walter Katte put up bridges. This English émigré had accomplished much in his career, including significant service with the US Military Railroads during the Civil War. He had never undertaken a job like this. No one in the world had.

Typically, Keystone's erecting teams used falsework to construct their superstructures. This temporary wooden staging in a riverbed or valley supported the men and parts during the short period required to assemble the trusses. Once they had built the falsework, the men could assemble their pin-connected bridges very quickly, as all the ironwork had been pre-fitted back in the shop.[1] But falsework was impossible at St. Louis. Its ceaseless river traffic would brook no obstacles of any kind.* For that reason, Captain Eads adopted a cantilevering method to support the arches during construction. Proposed and developed by Henry Flad, the technique entailed much new design work, laborious efforts to compute stresses and loads, and an interlocking chain of innovations.[2] This new stage in construction was as demanding as the first two phases,

---

* Two factors argued against falsework at St. Louis. That wooden staging in the river would have required weeks to build, obstructing steamer traffic, even if the work were undertaken one span at a time. Equally important, sudden rises in the river (all too common in the city) could carry off the falsework.

Union Pacific Railroad Bridge over Missouri River at Omaha, Nebr. Completed 1872.

The action here was four hundred miles from St. Louis, on the Missouri River at Omaha and Council Bluffs, where construction was well advanced on the Union Pacific bridge connecting those towns. This 1871 view from the Iowa (eastern) side of the river shows the first truss completed. Falsework supports the next three spans during their assembly. Further west, pneumatic piers stand ready for their trusses after crews build the temporary falsework, much of it in the river. Lucius Boomer's American Bridge Company used eleven Post trusses to cross the river and its broad flood plain here. Compared with Eads's pathfinding in St. Louis, these proven methods in design and construction cut overall costs and risks dramatically. *Union Pacific Railroad Bridge over Missouri River*, image PHYS 16980.tif, Union Pacific Railroad Museum, reproduced by permission.

the caissons to bedrock and the work to produce suitable steel components. This time the citizens of St. Louis watched a daily drama as the developing chords reached out above the churning river. No one could predict whether the method would work. On the other hand, everyone could understand the threats that might crash the arches during construction.

Success in this new chapter would depend very much on the cooperation and goodwill of the men at Keystone Bridge. But after two years of troubles with steel, Keystone's leadership had grown truculent. Jacob Linville had never wanted anything to do with Eads or his outlandish design. Three times, Carnegie had tried and failed to push the Captain aside. After all their aggravations, Carnegie & Associates clung to their promised completion bonus of $250,000. Nonetheless, the leadership at St. Louis Bridge had to cajole Keystone into fulfilling its contractual obligations.

Fits and Starts

With the cantilevering method, construction would start simultaneously on the two ends of a single rib, or more accurately the two beginnings. Once Katte's teams had installed the first three tubes, each developing chord would need some kind of support. Since falsework from below was impossible, Flad resolved to support the chords from above. Temporary wooden towers would sprout from each abutment and pier. As Katte's men added new tubes, cables would reach down from the wooden towers to bear the growing chords as they extended into empty space. If all went well, those lengths would meet at the midpoint above the river.

This cantilevering method was just as audacious as the pneumatic caissons. While doubt seldom assailed James Eads, Walter Katte may have felt differently. From its initial contract, Keystone had taken the liability to erect the arches; a supplementary agreement of March 5, 1873, renewed that obligation. Yet Keystone's Linville had grave and justified doubts that the arches could ever be completed.[3] Other pitfalls—equipment failure, powerful storm winds, or hidden faults in the steel—could threaten the chords long before Katte and Flad would confront the ultimate challenge that preoccupied Linville, namely, the difficulties inherent in placing the last connecting tube in an arch. Completion was far from assured, while acrimonious lawsuits or even foreclosure for St. Louis Bridge seemed nearly guaranteed.

While the spring of 1873 would bring real progress, this was not a brilliant new dawn. Instead, the painstaking efforts to cast, forge, and finish good steel grudgingly paid off over the year. More broadly, the complicated project struggled to advance on many fronts. During the preceding fall, the masons had largely completed the stone piers and abutments. According to a September 1872 story in the *East St. Louis Gazette*, the masonry "now presents an appearance that is grand to look upon."[4] Then winter

This August 1873 view from the St. Louis levee (looking downriver) shows Flad's cantilevering method to build the superstructure. Supported from above by temporary towers and cables, the steel tubes reached out from the St. Louis abutment. On the west river pier, work had also begun on the west and center arches. Workboats shuttled new tube sections out as needed; laborers swayed them up into place. In the distance, the East St. Louis Grain Elevator received wheat in rail cars and shipped it by rail and river routes. Photograph by Robert Benecke, in Woodward, *History*, plate 43.

stopped all work, froze the river, halted the ferries, and plunged the city back into isolation. The union depot organization existed on paper, nothing more. Thanks to the negotiations between Eads and Morgan in October 1872, the new St. Louis Tunnel Railroad Company had funds to begin building that important link. During fall excavations, the tunnel contractor William Skrainka encountered quicksand, which mired the digging and collapsed a wall.[5] Those problems in turn forced Eads and Flad to select a new route for the tunnel, requiring new authorizations from the city, which were granted only after the Tunnel Railroad paid a bribe demanded by a city official.[6] Ultimately, the brutal winter of 1872–73 shut down tunnel work. In November 1872, the president of St. Louis Bridge died. The board moved quickly to replace William McPherson with Gerard Allen, owner of the Fulton Iron Works and an investor in many St. Louis ventures. The work would go on.

In most American cities, trains ran in the streets, causing round-the-clock miseries year in and year out. Eads's tunnel would let St. Louis avoid that mayhem, although businesses along Washington Avenue endured months of disruption during the tunnel's construction. Detail from a stereoview by J. H. Fitzgibbon (1873), author's collection.

As warmth returned, the men atop St. Louis Bridge seized the moment. By April 1873, the reliable James Andrews had taken a revised contract to build the tunnel.[7] Tom Scott's old friend, Andrews had been the prime contractor for the bridge piers and abutments. He wasn't cheap; the tunnel was now budgeted at a million dollars.[8] By July, upwards of a thousand men were toiling through the heart of downtown. Where possible, the contractor chose the cut-and-cover method to build the tunnel, tearing apart the city's prime business district on Washington Avenue.

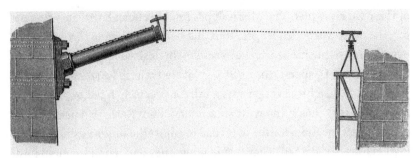

The first step in erecting the superstructure. The inaugural tube and skewback (*left*) were loosely bolted to the west abutment. A surveyor's transit (*right*) stood on a temporary wooden platform propped against the west river pier. While looking through the transit, an engineer shouted instructions to a crew positioning the inaugural tube. Once it was aligned, the crew fastened it permanently. Engraving from Woodward, *History*, 159.

In March 1873, a new contract established a fresh start for St. Louis Bridge, Keystone, and the new Midvale Steel Company, which had taken over from William Butcher.[9] At this time, Midvale adopted the open-hearth method to produce steel, with immediate improvements in quality.[10] Even so, Eads was forced to acknowledge a reality that had driven Butcher bankrupt: the sleeve couplings to join the steel tubes would not themselves be steel. The best wrought iron would have to be good enough.[11] And on March 13, Walter Katte led his team in installing the first steel components of the first arch. They started on the pair of chords making up rib D (the northernmost rib) in the span 1 (the west span). Seven years in the planning, this first step had to be perfect. Achieving the precise meeting point out over the river depended on correct placements for the first tubes. Working on and below the west abutment on the St. Louis levee, the men hoisted into place a wrought-iron skewback with its first segment of steel tube already screwed into place. This inaugural tube was one of four (two on the abutment, two on the west pier) that would become the starting points for building rib D. Outsized steel nuts held the three-ton skewback-tube assembly loosely in place.

Perched precariously five hundred feet away, an engineer (likely Flad) stood on a temporary platform propped upon the lower courses of masonry of the west river pier. The Mississippi's muddy churn lay twenty feet below the narrow wooden ledge. From that perch, he sighted through a surveyor's transit positioned at the exact height desired for the first inaugural tube in rib D, span 1, the one to be located where he stood beside the west pier. His sighting target was a mirror precisely fixed on the end of the inaugural tube on the west abutment, which Katte had placed in its approximate location. Katte's team had to adjust the tube's position so that it was correctly aligned. They would know they had found the right position once the surveyor saw through his transit the reflection of his own instrument in the mirror five hundred feet away. This elegant solution plainly showed the alterations required: left or right, up or down. To achieve that result entailed much shouting across the gap while Katte's men labored to adjust the heavy inaugural tube-skewback combination until its position registered perfectly through the transit. Katte and his team completed this first placement that day. To build the four ribs in each of three arches, they needed to place forty-eight of these three-ton inaugural tubes, using massive wrenches to tighten the steel nuts down on the forged anchor rods. These starting points would eventually transmit the arches' loads into the stone piers and abutments. Brute force and exquisite precision built the chords.

On April 7, crews began installing tubes in ribs C and B, the two center ribs of span 1. Then the work stopped again, awaiting materials.[12] This time the fault appeared to lie with Keystone. This pattern of fitful work broken by exasperating delay continued into August. The contractor had few reasons to hurry, given that the job entailed new and challenging tasks. Then Carnegie threatened to halt all work on site and at the Pittsburgh shops until the parties had arbitrated their financial disputes. He backed down after receiving a mollifying letter from William McPherson's replacement as president of the bridge company, but the languid pace of Keystone's men did not improve.[13]

Delays or outright stoppage amounted to dire threats for St. Louis Bridge. Every month, the company incurred over $23,000 in interest charges on the first mortgage bonds ($596,000 in 2023). In early August, Eads reported to his board of directors that the slow present was leading to a miserable future. If Keystone continued to work at its July pace, the last arch would close thirteen months later. That would not do. Besides the interest charges, other problems were mounting: forgone revenues, difficulties in cash flow, and the squandered patience of the entire city. While construction languished, the unfinished arches with their temporary towers and cables were particularly vulnerable in the event of another cyclone striking the city, a real if incalculable threat.

At a tense meeting in Philadelphia on August 14, Eads and Taussig thrashed out a new agreement with Keystone's officers.[14] The new plans required Keystone to double the number of temporary wooden towers and cantilevering cables then in use. For its part, St. Louis Bridge agreed to shoulder the additional costs of that apparatus. Once Flad's cantilevering infrastructures sprouted on all the piers and abutments, crews could work on the three spans simultaneously. The negotiators agreed that the new approach could result in completed arches by December 18, 1873, if Keystone would increase the number of machinists and the amount of tooling in Pittsburgh and the size of its workforce in St. Louis.

To motivate Keystone, the new contract included three incentives. That company would receive an additional $35,000 if it completed the arches by January 1, 1874. It would earn $30,000 more if the bridge was "ready for railway and highway traffic on the 1st of March 1874." On top of that, it could earn an early-completion bonus of $250 a day for each day *prior* to March 1 that the bridge was ready for traffic.[15] The new contract reflected raw facts of power and self-interest. By this point, Keystone's leaders could scarcely abide the Captain and his exotic bridge. For its part, St. Louis Bridge had no choice but to lay down sufficient enticements to drive Keystone to complete the work—even though it was already contractually obligated to do just that.

In New Directions

Even as James Eads sought fulfillment in the bridge, his restless mind and driven ambition turned to a new challenge. In May 1873, the Captain addressed the St. Louis River Convention. With thousands in attendance, the meeting included a dozen governors and more than one hundred US senators and representatives. This assembly of boatmen, merchants, shippers, and politicians wanted better navigation on the Mississippi. They especially wanted a deeper channel at the mouth of the river. Across the Midwest, on every rainy day acres of topsoil washed off the grassland prairies, now increasingly transformed into wheat fields, to travel downstream as silt to the Mississippi River Delta.* Once there, the muddy silt fell from the slowing current, clogging up the delta's many channels, creating sandbars, and blocking ocean freighters from reaching the docks at New Orleans. Speaking to this throng of influential men, Eads offered a solution, yet another novel conception of his singularly innovative mind. He would use the river to clear the river.

Then his health faltered again. On August 11, he wrote to Carnegie: "I have had several recent hemorrages [sic] from my lungs and am on my way for a trip to England + back for the benefit of my health and am anxious to sail at the earliest moment."[16] On the way to his New York departure, he stopped in Philadelphia to negotiate the new contracts of August 14 with Keystone's president, Jacob Linville.[17] Notwithstanding his illness, in early September Eads met with Junius Morgan in London. It could not have been otherwise. The banker needed a progress report, and Eads needed more money.

The 1869 prospectus had forecast that St. Louis Bridge would eventually offer a $2 million issue of second mortgage bonds. As conceived originally, those bonds would all be lovely gratuities to the stockholders upon completion of the bridge. Now the issue had become essential to raising the cash required to finish the structure. Since May 1871, Morgan had laid the ground for this second foray into the bond market.[18] Nonetheless, in their September 1873 meetings, he took a cautious approach, always mindful of what the market would bear. Eads's telegram back to St. Louis told the story: "Morgan thinks placing Seconds on market impracticable until completion bridge." He would, however, advance cash loans at the rate of £90,000 a month for three months, at the stiff interest rate of 10 percent. Morgan expected to make a market for the Seconds once the arches reached completion above the river.[19] That issue in turn would provide the cash to pay off the 10 percent loans, build the two decks, and finish the structure.

Behind these terms, Junius Morgan's patience was wearing thin. The charismatic Eads always spun a compelling tale. But work on the arches

---

* The Mississippi River Delta, located southeast of New Orleans, Louisiana, at the Gulf of Mexico, is three hundred miles downstream from a region located largely in the state of Mississippi known as the Mississippi Delta.

had barely begun, and completion could now come no sooner than two years after the original deadline. An intriguing question may have troubled Eads while he crossed the Atlantic: Did Morgan know that Thomson, Scott, and Carnegie had evolved from happy rainmakers to truculent contractors? Morgan likely knew something about those frictions, knowledge being essential currency in his line of work. The construction loans came with an ironclad stipulation. The first arch had to close by September 19, and the last by December 18, 1873, or he would cancel the loans.[20] In that event, foreclosure would soon follow.

Into the Void

Flad's cantilevering method to erect the arches was easily understood yet devilishly tricky to execute. The churning river below added to the challenges. After placing and bolting the inaugural tubes, gangs of men ventured out onto the developing rib to set up a temporary jib crane. The illustration on page 200 shows their work. After completing each tube connection, the jib crane and the men moved out gingerly like inchworms to do it all over again. In their early days on the west span, crews extended all four ribs simultaneously. The cross-bracing (between ribs D and C, C and B, and so on) strengthened the whole structure while helping to fix a precisely correct orientation to cross the river.

Support for the growing, cantilevered semi-arches required demanding work to position and tension a succession of cables. Once each chord had three tubes, the men set up temporary cables running from the west abutment to the upper chords in ribs C and B (known as third cables). When each chord had six tube sections, crews ran out new cables to the rib ends and tensioned them to support the load, replacing the third cables.[21] The trickiest challenges came with the twelfth cables. Actually made of iron bars, they ran from an anchorage on the levee up to a temporary wooden tower on the west abutment and then out to the twelfth joints in ribs C and B, a point in the void above the river, roughly 150 feet away from the abutment. Once these twelfth cables were lifted into place (a difficult job in itself), they carried the considerable and growing loads of the developing ribs as the men added new tube segments. To bear that additional weight, the men placed a temporary iron mast at the twelfth joint and strung a new eighteenth cable. Every step described here for the

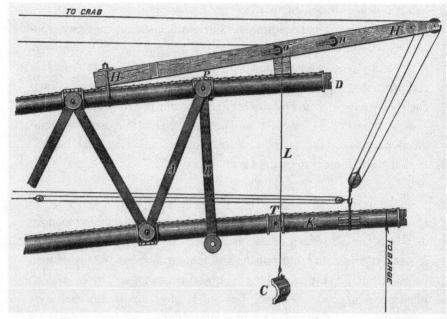

Assembling a rib. A gang of men set up a temporary jib crane (H) that served as the lifting point to bring a new tube (K) up from a barge in the river (each tube of twelve feet weighed two tons). After the laborers swayed the new section into its approximate alignment, they brought up the sleeve coupling (C) that would hold the new tube in place. Using wooden mauls, they hammered a turned steel pin through the two-part coupling. Then they bolted the coupling together to clamp down on the new tube section at point T. Finally, they would hinge brace bar B into place at point T, then tighten all the connections. With that work completed, the workers would bring up and place a new length of tube (not shown) to extend the upper chord at point D. Then they and the beam (H) moved out on the chord to repeat this work. Engraving from Woodward, *History*, 160.

semi-ribs from the abutment was executed concurrently for that span's counterpart, the growing ribs reaching westward from the west river pier. The spare image in the illustration on page 201 scarcely suggests the high drama imminent on that date, September 5: placing the last tube segments that would link up ribs C and B, creating the first complete and self-supporting arches of span 1.

Detailed planning and hard manual labor had brought the chords to this near culmination. By this time, Keystone had hired scores of workers. With the same essential steps now ongoing to complete all three spans, six

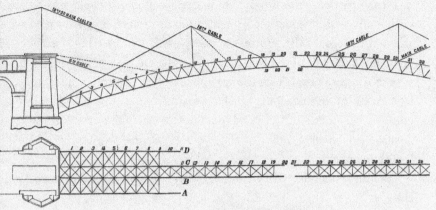

Fig. 29.—METHOD OF ERECTING THE ARCHES.

As the chords extended over the river, Katte's men set up successive cables to support their growing weight. As shown here in the plan view (*lower*), his teams initially built out all four ribs, then extended just ribs C and B (the two in the center). This eased the increase in the loads carried by the cantilevering cables and towers. After ribs C and B met up with their counterparts growing out of the west river pier, the arch would become structurally stable. This view shows the work as of September 5, 1873, without the final tube sections. On that date, no one knew whether they could be fitted into place. Engraving from Wood-ward, *History*, 173.

"tube raising gangs" worked simultaneously—all climbing about on the rough, loose planking that was their daily workplace. Other gangs fitted the tension rods and stays; more men did the riveting.* Upwards of two hundred men worked the daily shifts over the river. Some fell. In July, a plank in the center span broke, and three men tumbled into the river, one drowning.[22]

The tensioning of the cables created more drama. On September 5, the crucial twelfth cables each bore a strain of 176 tons.[23] As the ribs grew out over the river, tension in the cables needed to rise progressively, increasing the counterbalancing leverage as the gangs installed

---

*   Photographs show extensive riveting in brace bars and other iron components, but written sources shed little light on the men and methods that accomplished that essential work, much of it done in place, over the river.

new tube sections, couplings, and braces. To alter the cable tension at will, Flad and Katte had placed hydraulic rams vertically beneath the wooden towers on the abutments and piers.[24] By elevating the tower (and thus the apex of the twelfth cables), the engineers increased tension in those cables, lifting the rib ends. Here was another audacious, elegant bit of engineering. All the weights and strains had required extensive calculation, further efforts in the learning curve for Eads's design. The sheer number of those computations also increased the likelihood of errors.

Eads and Flad selected a young engineer, Theodore Cooper, to manage the erecting work done by Keystone and its gangs. Cooper had already demonstrated his merits as the company's quality control inspector at Midvale Steel. An entry from his daily journal portrays the challenges of his new job: "September 4: Raised 12 tubes. This brings the [west] side span within one tube at the top and two at the bottom member [whether rib C or rib B is unclear]. Set up 18th cable on C rib, and slacked off No. 6 a little in order to raise joint 16 and beyond for connecting tube-stays, the rib C being much lower (about 1.5 inches) than B. Very strong wind last night, but had no effect on the work except blowing off some loose planks from scaffolding."[25] To raise either end of rib C, Cooper could increase the tension in the twelfth cables. Filling in the gaps of ribs C and B (to unite the half arches in span 1) entailed similar care in tensioning and adjusting eight tube ends nearly simultaneously. The work called for a young man's constitution.

Another tricky challenge arose at this late stage. As the gap between the half ribs of arches C and B narrowed to just two tube sections, the chord ends perversely would not line up. Temperature caused the difficulty. The bridge has an east-west orientation, and the tubes and braces were all black. The hot afternoon sun of early September fell predominately on the south side of the structure, heating all that iron and steel unevenly. The metal most exposed to the sun, on the south side, expanded more than that on the north side. As a result, the tube ends began to turn away (heading upstream as it were) just as Cooper, Katte, and the Keystone teams planned to bring them into alignment. The only solution was not to attempt to close the gap on a hot sunny day. St. Louis had many in early September.

Still in Europe, James Eads clearly placed considerable confidence in Flad and Cooper. Indeed, he had to trust Katte and Keystone as well. Their work would make or break his deal with Morgan stipulating closure of the west span on or before September 19. On September 13, Flad, Cooper, and Katte tried to place the last tubes in ribs C and B. They failed. Flad ordered nearly two hundred tons of pressure in the main (twelfth) cables, but "the ribs were very stiff." The ends did not rise enough to align with the final tubes.[26] A mix of issues came into play at this point. Midvale had labored hard to make its steel with consistent qualities, especially in the modulus of elasticity. Yet the resilience in each tube segment varied somewhat, so that two outwardly identical chords did not behave identically under a given load. The variations would prove acceptable *after* the half ribs linked up into complete arches. Until then, they were a problem.

Another challenge in linking up was inherent in the design. From his initial report of May 1868, James Eads and his team of engineers had forecast the forces that compression would impose on the steel ribs in the arches. Because the "permanent load" would compress the tube lengths in the finished bridge, the tubes needed to be made somewhat longer than their dimension when under load. The precise ratio: 1.000363 times longer.[27] A small thing indeed. To Eads, such things mattered, as indeed they did for this arch bridge. For Jacob Linville or Lucius Boomer, designing any bridge to such impossible tolerances—a millionth of an inch—required equal measures of folly and hubris. For Henry Flad and Walter Katte, that miniscule amount became a major obstacle to closing the ribs and completing the arches. The permanent load would not come into play until the bridge was finished. Until then, each nearly completed chord was just a bit too long to readily accept its last tube segment.*

Murphy's Law also hindered placing the last tubes in the west arch. In making those components, the machinists and inspectors at Keystone had scrupulously followed Eads's specifications. As the west arch neared closure, however, Flad and Cooper saw that it needed to be 2.7 inches longer than drawn. Investigation revealed the source of the problem: the west pier had landed on bedrock out of place by just that amount.[28]

---

* Flad and Katte would eventually solve this problem by using the hydraulic rams and suspending cables to impose extra compression onto the half ribs.

Last Links

Under the August 14, 1873, contract, Keystone and St. Louis Bridge agreed, yet again, that Keystone bore full responsibility for closing the arches. Two days after Eads sailed for England, Keystone gave notice that it would not honor that commitment. To slither away from the obligation, Linville, Carnegie, and Piper claimed the whole contract was invalid. Their given rationale: their board had met on August 14 to ratify it but had failed to file proper legal notice of the meeting.

In an August 25 letter, Jacob Linville spoke frankly of the real problem: Keystone did not want the liability for closing the arches.[29] He promised to attempt completion by January 1, 1874, and Keystone wanted the bonus if it beat that date. But St. Louis Bridge would have to carry the responsibility for success or failure. Keystone's managers acted pragmatically here, if unethically. Eads and his team had mandated nearly every aspect of this unique bridge. No wonder that Keystone wanted the Captain's people to shoulder the execution risks: conundrums of difficulty, novelty, and unpredictability.

After their failed attempt of Saturday, September 13, Flad ordered the workforce to report on Sunday morning. He planned for a day of preparations—setting up parts, tensioning cables, taking measurements—with the consummating event to follow the next day. When Flad arrived at 6:30 a.m. on Sunday the temperature was a brisk 44 degrees. The cold shortened the semi-ribs, creating ideal gaps for placing the last lower tubes in ribs C and B. He would seize the moment. Laborers hurriedly lifted the tubes from a barge, even as a bright sun rose over the knots of excited workers. With every minute, the metal was heating and expanding. They slipped the final tube into the lower chord of rib C readily enough, but its counterpart for rib B fit partway, then became stuck. The engineers shouted for more tension on the cables, for men with sledges. But the lower tube in rib B simply would not take its final position. Fearful that rising temperatures would cause uneven strains, Cooper ordered the removal of both lower tubes while it was possible. The next day proved another disappointment: too hot, the gaps too short despite all their efforts.[30]

After Linville shucked the closing responsibility in late August, Eads wrote from London to his engineers with a back-up plan. He laid out designs for a special "closing tube." Machinists made up these new telescoping tubes with internal screw threads: a left-hand thread on one side and a right-hand thread on the other. It would work just like a telescoping shower rod.[31] At first, Flad chose to keep this solution in his back pocket. For three years he had sweated over the erecting plans and calculations. Not unreasonably, he wanted closure on his own terms. Sunday's low temperatures gave him an idea: he would bring cold to the bridge.

At 9:30 p.m. on Monday, September 15, fifteen tons of ice were delivered to the St. Louis levee and lifted up to ribs C and B. With temporary planks set on edge along the brace bars, the work gangs formed a makeshift trough upon the lower chords of those ribs. The men brought out ice all night long and well into the next day, sixty tons in all. It was a heroic effort, but nature held the whip hand. Before the work began, Flad had noted the temperature at 5:00 p.m. to be 98 degrees. In the shade. The night of ice and labor widened the gap by 1.625 inches, not quite enough. The work went on all day Tuesday, another scorcher. On the morning of Wednesday, September 17, Flad conceded to the sun. He ordered the telescoping tubes to the site. They worked almost flawlessly. The four special tubes (two for rib C and two more for rib B) plus their bracing were in place by 10:00 that night. Success, two days before Morgan's deadline. The *New York Times* described the closing as "one of the greatest triumphs of engineering skill the world has ever seen."[32]

The Panic and an Attack

In September and October 1873 two events, unfolding concurrently, shook the finances and the prospects of St. Louis Bridge. A day after the west arch closed, the leading investment bank in the country, Jay Cooke & Company of Philadelphia, shuttered its doors, unable to meet its obligations. Cooke had financed the Northern Pacific Railroad. That carrier's land grants and airy financing had, for a time, lifted the kites of many insiders and financiers.[33] Now the truth lay revealed. Another transcontinental railroad, one chartered to run two thousand miles through the trackless wilds of Dakota and Montana, was a road to bankruptcy. Cooke's failure spread to Wall Street, stock prices tumbled, and the New York Stock Exchange closed on September 20. Trading remained suspended for ten days. Spurred by Cooke's fall, banks called in their loans. Gold and currency

The Keokuk Northern Packet Line operated the Mississippi sidewheel steamer *Andy Johnson*, here on the St. Louis levee. Note its impressively tall stacks. Their height made for better combustion and hotter fires in its five boilers. They also contributed to a conflict with the new bridge in town. Photograph from Murphy Library Special Collections / Area Research Center, University of Wisconsin–LaCrosse, neg. #6889. Reproduced by permission.

grew scarce even in the nation's banking centers. This eastern distress burst the entire western railway bubble. By November, fifty-five carriers, including the Santa Fe, the Burlington, and the Kansas Pacific, had defaulted on their bonds.[34] The Panic of 1873 metastasized into a deep industrial recession that also depressed the entire western farming economy. Recovery would take five years.

At this climactic moment, the US Army launched an unlikely attack on the St. Louis Bridge. As the ribs of the west arch reached out over the river in the summer of 1873, a long-heralded phantasm appeared clearly, imminently on the cusp of becoming a real bridge. The Keokuk Northern Packet Company chose this moment to protest loudly to the Army Corps of Engineers and to friends in Washington, DC. Its complaint: the bridge was illegally obstructing navigation. A silent partner in the Keokuk firm was Secretary of War William Belknap, a powerful friend indeed. Wiggins Ferry soon joined the fight against the bridge.[35]

In the messy ways of American governance, a hodgepodge of jurisdictions shaped the western rivers and their bridges. St. Louis Bridge held charters from Illinois and Missouri, and it had received explicit authorization in a US statute of July 20, 1868. But Congress saw the Corps of Engineers as its essential instrument to shape the western rivers for the benefit of commerce. In an 1872 statute, the legislature charged the Corps with balancing the security of river navigation with the needs of railroads.[36]

From its creation in 1802, the Corps stood at the top rank of American engineering. In 1837, two of its lieutenants went out to St. Louis when the riverfront began to fill with silt, blocking steamboats from the levee. Robert E. Lee and Montgomery Meigs designed new wing dams to channel the river currents, clearing the sandbars.* By 1873, civilians—many of them graduates of new colleges in new fields like civil and mechanical engineering—wielded their own specialized expertise. Against those newcomers, the officers of the Corps still proclaimed themselves the country's engineering elite. Furthermore, their commanding officer loathed James Eads.[37]

On August 20, 1873, Brigadier General Andrew Atkinson Humphreys issued Special Orders No. 169, naming a board of five officers to convene hearings in St. Louis. Their charge: to "report whether the Bridge will prove a serious obstruction to the navigation of the Mississippi River." If the officers found against the bridge, they were to recommend "in what manner its construction can be modified."[38] It was a legitimate order and an outrageous maneuver. Colonel James Simpson convened the board in St. Louis on September 4. Its ranks including truly distinguished men; three had served as brevet major generals during the war. In the shrunken and beggared postwar army, the Corps was short-staffed and overworked.[39] Reverting to their permanent ranks as colonels or majors,

---

*   Meigs deserves a Lee-sized place in US history. Fiercely incorruptible, he served as the US quartermaster general during the Civil War, responsible for equipping the 2.1 million men of the Union army. It exaggerates nothing to say that Meigs made that army and victory possible.

the five officers selected by Humphreys knew how to follow orders, especially if their careers depended upon it.

Humphreys and Simpson did not need to tilt the board directly. Men like Major Gouverneur Kemble Warren had nothing in common with the self-taught Eads. Brother-in-law of Washington Roebling, this West Point graduate knew his engineering, his social affinities, and his self-interest. He was also something of a national hero. On the second day at Gettysburg, Warren, acting on his own initiative, had rushed his men to secure a Union position on a hill called Little Round Top. That desperate strategic move, made under continuous enemy fire, had preserved the Union line and thus blocked a Confederate victory.

The board met for two days in early September, showing stark bias favoring the steamboat interests. The officers declined to hear any testimony on the statutes and charters that authorized the bridge and specified its clearances. With Eads in London, it fell to Taussig and Flad to mount a defense. Major Warren denied their request for additional testimony, saying that he had already decided the case and that a thousand witnesses favoring the bridge would make no difference.[40] The ostensible problem: with their tall smokestacks, larger steamboats could not pass beneath the bridge. It came as no surprise that the officers, with their purpose preordained and their proceedings slanted, found that the bridge did obstruct navigation. Their action was not without precedent. In 1872, the Corps had ordered and secured alterations to a massive bridge then under construction over the Ohio River. Keystone Bridge had made those changes at a cost of nearly $500,000 ($12.7 million in 2023).[41]

Raising the height of the arches at St. Louis was impossible, so the board advocated a bypass canal. A thousand feet long and two hundred feet wide, this expensive ditch would cut through East St. Louis and its rail lines. Steamers that could not pass beneath the arches would take this detour when the need arose. Rail traffic to and from the St. Louis Bridge would necessarily cross a new drawbridge over the new canal. When steamers entered the canal, the drawbridge would open, halting the trains. The board claimed that "it will only be in exceptional cases that boats will desire to pass through this draw."[42] Why capable officers offered such nonsense seems nearly inexplicable.

The historians Ralph Gilbert and David Billington offered a compelling answer: Eads's cocksure promotion of an innovation fifteen hundred miles from St. Louis had ruffled high-ranking feathers.[43] During the rebellion, the navigation channels through the Mississippi River Delta in Louisiana had become choked with silt. To solve the problem, the Corps had proposed to build a new bypass canal at a cost of $7.4 million ($128.1 million in 2023).[44] In his May 1873 address to the St. Louis River Convention James Eads had offered a radically different solution: the creation of a deepwater channel from New Orleans through the delta to the sea. He planned to use the river's current to flush the silt out to the Gulf of Mexico. With his proposal for new jetties to concentrate the current's force, the Captain had made an opponent of General Humphreys. Adding salt to their conflict, Eads had publicly criticized the army's planned canal as an expensive boondoggle.

For the professional fighters of the Corps of Engineers, the unfinished St. Louis Bridge became a handy cudgel in battles unfolding many miles away in Louisiana and Washington, DC. The officers did not really need or want any solution for the steamboat interests of St. Louis. They sought only to wound James Eads. As the board took its slanted testimony, the arches in the west span advanced to the precarious state of September 5 (see the illustration on p. 201). Work had barely begun on the center and east spans. Money was again tight. Another delay, even just a few months, could pull down the whole financial structure of St. Louis Bridge, especially after the Panic hit that month.

The board finished its work in early October. General Humphreys signed the report recommending a bypass canal through East St. Louis, and Secretary Belknap approved it on October 10, 1873. Back in St. Louis, Henry Flad declined to take the army officers very seriously. Interviewed by the *St. Louis Democrat*, he offered these responses: Perhaps six large steamers engaged in routes on the lower river could not pass upstream under the bridge; all other vessels would travel unimpeded. In any case, "a few years will play the big boats out," and he predicted that they would be replaced by barges and towboats. He felt confident that Congress would neither require nor fund a canal. Flad believed that if events proved him wrong, canal construction would delay the bridge's opening by two years or more. In that case, all St. Louis would rise up in anger, and "the popular feeling would drive the steamboatmen who started the thing out of town."[45]

Flad's public nonchalance contrasted with the private concerns of William Taussig. As the top financial officer for St. Louis Bridge, Taussig had plenty to worry about that October. He hurried to New York, where he met with Pierpont Morgan and James Eads, just off the steamer from Britain. Morgan shared Taussig's worry that the Corps's verdict could have a "disastrous effect" on the London market for bridge company securities.[46] Undaunted, Eads again resolved to drive on. He too had friends in high places.

On October 27, 1873, Taussig and Eads arrived at the White House. Both men had met Ulysses Grant in earlier chapters of their lives. As a judge on the St. Louis County Court before the war, Taussig had turned Grant down for a job. In the first important Union victories of that war, Eads's ironclads had supported Grant's attacks on Forts Henry and Donelson. Recent events provoked worries, however, because the visitors had supported the Liberal Republican challengers to Grant in the election of 1872. President Grant greeted the St. Louis men warmly in the Cabinet Room, asking Taussig, "How are you, Judge?" Noting the honorific, Taussig hoped that Grant did not hold the past against him. Grant laughed, replying, "Oh no, you see how much better it is than it might have been."[47]

Taussig and Eads described the Corps vendetta against the bridge, all of it news to Grant. The president then summoned Secretary Belknap, who arrived within minutes. Under Grant's questioning, Belknap admitted that the bridge complied with its legislative authorizations. Furthermore, his predecessor as secretary of war, William Tecumseh Sherman, had fully endorsed the project. After a few moments of silent reflection, Grant dismissed Belknap, saying: "If your Keokuk friends feel aggrieved, let them sue the bridge people for damages. I think, General, you had better drop the case." The president's decision was final. The Corps had picked a fight, then lost it. In November 1873, Grant visited St. Louis, venturing on a cold and blustery day onto the unfinished west and center arches. He approved of all that he saw.[48]

Two Last Links

The stormy fall of 1873 culminated with completion of the center and east spans. Working so closely in these demanding conditions, Flad and Coo-

The arches nearing completion, December 1873. A Wiggins ferry awaited passengers on the Illinois shore. Photograph by Robert Benecke, in Woodward, *History*, plate 44

per settled into a harmonious relationship with Walter Katte and his gangs of erectors from Keystone. The work of raising, placing, and bracing the steel tubes became routine even as the whole project took on a much more dramatic appearance. By October the men were completing ribs D and A in the west span. At the same time, towers and cables on each of the piers and abutments were reaching to support construction of ribs C and B in the center and east spans. Every day, onlookers saw a taut drama on the city's doorstep, the unfinished ribs extending out into thin air. Pressing the work concurrently on all the arches largely balanced the massive weights that bore down each pier, an important consideration. Furthermore, the pace was essential in meeting Morgan's stipulated dates for closing the center and east arches.[49]

The urgency caused men to fall. On December 2, it was Theodore Cooper's turn. Working on span 3 (the east arch), he tripped and plummeted. As he fell, his rational nature took over. Contemplating the distance and the probable force of the impact, he tucked himself into a ball to lessen the likelihood of injury. After falling ninety feet (he later measured), he plunged deep into the river, came up gasping, and struck out for the eastern shore. Surfacing, he still gripped his pencil. A boat rowed to his rescue, and he was back at work within hours.[50]

Overseen by Cooper, the erecting gangs closed the inner ribs of span 2 on December 16 without trouble. Like so many aspects of this project, filling in the last gaps of span 3 proved difficult. With the eastern half of the arch too low, Cooper and his men resorted to every stratagem they had. At 10:00 p.m. on December 17, the wind shifted from south to north, bringing a cooling and contracting breeze. As the men worked on the scaffolding by torchlight, the central tubes in span 3 finally slipped into place at 7:40 the next morning.[51] It was December 18, the date of Junius Morgan's deadline. When the news of this triumph spread, factories and river steamers tooted their whistles in joy. Finally, the city had its bridge, seven years in the making. Flad's audacious cantilevering process had proven its worth. At noon, "a party of ladies and gentlemen walked out [on scaffolding] to the middle arch and hoisted the Stars and Stripes." More whistling and cheering marked the moment.[52]

Two days later, good news arrived from London. Rather than waiting until full completion to market the second mortgage bonds, J. S. Morgan & Company placed the $2 million issue on the market on December 20, 1873.[53] The bonds sold readily, pushing total capitalization to $10 million. Those British investors might have held on to their money if they could have read the minutes of the bridge company's Executive Committee meetings that month: no funds in the treasury to pay current bills (December 6), demand all unpaid assessments on stock (December 12), delay remittances on all bills from vendors (December 25), and pressure the New York stockholders to take a new issue of preferred stock (December 25).[54]

Finishing Touches and a Bitter End

With the arches completed, Keystone's men needed four months to build the rail and roadway decks, rivet the wind truss into place, paint the structure, add the sidewalks, and build the approaches in St. Louis. They overcame significant problems. To make the half arches line up, Flad and Cooper had forced some of the tube connections by putting extra tension in the cables. As a result, the rib crowns in span 1 were "several inches" lower than designed.[55] Keystone had to cut down some parts of the wind truss to make it fit.[56]

A notable turning point came on January 31, when Keystone's men removed all of Flad's wooden towers and cantilevering cables. Another milestone occurred on April 15, when Keystone's managers laid on the traditional celebratory dinner for the laboring men who had built the bridge. The bosses had cause to rejoice as well. As defined by its contract, Keystone had achieved substantial completion fifteen days ahead of the deadline, earning a bonus of $15,000.[57] Many details remained unfinished on both decks, but the end was in sight.

With the great structure nearly completed, Andrew Carnegie regained his characteristic exuberance. He wrote to Junius Morgan on February 9 to describe a special side deal he had negotiated. Outside Keystone's direct obligations, Carnegie now had his own bonus/penalty contract with the bridge company. He would lose a thousand dollars a day if the bridge was not ready for rail and highway traffic by June 1. Conversely, he would receive that sum in a bonus for each day of completion before that date. He wrote Morgan that he expected "to take *at least* 25,000$ [$681,000 in 2023] out of that bargain." Carnegie concluded: "Allowing one year to enable the Railway Companies to get their business fully changed from Ferry to Bridge, the expectations of the most sanguine will be realized."[58]

By April 15 the roadway deck was finished. After clearing the decision with Keystone's representatives, Taussig announced in the St. Louis newspapers for Saturday, April 18, that the bridge would welcome visitors the next day. Finally, the public would cross and inspect the new engineering marvel. At midnight, a messenger brought sour news: Keystone had unilaterally canceled its agreement about visitors and would retain full possession of the structure.[59] To ensure its control, Keystone sent workmen to tear up a section of the wood-decked approach road. As the day dawned, that squad stood guard, preventing anyone from venturing onto the bridge. It was a mean-spirited gesture toward the men of St. Louis Bridge, indeed toward the city as a whole. Nature delivered some payback, however. Gray clouds and cold driving rains on Sunday and Monday discouraged nearly all the curious, while drenching Keystone's guards.[60]

Advised by their lawyers, Keystone's Carnegie and John Piper had decided to retain the bridge to improve their leverage in negotiating a final settlement of monies owed. That thicket of claim and counterclaim had grown nearly impenetrable over the years of delays, alterations, and renegotiations. Anyone reading the terms of the original negotiations of February 1870 between Eads and Carnegie can see that this bitter end was nearly ordained from the start by the ambiguous clauses in their agreement. Eads's innovations, his perfectionism, the delays, and the troubles with steel all then took a bad situation and poisoned it. Anyway, Keystone had never really wanted to build the Captain's bridge. Now that the deed was done, its owner-managers were determined to secure full payment for all their work and all the aggravations. Resolving who owed what became the preoccupation of arbitrators and lawyers for years thereafter.

It is impossible to trace all the claims and counters. Some highlights: Jacob Linville believed that Eads's capricious demands had increased Keystone's costs by $500,000; he had begun documenting that figure in 1872.[61] In turn, St. Louis Bridge maintained that Keystone owed it $241,000, mostly for excessive billing on fixed-price contracts.[62] Under his personal bonus agreement, Carnegie claimed substantial completion effective April 14, although this railroad bridge lacked rails until the end of the month.[63] An arbitrator selected May 2 as the completion date, awarding Carnegie a bonus of $30,000.[64] St. Louis Bridge paid that debt with a promissory note. When Carnegie tried to collect the funds six months later, there was no money in the account.[65] Here was a practical problem: the company could not pay what it did not have.

The thistles at the center of the thicket were the claims involving Andrew Carnegie & Associates, as distinct from Keystone. Early on, Thomson, Scott, and Carnegie had received stock in compensation for their services in placing the first mortgage bonds with Morgan. (Unlike other insiders who were gifted with stock, the trio had played a key role in making that connection, justifying this compensation.) When St. Louis Bridge decided to assess all its stockholders to raise cash, the associates declined to pay up. The arrears on their calls totaled $179,750.[66] On the other hand, the bridge company owed $250,000 to the trio as *their* completion bonus. They did receive payment, after a fashion, in the form of "bonus" stock rather than cash.[67] This no-cost stock might soon prove of no value, however.

Carnegie had extracted their due for the associates, but he left a bad taste with Taussig. The day-to-day manager of St. Louis Bridge wrote resentfully to Carnegie that he had received a scant salary while protecting the interests of the company. "My stock is a trifle to yours and cost me, every dollar of it, hard earned cash money. Yours costs you nothing." Taussig felt that he had been "very badly treated." He complained that the union depot project was again languishing, adding that "Scott didn't even answer my letter. . . . This whole business [illustrates] the careless and torpid manner in which the Penn. Interest attends to its affairs."[68]

Still, not all was sour. In the spring, the Baltimore Bridge Company completed the eastern approaches to the bridge, an iron viaduct through East St. Louis.[69] By late May, St. Louis Bridge had control of its $10 million property. On Sunday, May 24, upwards of twenty thousand happy visitors made the pilgrimage to and over the river. The modest admission fee counted for something with Taussig.[70]

The tunnel was nearly finished by June, so Washington Avenue no longer evoked the siege of Paris. Its cost came to $1.066 million, with construction sustained by loans from Junius Morgan. In a letter to the banker, Taussig described the tunnel. Lit by gas jets, it "presents a very cheerful and pleasant appearance."[71] At 4,400 feet in length, it ranked among the longest tunnels in North America. A page 1 story in the *New York Times* described in glowing terms the many benefits to flow from this route beneath downtown. Without the tunnel, trains to and from the new bridge would have had to run in city streets, as they did on New York City's Park Avenue, Tenth Avenue (nicknamed "Death Alley"), and many other arteries. If St. Louis had emulated Manhattan, the rail traffic would have "killed its yearly hundreds, would have smashed buggies and wagons by the score, would have blocked up the roads with long lines of freight-wagons, and would have made the St. Louisans gnash their teeth generally at the mad ambition that prompted them to become a railroad centre."[72]

As befitted the occasion, James Eads took the central role at the bridge's grand opening ceremonies on July 4, 1874. Few in the audience knew that he was then fully committed to his next project, improving navigation through the Mississippi River Delta. To Eads, the St. Louis Bridge was both a proud accomplishment and old news. On July 12, he wrote a leave-taking letter to Junius Morgan. This remarkable document offered equal measures of self-serving justifications and score settling. Eads wrote that "the delay . . . of the railroads in making their connections with the Bridge, which I have always expected, is now occurring." He claimed that "the bridge will very soon prove that it is a *great financial* success." According to the Captain, the cost overruns and delayed opening had resulted from "the unjust demands of the Keystone Bridge Co." Furthermore, "the fear of having the influence of the Pennsylvania Road turned against the [bridge company] has, I believe, cost our company not less than a million and a half of dollars."[73] While that estimate was almost certainly inflated, Eads was correct about the obstructionism, while omitting the many problems of his own making.

Eads was moving on. Morgan would now have to plunge deeply into bridge matters—finance, management, operations—to make all this work. His reputation hung in the balance, alongside millions of his own money and more millions from his investors. The bridge stood completed. Engineering its financial viability was another matter entirely.

CHAPTER 9

Foreclosure and a Pool

ON OPENING DAY, July 4, 1874, William Taussig feared "impending bankruptcy" for St. Louis Bridge. A week later, James Eads predicted to Junius Morgan that the company would be a "great financial success." Their split opinions partly reflected their natures and labors. A sober daily manager, Taussig knew where every penny went. Conversely, the expansive promoter never let dollars interfere with his visions. As matters unfolded, both predictions proved true.

In July and August 1874, the people of St. Louis flocked to their grand new public space, drawn by cooling breezes off the river, panoramic vistas of the city and the Mississippi, and the chance to see and be seen. On summer nights the bridge company put on concerts at midspan, popular gatherings costing only a dime for the pedestrian toll ($2.73 in 2023).[1] In August the first eastbound freight crossed the river, a boxcar laden with wheat from the Central Elevator.[2] Sporadic freight shipments followed. In early September, thirty stock cars carrying livestock traversed the tunnel and bridge to reach East St. Louis. The cattle may have met their maker in a slaughterhouse at the new National Stock Yards, which had opened a year earlier, anticipating the bridge.[3] The press also reported a meeting that September of local passenger agents, midlevel railroad officers charged with managing services for passengers. They sought to craft the arrangements required to let people cross, just as the cows had.[4]

If the meeting was accurately reported, nothing came of it. For one thing, only a single carrier connected at each end: the Missouri Pacific to the west, the Vandalia line (a PRR affiliate) to the east. The railroad long identified with Eads, the North Missouri, had no connection with the bridge. Passengers from the East stepped off trains in East St. Louis—and onto Wiggins ferries. Their short ride provided superb views of the new bridge, which the *New York Times* had described as "the eighth wonder of the world."[5] Newcomers to the city could only guess why their trains were not crossing the brand-new St. Louis Bridge. A snarl of issues kept most rail traffic off the bridge in that first year. Eads's letter to Morgan was correct in at least one regard: commerce would need time to flow into new routes. But the bridge company had to overcome far more than inertia.

Railroads: Indifferent or Opposed

The western railroads' chronic shortfalls in revenues and liquidity grew worse after the Panic of 1873. The North Missouri had entered receivership even before that general collapse. Within two years, the Missouri Pacific and the Atlantic & Pacific also became wards of the courts. With red ink filling their ledgers, the western lines lacked the funds to build direct connections to the tunnel or to the MoPac tracks in the Mill Creek Valley. The history of broken promises did not help. Since 1867 Eads and Taussig had touted the bridge to every top railroad officer in St. Louis, heralding services that proved chronically late to show up. At this juncture, the carriers had little reason to hurry in building facilities that would mostly benefit the bridge company and its London bondholders.

Some of the eastern carriers had more money, but most had fewer incentives to cooperate with St. Louis Bridge. Their freight yards in East St. Louis adequately served their needs. True, consigners of freight paid charges to Wiggins Ferry and the St. Louis Transfer Company. This was perhaps unfortunate for the shippers but of little concern to the carriers. If the Alton or the Ohio & Mississippi built track connections to the bridge, their freight trains could run directly into St. Louis. Service over the river would then require new yards, sidings, switching engines, locomotive terminals, freight houses, clerks, maintenance staff, and laborers—all in Missouri. The eastern carriers had already provided those facilities and personnel at considerable expense in East St. Louis, Illinois. Passengers from the East would appreciate service directly into the city. Unfortunately, there was no station.

William Taussig had labored tirelessly to create a union depot for all the city's railroads. Much of the land had been purchased, and the city had condemned or vacated streets in the path of progress.[6] Five carriers had guaranteed to use the station and service its securities. In April 1874, the bridge company had signed contracts with those railroads, authorizing their trains to run over the bridge.[7] All for naught. In a meeting of representatives from five eastern lines on July 9, 1874, the PRR's men pushed through an agreement to boycott the bridge until the depot was a settled fact on the ground.[8]

Why Tom Scott chose to humble St. Louis Bridge is a mystery. Edgar Thomson had died in May, and Scott had immediately risen to the PRR presidency. Any number of calculations may have motivated him. Carnegie & Associates believed themselves shorted, even swindled, out of the proceeds they had counted as their due from building the bridge. Perhaps Scott simply was nursing a grudge. Or he may have wanted the bridge for the PRR. If St. Louis Bridge were driven to foreclosure, new owners could acquire the structure at a fraction of its $12.5 million cost.[9] The historical record leaves no doubt that Scott stood opposed, and his opposition mattered. His PRR had dominance in the industry, connections to other lines, political pull, and access to capital; in short, it had the raw power to impose nearly any stance or policy it chose. And its allied carrier, the Vandalia line, provided the bridge with its only eastern connection in the early days.

Over at the tunnel's western mouth the situation was worse. For nearly two years after the bridge opened, the Missouri Pacific provided its sole western connection. The MoPac took any westbound freight emerging from the tunnel, receiving fees to move it onward. Eastbound traffic was a different story. Whenever possible, the MoPac directed its own freight cars or those of carriers moving across its lines to its own ferry service over the Mississippi at Carondelet.[10] Desire for profit and duty to shareholders drove those choices. A long-standing personal grudge between Eads and the MoPac president, Captain Joseph Brown, did not help matters. Furthermore, Brown was then mayor of St. Louis, making him an especially challenging adversary.[11] In the past, the city of St. Louis had supported the carrier with massive infusions of public funds. Now the railroad and the mayor were undermining the most important and expensive innovation in the commercial life of the city and the region.

After the associates withdrew their support for the tunnel and the union depot in November 1871, the local leaders of St. Louis Bridge had struggled to craft an alternative model for their business. Their inexperience, meager capital, and no small measure of naïveté contributed to their difficulties. The expansive vision of Eads's 1868 plans far outmatched their own capacities to execute it.

Originally, Eads, McPherson, Taussig, and the investors envisioned the bridge as a toll highway, open to all railroads on equal terms. The bridge company prepared a standard-form contract that detailed those terms.[12] Every carrier that signed would receive equal access to the bridge, charged at the same rates. The contract stipulated that the roads would build their own track connections, run their own trains, and give all their river-crossing business to the bridge. That clause aimed to freeze out Wiggins. St. Louis Bridge would limit its own operational roles to scheduling, dispatching, and signaling. Passenger trains would receive priority over freights, scheduled trains over extras. It all seemed logical on paper. The carriers had nothing to lose in signing that paper. By July 1874, it had become clear that they had little to lose by abrogating their obligations.

The railroads mounted a legal argument for their inability to fulfill their contractual duties. The eastern lines claimed that their charters provided no authorization to operate in Missouri; the western carriers said the same of Illinois. If the claims were true, the bridge company had incompetent legal counsel (for failing to foresee this problem).[13] If untrue, it had treacherous railroad partners. In any case, one fact dominated everything: throughout its first year in operation, the officers of railroads serving the region would not lift a finger to help St. Louis Bridge.

Bridge versus Ferry

Railroads were not the only problem. The Wiggins Ferry Company had joined the Corps of Engineers to battle the bridge, and the army had been routed at that White House meeting with President Grant. Wiggins next challenged the bridge company directly. The ferry monopoly had proven lucrative to its owners since its charter by Illinois in 1819. Aside from its fees for moving cargos and people, Wiggins owned outright most of the

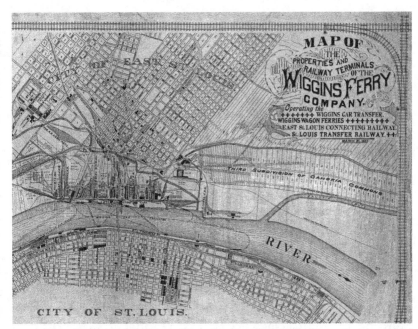

This 1897 map includes the main features of East St. Louis, Illinois. Rail yards and the docks of the Wiggins Ferry Company lined the river. Good rail connections tied the National Stock Yards (*top left*) to the bridge (*center*). Dotted lines in the river show the Wiggins ferry routes. Detail of a map dated March 30, 1897, from the Herman T. Pott Inland Waterways Library at the University of Missouri, St. Louis.

waterfront and many of the rail yards in East St. Louis, real-estate holdings worth more than $2.5 million in 1874 ($68.1 million in 2023).[14] Its powerful owners had no intention of surrendering the business.

Instead, they went on offense. In June 1874, Wiggins bought the *D. W. Hewitt,* a tugboat that St. Louis Bridge had used during construction. Wiggins converted it into the first ferry outfitted expressly for foot passengers. Until then, pedestrians had had to seek safety amid the jostling horses and laden wagons on the ferries, no easy feat. The "elegantly fitted" *Hewitt* crossed to Washington Avenue in just three minutes. Passengers paid a nickel each way, while avoiding the effort and time required to walk up to and over the bridge.[15] An accounting from 1875 found eight ferries and towboats in the Wiggins fleet.[16] With multiple landings on each shore, Wiggins offered convenience to foot passengers in a city short on mass-transit options.

To protect its freighting business, Wiggins placed a barrel of whiskey on its boats, giving free drinks to the drovers and teamsters who crossed. The bridge company countered with free ice water at midspan.[17] St. Louis Bridge opened the war over freight rates, charging just five cents per hundredweight versus nine cents on the ferry. Wiggins immediately undercut the bridge rates, while the St. Louis Transfer Company also slashed its fees.[18] Wiggins offered ferry services to all comers. The transfer company provided the drayage, invoicing, and delivery services required to move freight from railcars in the East St. Louis yards to consignees in the city: homeowners, merchants, or the western carrriers that sent the goods onward. Most shipments bore a fee from each company.

The transfer company could have shifted its cross-river route from Wiggins to the bridge. Instead, it remained allied to the ferry company and fought to hold its own business. Shares in the closely held transfer corporation were highly valued and rewarding to many in the local business elite. Equity holders included officers of the Alton and Ohio & Mississippi railroads. The railroad men saw no reason to alter arrangements with Wiggins and the transfer company, dealings that enriched them.[19] In point of fact, the status quo worked. Once the rate war broke out, shippers had less motivation to select the bridge. About 50 percent less.

Furthermore, Wiggins did a large and growing business in moving railcars (loaded or empty) on purpose-built ferries across the river. Its two landings on each shore connected to four railroads, twice as many connections as the bridge could claim. In the year ending in July 1873, Wiggins ferried 18,755 cars across the Mississippi. In 1878, more than 101,000 cars made the trip by boat.[20] As those figures suggest, the railroads had a viable alternative to the bridge, except when the river froze.[21]

Overcoming the habits of established networks was the ultimate challenge. While design and construction had raised difficult issues, now the bridge men confronted problems of an entirely different character and magnitude. To change the flow of vast and complicated streams of commerce would require thousands of shippers, freight agents, and merchants—in St. Louis and around the country—to think and act anew. Crafting a new approach for passengers, in senses figurative and literal, was comparatively easy. Once a suitable depot came into being, their direct arrival by rail into St. Louis could follow immediately. Shifting freight in

The bridge entrance at Washington Avenue and Third Street, ca. 1877. A heavily loaded wagon (*center*) will soon cross the river. A horsecar (*right of wagon*), sheltered by the tollhouse roof, awaits its passengers before departing for East St. Louis. Washington Avenue was paved, although well encrusted with manure. The tunnel ran beneath the scene shown here. Unknown photographer, MHS, N 39492.

all its varieties into new distribution channels would take time. And time would prove scarce.

Building Anew

Even before the bridge opened, its managers perceived these growing obstacles and rising headwinds. Facts are stubborn things. They had to craft new business models even as they struggled with Keystone and steel, negotiated with railroads, completed the tunnel, searched for fresh capital, and reeled from the Panic and depression. In April 1874, the bridge men took a pivotal step by forming the Union Railway and Transit Company. This new corporation would build and own freight terminals, purchase locomotives, create switching facilities, hire workers, and, crucially, run trains across the bridge and through the tunnel. By taking this step, the bridge men put themselves in the railroad business. When it began operations during the summer of 1874, theirs was the most unusual railroad in the country.* It owned no main line; it simply provided railroad services to the bridge and tunnel companies for negotiated fees. Hauling the cars of other railroads, Union's trains ran on tracks (the bridge, tunnel, and approaches) totaling three miles, had two corporate owners, and had cost upwards of $15 million (or $418 million in 2023).[22]

At the start, Union's inadequate staff, equipment, yards, capital, and experience all showed. On June 8, 1874, nearly a month before the official opening, "a small party" of bridge company officers, engineers, and invited guests boarded a two-car train in East St. Louis. The locomotive and cars were borrowed from area carriers. Unfortunately, the neophyte railroaders of St. Louis Bridge failed to realize the implications of taking a passenger car from the Ohio & Mississippi. Originally built to the wide gauge of six feet, that carrier converted its tracks in 1871 to standard gauge (4 feet, 8.5 inches). To save money, it simply put new standard-gauge wheel sets under its old wide cars. The O&M coach ran over the bridge without difficulty. When it entered the tunnel, its wooden sides began to scrape and splinter on the walls, darkening the celebratory mood aboard. Deeper fear set in when the train, stuck in the tunnel, began to fill with smoke from the coal-fired locomotive. After too many anxious minutes, the engineer managed to back up, returning his train to daylight and fresh air.[23]

On the grand celebratory day, July 4, the men running St. Louis Bridge again revealed their inexperience. Working with its friends on the Vandalia line, the company loaded five hundred "distinguished guests" aboard a train of fifteen palace cars, hauled by three coal-fired locomotives, intending to run it from East St. Louis over the bridge and through the tunnel. But the train halted before its cars had fully exited the bore. It had to stop because the company had failed to secure city permission to run its tracks far beyond the tunnel's western mouth. The two locomotives up front caused no difficulty, but a third engine was coupled to the rear, so it

---

* The bridge men needed Union because their existing corporate structures—the Illinois & St. Louis Bridge Company and the Tunnel Railroad of St. Louis—lacked any legal authority under their charters to operate trains. Union was actually two companies, one chartered in Illinois, the other in Missouri.

remained under ground. For "ten or fifteen minutes" its smoking exhaust threatened the honored guests. As Eads later confessed to Junius Morgan, the air in the tunnel and cars soon grew "close and unpleasant," and passengers became "sick and faint."[24] After too many anxious moments, the special returned to the bridge and clear air.[25] By August, four coke-fired steam locomotives were heading to St. Louis from Philadelphia's Baldwin Locomotive Works.[26] If not quite the "smokeless" locomotives that Eads had promised, they appeared to offer an improvement. In the years to come, passengers would often find the tunnel air unpleasant, if not life-threatening.

New responsibilities kept piling up. By November 1874 William Taussig had pushed through construction of a temporary depot, a few sidings, and a freight house at the tunnel's west end. The railroads gave little note or business to these facilities.[27] For the men who had guided the bridge project through its many travails and triumphs, all this was maddening. Among that cohort of backers and allies, Junius Morgan ranked at the top.

Junius Morgan in America

Despite these troubles in St. Louis, the bridge company stood tall in far-off London. On September 16, 1874, fully a year after the Panic took down Jay Cooke, some equity shares of St. Louis Bridge traded there at 103.75.[28] In December, its first mortgage bonds found buyers on the London exchange at their par value.[29] That strength reflected their attractive interest rate and Junius Morgan's impeccable standing with British investors. Protecting that high regard had to be his top priority. In September he had traveled to the United States. Typically he made an extended visit each year. The trips allowed for meetings with his American partners, offered fresh insights into the US financial scene, and included firsthand updates with clients. On his varied agenda, St. Louis Bridge took a worrying primacy.

J. S. Morgan & Company had already sustained the bridge with a Mississippi-sized flood of capital. In all, these debt obligations totaled $10 million, and they paid interest at rates ranging from 7 to 9 percent. Note, however, that while the bridge and tunnel companies had only received 60 to 80 percent of par value for these securities, they would have to repay the full par value, $10 million plus interest, to Morgan's bondholding clients. In September 1874 he had mounting cause for concern on that point.

During his American tour, Morgan periodically updated his partners back in Britain. From New London, Connecticut, on September 1, 1874, he wrote that "the delay in opening rail traffic on [the] Bridge is provoking, but it is now . . . being arranged at an early date." Meetings in New York City that month drained his confidence. He confessed in a letter of September 23 that "there does not seem to be anyone in the business with brains enough to organize some plans" to manage the financial obligations over the next twelve to eighteen months while developing the rail business. "I had thought we had amongst the shareholders & directors people able & willing, but [I] find that all they are competent for is [to] take care of themselves. . . . I am thoroughly despondant." An October 20 letter recounted three days of meetings in St. Louis and sketched the bad situation: "The competition with [the] ferry co is severe [and the railroads] are taking advantage of it to force lower terms from the bridge. Scott was there with me, and I tried hard to get him to order his trains over." Morgan concluded that Scott would seek board approval "for the rates I required." All hung on the PRR's president.[30]

As time passed, Morgan offered further insight into the mess in St. Louis. His October 30 letter from Philadelphia was stark: "You have no idea how wretchedly, engineering excepted, all the bridge affairs have been managed." Four days later, he explained the wretchedness. "The management of [bridge company] affairs has been as bad as was possible. Which you will hardly believe when I tell you that the present floating debt is $1,100,000 [$30 million in 2023]."[31] "About $600,000 is secured in one way or another, but the security in all cases is not adequate." A letter from Hartford on November 20 noted more talks with Scott days earlier, but the PRR president still would not order his affiliates to use the bridge.[32]

Morgan had ample cause for disgruntled ruminations during his return voyage to Britain that December. He had finally seen the great bridge in October, and he had finally perceived the real character and present situation of his debtors and their associates. Scott's bland intransigence must have been especially maddening. At least Morgan now saw the facts and players for what they really were. A telegram of February 27, 1875, sent from his New York partners at Drexel, Morgan & Company to the London house offered no surprise. "Satisfied foreclosure inevitable."[33]

A Postmortem

This descent into foreclosure provokes a host of questions.* After all, St. Louisans had fervently desired this connection to the national railway network for a decade or more. The upper deck welcomed all manner of local traffic. But the bridge company became a foreclosed ward of the court ten months after its grand opening. Certainly Scott's vendetta, the war with Wiggins, and the depression all bore down on St. Louis Bridge. But deeper inquiries seem warranted. Were the company's top managers and directors guilty of malfeasance? Does the foreclosure shed new light on the Morgans? On its face, it would appear that they had little business, and worse judgment, choosing to finance such a dance hall.

It is impossible to know precisely when Eads, Taussig, or Junius Morgan realized that the bridge company would default on its debt obligations. Carnegie had seen it looming back in 1871. On April 14, 1875, the trustees of the fourth mortgage bonds petitioned the US Circuit Court for the Eastern District of Missouri to foreclose on the bridge company.[34] Under the court's oversight, the trustees—Pierpont Morgan and Solon Humphreys—took responsibility for running the bridge company, preserving its assets and revenues to benefit the London holders of the first mortgage bonds.

---

* Just as Scandinavians have many words for snow, Americans of the Gilded Age needed a broad vocabulary to describe financial failure. A note *went to protest* when Acme Bridge failed to honor (pay) one of its own debt obligations. If the parties could not work out an agreement on that specific debt, Acme had then *defaulted* on its obligation. If Acme had bondholders, they would be wise to *foreclose* on the company at this point to ensure their own primacy over other creditors. That step made Acme a ward of a court, which would appoint *receivers* to manage its operations during a *receivership*. If the receivers failed to reach agreements with creditors (to pay some or all of the debt), the court would declare *bankruptcy* for Acme, and the receivers could oversee the *liquidation* of its assets. As an alternative, judges, receivers, and creditors often preferred to give such firms another opportunity to turn a profit. In that case, the receivers sought court approval to install new managers, restructure Acme's debts, pay its preferred creditors, and allow Acme to start anew. Such a *reorganized* firm had a new corporate name to signal that break with the past.

Liquidation of the assets would prove impossible; perhaps the trustees could make the venture pay. The *Railroad Gazette* offered this backhanded verdict: "There was hardly ever so costly a new work in this country that had so little trouble in obtaining capital abroad, notwithstanding the fact that only a very small fraction of it was provided by stockholders."[35]

True enough, although the *Gazette* might have noted that those stockholders now stood to lose every cent of the nearly $4 million that they had eventually paid in for those shares. June brought a bitter moment when the sheriff carted away from the office on Main Street the desks and files, even the water cooler.[36] In October, the credit reporters at R. G. Dun & Company wrote that "the original stock is not regarded as of any actual value although some shares have lately been sold at 2 or 3 percent of their face."[37]

Cash-flow problems were the proximate trigger for foreclosure; the ultimate problem was miserable earnings that traced directly to the railroads. In its first full year of operations, an average of just sixty-four freight cars a day crossed the bridge. With its dual tracks, it could readily have handled trains with sixty-four cars every five minutes. Nearly all the carriers had lagged in building their own rail connections, a delay largely owing to Scott's boycott and the price war with Wiggins. February 1875 brought a revealing if temporary change. The Mississippi froze solid, so traffic had to take the bridge or go nowhere. The Union Transit Company's locomotives hauled 5,300 cars over the river that month, a daily average of 189.[38] The frigid February weather proved that its customers were conspiring to kill the bridge company by premeditation.

Two months after the foreclosure, the new Union Depot opened on Poplar Street, its eleven tracks occupying city blocks from 10th to 14th Streets. The first passenger train to the new station crossed the bridge on June 13, 1875.[39] The master schedule for the depot, Time Card #1, bore the same date and listed sixty passenger trains coming and going daily plus nine scheduled freights.[40] The business had turned a corner, a few months too late.

In June, J. S. Morgan & Company sent out a "Circular to Bondholders," which put on a brave face. The bank attributed the foreclosure to construction delays (due to engineering difficulties), the Panic, and the railroads' broken promises. The circular included better news: two hundred passenger cars were now crossing each day, and "London parties"

UNION DEPOT.

Once it finally opened, the St. Louis Union Depot proved a fine facility, emi-nently suited to the needs of travelers. Passenger trains from the East still stopped in East St. Louis because those carriers kept their repair and servicing facilities in Illinois. The locomotives and crews of St. Louis Bridge (really its Union Transit subsidiary) handled train operations between the Relay Station in East St. Louis and this depot. A new business district grew to surround the station, oriented to the rails, not the river. Image from *Souvenir of St. Louis*, ca. 1882.

had invested $250,000 in new money in the Union Transit Company. The circular closed by urging the bondholders "not to take alarm or lose confidence in this undertaking" as J. S. Morgan & Company retained full faith in the securities and the business.[41] Morgan was as good as his word; indeed, he had financed the loans to Union Transit.

But the circular told far less than a complete story. Most importantly, Eads's broad vision for a complete transport infrastructure for the region far outreached his company's capacities to execute it. Nor did his erstwhile allies, the managers of the area's railroads, appear to care very much about that vision

Furthermore, many businessmen in St. Louis saw a silver lining in the foreclosure. William Taussig would fault the merchants' ingratitude to the "enterprising and public-spirited citizens who risked their money and induced so much foreign money to be risked in aid of their great undertaking."[42] Many merchants, however, had come to a starkly different conclusion. The bridge company would clearly need to charge high tolls to pay off its heavy debt-servicing costs. As the merchants saw it, the commerce of the entire region *"should not* be taxed" by those interest charges.[43] Such a course would make all St. Louis's industries and every one of its workers unable to compete with those in rival cities not choked by the grip of a monopoly.* An analyst for the Merchants Exchange looked at the bridge and wondered whether the business community had *"an elephant on their hands."*[44] Foreclosure could trim the debt burdens on bonds and erase all dividends to stockholders. This in turn could lead to lower rates charged by St. Louis Bridge—a welcome result for businesses and consumers throughout the region, if not for the insiders who had promoted the elephant and the investors who had paid for it.

Eads's determination to force innovation proved exorbitantly expensive, another factor behind the foreclosure. Driven by his perfectionism, the Captain pushed up costs without apparent limit. Not surprisingly, he was a bit touchy on this topic. For example, in 1875 he placed the cost of the bridge at $6,680,331, which he claimed was only 20 percent more than his original estimates. In fact, it was 39 percent higher than his 1868 forecast.[45] And that total just covered the bridge itself. Nearly $6 million more was swallowed by interest charges, commissions on bonds, charters, legal expenses, real estate acquisition, and other incidentals. In all, Eads himself put the final tally at $12.5 million as of 1875.[46] More interest charges would pile up in decades to come. Then there was the million-dollar tunnel.[47]

---

* These concerns about competition provoke a hypothesis. St. Louis merchants correctly feared that their shipments (in or out) would bear high charges to pay off the bridge company's burdensome debt. If Wiggins could be driven out of business (as seemed likely), those charges would arise from the bridge's monopoly over the rail traffic carrying St. Louis commerce. That market control ultimately traced back to the Illinois legislature. Perhaps Illinois legislators foresaw that this gift to the bridge company would amount to a long-term burden on the commerce of Missouri's leading city. If so, Chicago's representatives had ample cause to vote in favor of this Trojan horse.

Judging Their Rectitude

Fraud also played a sizeable role in pushing up the costs and driving the bridge company to foreclosure. The looting came in many forms. Overpayments for materials were easiest to hide. Consider the contract for stone. Tom Scott's friend James Andrews was the prime contractor for the masonry piers and abutments. But he bought stone from the bridge company, not directly from quarries. That arrangement gave bridge insiders the opportunity to skim a lucrative markup. Furthermore, back in 1868 Andrews had agreed to share the stone contract with Scott.[48]

Stock proved ideal for frauds in many forms. The company had charter authorization for $4 million in share capital. Under its February 1869 prospectus, it issued $3 million in stock to its initial investors. At that time, it assured those men they would only have to pay in 40 percent of par, which most did. A year later, when its prospects still looked gilded, the bridge company plucked from its vaults a pile of its unissued shares, giving $800,000 (par value) to a small group of insiders. These bonus shares went to members of the original investing pool of the previous year.

Here its actions veered into fraud. Ostensibly the bonus shares went to insiders for their helpful role in placing the first mortgage bonds with Morgan in London. This was true for Carnegie & Associates but untrue for other insiders who received bonus shares. Also untrue was the accounting in the bridge company's books claiming that bonus shareholders had paid in cash for those shares at the customary 40 percent assessment just as other stockholders had. The books lied; the bonus shares were free. The stock frauds had three larger results. The company had less cash than it claimed on its balance sheet, presenting a deceptive portrait to its bondholders. When troubles with steel and other problems soured the company's prospects, these knowledgeable insiders could sell those shares to unwitting buyers. And creditors were ultimately defrauded by this chicanery.[49]

When delays and cost overruns began to overwhelm St. Louis Bridge, its principals quickly turned to protecting their own self-interest, exacerbating the company's troubles. James Eads offered a breathtaking excuse for his own failure to pay the assessments on his shares. At the time of his resignation in July 1874, he owed $100,000. Writing to General Manager

Taussig, Eads claimed that he could have sold his shares years earlier (dodging that liability) given his inside knowledge of the upcoming levies on the stock and of the company's cloudy future. But "the necessity of sustaining the faith of *all* parties pecuniarily interested in [the venture's] ultimate success was of vital importance."[50] So he had neither sold his shares nor paid his assessment.

By July 1874 the bridge was finished, but its cash-flow struggles had grown desperate. Eads now claimed that he should not have to pay the assessment in cash. Having duped his fellow stockholders (and preserved his own capital) two years earlier, he now cast his choices as laudable actions to serve the greater good. Here again we see the era's business ethics, characterized by the historian Richard White: "Good men lied and manipulated information to maintain or increase values. . . . Men of character considered themselves the final judges of their own rectitude."[51]

Other men in the St. Louis business community reached their own conclusions about James Eads. Recollect the Eads-led investor group that took over the State Bank of Missouri in November 1866, transforming it into the National Bank of the State of Missouri. In June 1877 the bank shuttered its doors, its liabilities far outstripping its assets. Eads and his friends still sat on its board. A newspaper story that month claimed that the bank's stockholder-owners would "not only lose their stock, but are liable to an equal amount besides to make good the deposits." Eads was liable for roughly $250,000.[52] To head off an indictment for fraud, he negotiated a cash settlement.[53] His close engineering associate Shaler Smith privately offered his own guilty verdict: "Eads, smartest of the smart, slipped out from under . . . and the arch beguiler went his way rejoicing."[54] This would stand among the biggest bank failures in the country in the depression following the 1873 Panic.[55]

Ample servings of impropriety, as we construe it today, were nearly baked into the bridge project from the start, given the cloaked insider deals, asymmetric information, conflicts between personal and corporate profits, and comingling of private enterprise with public purposes and state powers.[56] As a financial debacle loomed, insiders like Eads, Scott, and many others cared only for their own advantage, heedless of obligations to one another, to unwitting stockholders in St. Louis and New

York, or to the London bondholders. Their behavior appalled Junius Morgan. During his US visit in September 1874, he saw their financial chicanery up close. He wrote back to London: "Any man or set of men who could act in such a manner are not worthy of our confidence."[57] By that insight, he insinuated that they were also unworthy of the millions in British gold that he had channeled to the project.

Morgan might have looked deeper sooner. The stockholders in St. Louis and New York, at least those who met their assessments and did not jump early from the sinking enterprise, would lose the money they had paid for their equity. Two years before it did sink, the company crafted a deal allowing shareholders to convert 25 percent of their stock's par value into second mortgage bonds. By this alchemy, holdings destined to be worthless stocks instead became debt securities traded in London and underwritten by J. S. Morgan & Company.[58] And the Morgans were the agents of this transformation.[59] Most of the costs—fraudulent, exorbitant, or essential—would eventually fall on the people and commerce of St. Louis. Striking an accurate balance in that ledger would require many decades.

Who Wants This Bridge?

The Omnibus Bridge Act of July 1866 authorized eight bridges over the Mississippi and one over the Missouri River. The St. Louis Bridge was the last one completed. A month before its grand opening, James Eads was in Chicago to attend the annual convention of the American Society of Civil Engineers. His protean mind had new ideas on bridging. In a presentation titled "Upright Arched Bridges," Eads claimed to have revolutionized bridge engineering. Again. Unlike the structure just completed in St. Louis, his latest approach placed the structural arch *above* the deck. With his cocksure confidence, he claimed that this new paradigm for metal bridges "must ultimately supersede every other one now in use."[60]

The presentation did not go well. One commentator, Willard Pope, characterized the paper as "a bold and vigorous attack upon the labors and opinions of the great majority of modern bridge engineers." The conventioneers' comments were divided between condemnation of Eads's new approach and criticism of the just-completed St. Louis Bridge. Many decried its expensive price tag. Ashbel Welch declared that "that engineering is best which most fully answers its purpose at the least cost." The country's leading authority on stresses in engineering, Samuel Shreve, went further, arguing that the bridge at St. Louis was notable only because it was in danger of imminent collapse. He claimed: "The great importance of immediately strengthening the ribs of the St. Louis bridge can no longer be ignored."[61]

We don't know what led Shreve to this howling gaffe. Perhaps he failed to understand the calculus that reckoned the actual stresses. Or perhaps he understood Eads all too well. Like every engineer in the hall, Shreve knew that the Captain had contempt for orthodox thinking and incremental design improvements. Most engineers clung to both, perhaps wisely, given the difficulties and costs inherent in long-span bridges. Not surprisingly, many of those professionals responded with contempt for Eads.

Like the Mississippi itself, time would now flow on around the bridge. Over time, views about this accomplishment would evolve. James Eads had succeeded in the difficult challenges of building a pathbreaking structure. It remained to be seen whether or how Pierpont Morgan would manage to make it profitable.

The Morgans Take Charge

The depression that followed the Panic of 1873 was a tough time to run a railroad, yet that had become the Morgans' challenge. To turn St. Louis Bridge around, the bankers had to throw themselves into daily management issues and larger matters of business strategy and development.[62] They had to achieve what James Eads had promised, while overcoming obstacles that he had not faced. They worked remotely, from New York City and London. And they worked against a backdrop of depression, deflation, general labor unrest, and widespread railroad restructuring.[63] Turning failure into financial viability would require five years of hard work.

In the first year of operations, rail traffic for the bridge and the tunnel was largely choked off by Tom Scott to the east and the MoPac to the west. The company's meager earnings derived mostly from customers on the upper deck. Pedestrian tolls held steady, as did the earnings of the horse-drawn streetcar line. Steady but unremarkable. Wheeled traffic included private carriages, omnibuses, express wagons, and coal vans. That business doubled with the river's winter freeze, then tumbled with the arrival of spring. The price war with Wiggins caused a lot of pain at St. Louis Bridge.[64] Absent that conflict, however, the upper deck revenues alone could never have made the company profitable given its huge burdens in debt service.

The war continued for two years. By October 1875 Wiggins had hammered down its charge for moving a loaded freight car over the river to two dollars; the return trip for the emptied car was free. By contrast, the tariff for the bridge and tunnel came to five dollars.[65] While the contest benefitted the public and merchants, the bridge company was losing money. Probably Wiggins was too. The two companies began to negotiate a truce in the summer of 1876, signing an agreement on April 23, 1877.[66] Apparently, Wiggins came to realize that its real opponent had become J. S. Morgan & Company, the power behind St. Louis Bridge. Morgan wouldn't, couldn't give up. Furthermore, as the economy began its halting advance out of depression, rail traffic would inevitably rise. With that growth, the bridge company would harvest economies of scale unavailable to the labor-intensive operations at Wiggins. It was time to reach an understanding.

Pierpont Morgan's very name would become a synonym for the elements of that deal. From June 1877, Wiggins and St. Louis Bridge charged the same tolls for each class of service, from loaded box cars to hogs in transit. All earnings went into a common pool. Until net revenues reached $400,000 a year, the bridge and tunnel companies received 75 percent of the total and the ferry took 25 percent. That scale slid progressively; when the proceeds passed $1 million, Morgan's clients would take 97.5 percent of the pooled total. In sixteen sections the pooling agreement spelled out details for auditing the books, maintaining the properties, and preventing cheating. Its rationale appeared in section 15: "to put an end to a mutually destructive competition."

Antitrust law did not exist in 1877, while deflation and profit-killing competition appeared at every hand. Over time, *Morganization* became the term of art to describe how Pierpont Morgan forced his debtor railroad clients to negotiate pools or mergers, ending their rate wars to assure steady returns to capital.[67] To ensure that these agreements were honored, a Morgan ally or partner joined the board of the Morganized company. Solon Humphreys served that function at St. Louis Bridge, taking its presidency. Pierpont Morgan could impose these policies because his associated banks were the best gateway to more capital. He would offer no apology for imposing oligopoly or monopoly on railroad companies or the commerce of St. Louis. His goal was profitable stability, which he claimed would benefit laboring men as well as capitalists.

The pool succeeded. The bridge and ferry companies still had separate legal status, while operating jointly. Within months, both firms enjoyed rising earnings.[68] The bridge company's larger slice of the growing revenues reflected pragmatic realities. The ferries did not cross at night, and winter ice could stop them for weeks or months. Trains ran around the clock. With time and economic recovery, St. Louis Bridge would dominate the trans-river business and the region.

Yet Wiggins would endure, as did successive pooling agreements. In 1878 it built its own switching railroad along the Illinois riverfront, tying all the carriers terminating there to docks that transferred cars to or from the Wiggins ferries crossing the river.[69] Many railroad and shipping customers preferred the ferries' connections and convenience. James Eads and his banker-investors had presumed that the new would beat the old, that the bridge would kill the ferry. Not so.[70]

Thanks to the pool, however, St. Louis Bridge had turned a corner by May 1877. Every day, upwards of sixty-four scheduled trains (hauled by Union Transit's locomotives) carried passengers over the bridge and through the tunnel, to or from the East. The western lines funneled more traffic to the Union Depot. Freight traffic on the bridge grew by 77 percent in the fiscal year, to nearly 80,000 cars.[71] Within months, nearly all that traffic halted.

The Battles of 1877

In their struggles to remain solvent in the sputtering economy, the four big trunk lines agreed in May 1877 to fix their rates and pool their revenues on freight traffic from the Atlantic Seaboard to Chicago.[72] Apparently, the managers of the New York Central, the Erie, the Pennsylvania, and the Baltimore & Ohio also forged a secret agreement to cut wages across their systems. By June, isolated groups of workers across the PRR's vast system

had begun to protest over wage cuts and safety concerns, actions that local managers mostly dampened down.[73] The Baltimore & Ohio then emulated its larger and more profitable competitor, imposing systemwide
wage cuts of 10 percent, effective July 16. In a well-shaken mix of hubris
and condescension, on that date the B&O president, John Garrett, also announced a 10 percent stock dividend.[74] His actions sparked the worst
labor unrest in the nation's history to that time.

After three years of accumulated wage cuts, speed-ups, and other prov-
ocations, the strikes of July 1877 quickly became wildcat actions that
shut down all the eastern trunk lines and most freight traffic in the Midwest. The strikers were desperate men. Few had effective union representation. Even fewer had any savings to feed their families or pay the
rent. And with unemployment above 20 percent, the strikers and their
bosses knew that replacement workers—scabs—were hungry to take
their jobs. Despite the odds against them, work stoppages quickly spread
as far as Omaha, Galveston, and St. Louis.[75]

On July 19, Junius Morgan cabled John Garrett, an important client and
a close friend: "Our sympathies with you. Hope you will hold firm, accepting no decision but unconditional surrender." Garrett understood the
message: the London bankers wanted all the carriers to show a united
front against labor's demands.[76] His reply aimed to reassure Morgan:
"Thanks. Fully determined to maintain discipline absolutely. President
Hayes promptly furnished troops. We look for Early restoration of control
unconditionally."[77]

This deployment of federal soldiers rather than state militias in a
labor dispute was unprecedented. To give his act a figment of legality,
Hayes declared the strikes on the B&O to be "unlawful and insurrection-
ary proceedings."[78] The new president lacked a sense of irony. A few
months earlier, he had withdrawn the last army units detailed to the
South to protect freed slaves from escalating assaults and intimidation
after the all-too-real rebellion there. With his new proclamation, Hayes
invented a faux insurrection in the North, then detailed the US Army to
quell it.

Within days, railroad workers in St. Louis also walked off the job. A
meeting in East St. Louis on Saturday, July 21, drew six hundred men from
many railroads and diverse trades.[79] Reporting on these developments, the

New York Times predicted that employees of Union Transit "will probably be the first to join in the strike on this [west] side of the river, they being much incensed at the recent action [wage cuts of 10 percent] of that company."[80] Mass meetings on July 22 culminated with a call to stop all freight trains, effective midnight that date, on five major eastern lines terminating in East St. Louis.[81]

A day later, the strike in St. Louis shut down work across much of the city. A newspaper report claimed that five thousand railroad workers had taken control of the Union Depot and nearby yards, joined by three thousand sympathizers from other occupations. Their resolute, orderly determination won "hosts of friends and sympathizers." Nonetheless, rumors spread of a rabid element that might burn the eastern approaches of the bridge or dynamite a span. Perhaps they would set off charges to collapse the tunnel. Or perhaps all this fear-mongering was the work of labor's opponents. In any case, William Taussig reinforced his force of hired guards at all vulnerable points.[82]

On Monday July 23, workers on the MoPac struck for wage restorations. All freights east and west halted, while strikers ensured that passenger trains carrying the US mail continued without interruption. Stopping the mail could provide an excuse for federal troops to intervene in St. Louis. To preserve public order, the strikers and the sympathetic mayor of East St. Louis shut down the saloons there. Late on that Monday, two developments raised tensions even further: two companies of US Army troops out of Fort Leavenworth, Kansas, arrived in St. Louis; and the Workingmen's Party led a "noisy procession" of thousands through the city streets.[83]

On Tuesday, July 24, the job action against the area's railroads became a general strike, the first in America. After a mass meeting of ten thousand that night, led by the Workingmen's Party, the movement sought an end to child labor and the adoption of an eight-hour workday. The *Missouri Republican* called it "a labor revolution."[84]

The Morgan partners took a dark view of all this social upheaval. On Wednesday, July 25, the bankers at J. S. Morgan & Company received a worrisome update in a telegram from their American partners: "Affairs continue serious very. New York quiet but military remain under arms at armouries. What at commencement [was a] railway strike has now become general labor riots. Traffic Ill. & St. Louis Bridge Co. stopped." The New York partners added news that highlighted the dire trajectory of events: the bridge company's receivers had appealed to the governors of Illinois and Missouri to call up their state militias "for protection to passage [of] freight trains to Pittsburg, Albany, Hornellsville, [and] Martinsburg."[85] Pierpont Morgan and Solon Humphreys wanted the soldiers and their guns to intimidate the striking workers into submission.

In secret, the big men in St. Louis politics, industry, railroads, and banking were also organizing, making plans to crush the strikes. Supported by the governor and the mayor, an armed Committee of Public Safety raided the headquarters of the labor movement on Friday, July 27, arresting its leaders. The committee received its rifles from the state armory, its pistols from the US Army's Rock Island Arsenal in Illinois, and its reinforcements, three companies of US infantry, from Fort Leavenworth.[86] Resistance surely was futile. Soon, 150 strikers and sympathizers were in jail.[87] This decisive action collapsed the workers' cause overnight. Over in East St. Louis, the Illinois governor arrived on Saturday, July 28, joined by eight companies of US Army soldiers. There too, labor's militancy immediately evaporated in the face of overwhelming force.

In a backhanded way, the strikes revealed the central role of the infrastructures proposed by James Eads a decade earlier. Seeing their strategic value, the strikers had taken control of the Union Depot in St. Louis and the Relay Depot over in Illinois. At that stop, trains to and from the east changed their locomotives with the engines from Union Transit that pulled the cars across the river. By early August, the old order was restored across the country. Thanks to the durable alliance that wealthy men like Morgan and Garrett had forged with the state, capital had enlisted armed force to crush the militancy out of labor.

A Fresh Start

In October 1877, Junius Morgan made his customary fall trip to America. Large bond deals with the Baltimore & Ohio and New York Central were on his mind. So was St. Louis Bridge. For two years the company had operated under foreclosure. This protected the bondholders' priority over the claims of lesser creditors while buying time to work out agreements with Union Transit, Wiggins, and the railroads. Now the time was nearing to reorganize, to create a new company to take the venture out of court and off the bankers' hands. While in New York, Junius discussed its "reconstruction" several times with Solon Humphreys.[88] He reported back to London that Colonel Scott was coming up from Philadelphia to discuss "bridge matters."[89] Morgan's partners were clear: the time had come to develop a reorganization plan for St. Louis Bridge.[90]

From his federal courtroom in St. Louis that October, Judge Samuel Treat signaled his impatience, two years after approving the foreclosure. St. Louis Bridge was finally making money, but profits were insufficient to pay the interest due on its first mortgage bonds, let alone the other securities. The London bank was loaning the company the funds necessary to pay the semiannual interest on those bonds.[91] But the judge was tiring of this temporizing measure. The company needed to stand on its own, which meant a restructuring. Developing that plan consumed a year of negotiations between the parties in London, New York, and St. Louis.[92] The primary issue: the debt burdens on the bridge and tunnel companies needed trimming.

A consortium of London investment bankers had joined J. S. Morgan & Company in extending financing to St. Louis Bridge during the receivership. Once the broad outlines for reorganization were settled in London, Charles Branch, of Foster & Braithwaite, traveled to St. Louis in December 1878 to guide the foreclosure sale. Reporting back to London, he wrote: "This has been a very busy week, one of the hardest and most anxious I ever spent. From breakfast time until midnight the lobby of this hotel and some of its bedrooms have been entirely occupied by innumerable conferences about [the] Bridge, Tunnel, Ferry + Transit [companies]."[93]

On December 20, 1878, Branch had the leading role in an auction on the steps of the St. Louis Court House. He proved to be the sole bidder for the assets of the old company, which sold for $2 million. The fastidious Branch found it all a bit unnerving. "There was quite an excitement about the Court House at the sale, and there was a crowd of uglier looking roughs than I ever saw together since the days when they hanged men publicly at the Old Bailey." He had ample cause to worry about the crowd given that he had brought $50,000 in gold double eagles, earnest money to cement the sale.[94] In all, those $20 gold pieces amounted to 184 pounds of specie, quite a burden to haul around St. Louis.

The ventures emerged from reorganization on March 28, 1879, as new corporations: the St. Louis Bridge Company and the Tunnel Railroad of St. Louis (the companies remained legally independent).[95] Under the new capital structure, the stockholders in Eads's old company lost the full value of their equity stakes, nearly $4 million. Holders of the senior debt in the old bridge company did well under the plan, receiving equivalent first mortgage bonds in the new corporation, which paid the same rate of interest.[96] Protecting those elite London investors was Junius Morgan's primary concern and obligation.

The Morgan banks took a sizeable cut in the fees and interest due them in New York and London. This amounted to an implicit concession that they had misjudged Eads's venture, a pretty incontrovertible conclusion by 1879. Crucially, the reorganization committee in London retained a new class of common stock "in trust."[97] Ownership of the bridge and tunnel companies conveyed with the common shares. That clause spoke to the bankers' ultimate objective. Having settled equitably with the bondholders, with the past, they could now shop for an American company to take over the business and make a new future.

The plan made the best of an unfortunate situation. The old company had chewed through stocks and bonds with par values totaling $13 million ($354.3 million in 2023), not including its floating debt or the tunnel's financing. The new St. Louis Bridge Company would issue securities with the same par value, but the plan shaved its debt-service costs for the first three years. In reality, the reorganization amounted to a fastidious trim, not a serious haircut. The new capital structure changed little for the merchants and shippers of St. Louis, and its terms meant that their fears in 1874 would prove true. Every participant in the commerce of the city would pay burdensome tolls to the benefit of far-off investors.[98]

Table 1 gives a statistical view of St. Louis Bridge between 1875 and 1881. Many developments drove the numbers up. The pool with Wiggins and general economic recovery were of primary importance. Time (to enable altered flows of commerce) and new track connections contributed their share. An analysis in July 1881 forecast steady growth for the city of

Table 1   Measures of Success, St. Louis Bridge and Tunnel Railroad, 1875–1881

Year	Gross Earnings ($)	Net Earnings ($)	Loaded Freight Cars	Passengers
1875	287,528	0	16,364	0
1878	690,604	219,598	81,227	667,294
1881	1,128,627	453,040	140,128	1,083,892

Source: Data from St. Louis Bridge, *Report*, 1881, 5, 33.

Note: These numbers show how Junius Morgan's faith in St. Louis Bridge eventually paid off. Gross earnings, included all upper-deck (roadway) revenues, approximately 18 percent of the totals shown here. Net earnings excluded interest and dividend payments due on securities. Counts for freight cars and passengers combined east- and westbound. Annual data are based on the company's fiscal year, ending on April 30. Daily passenger-train service over the bridge began in June 1875.

St. Louis thanks to its ties to the Southwest. The *Railroad Gazette* predicted that the bridge would eventually vanquish the ferry, so "it seems sure of monopolizing the whole transfer business eventually."[99] That year the company finally earned profits sufficient to pay the interest on its bonded debt. The Captain's venture arrived at that milestone a decade late, but just in time to interest Jay Gould.

Enter Gould

In 1878, Cornelius Garrison controlled two of the three railroads that crossed Missouri, the Missouri Pacific and the North Missouri. Few men had enjoyed a more colorful career. In the 1840s, Garrison had built, owned, and captained Mississippi steamboats. Then he had moved to gold-rush California, becoming mayor of San Francisco. His reputation for fast dealing and sharp cardplaying led one admiring Californian to say that it took "twenty men to watch him."[100] By 1878 Garrison had become the leading railroad magnate of St. Louis. His North Missouri put much traffic over the bridge, while his MoPac still directed its business away from the crossing, as it had when his brother, Daniel, ran that line.[101] Late in 1878, Garrison also bought equity control of the Wabash, a line running from St. Louis to Toledo, Ohio. It was comparatively easy for Garrison to acquire a controlling stock interest in these carriers, thanks to the depression and the plunge in their equity values. A few months later, in April 1879, the 70-year-old Garrison sold out to the most vilified tycoon of the age.[102]

In just three years, 1879–81, Jay Gould amassed control of railroad lines with ten thousand miles of track, creating—seemingly overnight—one of the largest rail empires in the country.[103] Throughout his business career, railway competitors and newspaper editors portrayed Gould in scathing terms. In the eyes of Joseph Pulitzer, publisher of the *St. Louis Post-Dispatch*, Gould was "one of the most sinister figures that have ever flitted bat-like across the vision of the American people."[104] Such vivid imagery sold a lot of newspapers. For his part, Gould ignored the public catcalls, focusing in these years on creating and extending rationalized railroad systems.

St. Louis Bridge could serve as the central link to unite Gould's network. Starting in 1879 at St. Louis, he acquired operating control of the Missouri Pacific, the old North Missouri, the Wabash, and the Iron Mountain. That last carrier ran from St. Louis to Texas. He also picked up Tom Scott's Texas & Pacific, and he pushed a line to New Orleans. The MoPac was central to this new empire. The Wabash proved equally important. Running from Toledo, Ohio, to Council Bluffs, Iowa, and Kansas City, Missouri, by 1880, it was the only railroad to operate on both sides of the Mississippi. Gould acquired control of these properties at low prices, thanks to the general depression in western railroad stocks after 1873. Their amalgamation added value to each part.[105]

To anchor the network, Gould turned his focus to the bridge, opening talks in December 1879 to lease the company while Junius was again visiting New York City. Morgan telegraphed his London partners: "We think very favorably Illinois & St. Louis Bridge proposition. . . . Lessees by controlling both Wabash on one side and Missouri Pacific . . . [and North Missouri] on the other together . . . are practically masters of the situation."[106] But glitches blocked a final agreement for twenty months, leaving William Taussig to run the business on behalf of the Morgans and their investors.

In June 1881, representatives for the Gould roads again sought to close a deal with Drexel, Morgan & Company in New York to lease the St. Louis Bridge Company and the Tunnel Railroad. The rental terms would mean

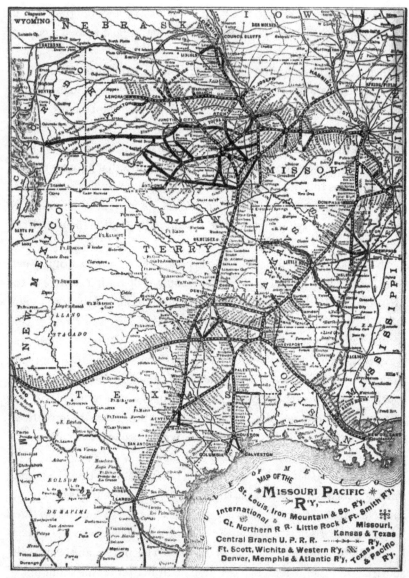

This 1887 route map for the Missouri Pacific and associated lines shows some of Gould's triumphs in system building. His roads tied St. Louis to Council Bluffs, Iowa; Lincoln, Nebraska; Kansas City, Kansas; and Pueblo, Colorado. The system also reached into Texas and Louisiana. St. Louis appears at right, the junction of the east-west MoPac main line and the north-south route of the Iron Mountain. Map from St. Louis Merchants Exchange, *Annual Statement* (1887), 80.

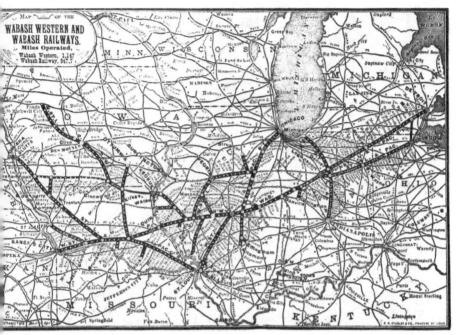

Stretching from Detroit and Toledo in the east to St. Louis, Kansas City, and Omaha to the west, the Wabash did fine long-haul business. The route to Chicago added value. St. Louis and its bridge were the lynchpins uniting the MoPac and Wabash systems. The city appears here at the bottom point of a triangle formed by three Wabash lines. Map from St. Louis Merchants Exchange, *Annual Statement* (1887), 86.

a guaranteed income stream for those companies to pay the interest due their British bondholders. By leasing, Gould avoided the capital costs and complications of an outright purchase. Upon learning this good news, the London bankers wrote to thank their American partners. "Congratulations on success negociations. Very satisfactory. Think Jay Gould exercised usual sagacity. Far better business [for the] roads than [for the] bridge owners. Many thanks. All your zeal is greatly appreciated."[107] If the deal went through, Gould would take a fine upside when bridge traffic and revenues grew.

Then another glitch nearly collapsed everything. On July 2, Pierpont cabled to London that Gould wanted a new clause in the lease agreement. The St. Louis Bridge Company, not the lessee railroads, should bear all responsibility and costs if "the arch" suffered catastrophic failure or

The cause of Gould's worries, Lucius Boomer's Omaha Bridge lost two of its spans to a prairie cyclone in 1877, five years after its opening. Ferries provided a temporary link while the railroad rebuilt this key crossing. This structure used the same elements that Boomer had proposed a decade earlier for St. Louis: Post trusses on iron pneumatic piers. *Union Pacific's First Bridge*, image BF14-236, KMTV/ Bostwick-Frohardt Photograph Collection, permanently housed at The Durham Museum, Omaha. Reproduced with permission.

destruction. On reading this, the London bankers grew alarmed and suspicious. A partner immediately replied to New York: "Jay Gould's insistence upon point so little likelihood occurring . . . raises doubts about his sincerity. . . . If he wishes withdraw, sooner we know the better." Pierpont wrote a soothing reply "Think Jay Gould sincere, but influenced by experience Omaha Bridge [where in 1877] two spans disappeared by lightning, never heard of."[108] In fact, a cyclone had ripped out of the grassland plains and blown down two of Simeon Post's 250-foot-long trusses, breaking the Union Pacific crossing of the Missouri River. After Pierpont's assurances, J. S. Morgan & Company accepted the last-minute clause, and the deal was done. Gould himself signed the lease on July 25, 1881.

With that consummation, the Morgans achieved all their immediate and long-term goals. Through seven years of work, they had laid a profitable foundation under St. Louis Bridge. In the 1879 reorganization they had protected the bondholders, their infrastructure, and its future. And they had made money on all of this: fees as receivers, interest on their loans to St. Louis Bridge and Union Transit, charges to set up the new firm, and fees to issue and service its bonds. The continued pool with Wiggins Ferry ensured sufficient profits to pay all those charges. For 1882 (the first full year under the Gould lease) the net profit on operations reached nearly $850,000 ($25.8 million in 2023).[109] St. Louis Bridge thus had ample reserves to pay the guaranteed interest due that year on the new first mortgage bonds ($350,000), plus $124,500 in dividend payments on the new first preferred stock.* The Morgans earned their reputation by taking good care of their companies, investors, and allies.

The bridge company's record of substantial earnings for many decades after 1880 ratified Junius Morgan's initial assessment of 1870 that the venture "will be a financial success."[110] At that time, Morgan had believed that St. Louis Bridge held three invaluable assets: the support of Thomson and Scott, the rail-bridge monopoly granted by Illinois, and the solid growth prospects for the city of St. Louis. As matters unfolded, the exclusive charter right and the prosperity of the city lifted and sustained the business despite Scott's betrayals.

Rethinking Failure

The expression that "failure is in the eye of the beholder" merits a corollary. Given sufficient time, failure may become triumph. To most civil engineers attending their 1874 convention in Chicago, the St. Louis Bridge was at best a dubious structure on its technical merit and a failure on economic grounds. Its foreclosure a year later ratified their skepticism. These men shared an unshakable faith in pin-connected iron truss bridges. The Boomer-Post design proposed for St. Louis reflected that orthodoxy, as did its near relation at Omaha, also designed by Simeon Post and erected

---

* The first preferred stock of the new company had gone to holders of the old company's second mortgage bonds. In 1872, some of those insiders had received those bonds in exchange for a portion of their stock in the old firm (described earlier in this chapter). It paid to be on friendly terms with the Morgans.

by Lucius Boomer. Then a Nebraska cyclone blew those certainties away. By 1881, Eads's massive steel arches provided solid reassurance to Pierpont Morgan and Jay Gould, placing a new economic foundation under the business. Time and events bolstered St. Louis Bridge. It took a bold or a reckless man to now argue that the best engineering "answers its purpose at the least cost."* The Captain's novel design for St. Louis had earned fresh reappraisals and a new start.

---

* The Omaha bridge collapse was no anomaly. The original Rock Island Bridge (1856) burned the year it opened. Its wooden replacement was heavily damaged in an 1866 cyclone, replaced by a wooden crossing two years later, then by an iron bridge in 1872. The Hannibal Bridge (1871) lost a 250-foot iron span in 1876, hit by a barge during high water. Later that year, flood conditions overwhelmed a pier of the same bridge, collapsing two spans into the Mississippi. In 1879, a 318-foot span of the St. Charles Bridge collapsed into the Missouri River after a freight locomotive derailed while crossing; another span of the same bridge fell into the river after another derailment two years later. At the civil engineers' convention in 1874, Willard Pope had condemned Eads's arch bridge in St. Louis, advocating instead conventional, pin-connected iron trusses. In 1887, a cyclone hit Pope's orthodox Hannibal Bridge over the Mississippi, collapsing its iron center span.

CHAPTER 10

Successes across Time

CAPTAIN EADS PERSEVERED against long odds to build his bridge, then Pierpont Morgan made the venture profitable, attracting Jay Gould to take control. By that time, Eads had taken up other challenges, but the strength and breadth of his vision continued to shape St. Louis Bridge, the city, and the region for many decades. The bridge was the central piece in a mosaic of transformations across a half century or more. Some continue into the present. It had such broad influence partly because Eads proposed far more than the structure. As he sketched in 1868 in his first report as chief engineer, he integrated the bridge into plans for rail, road, river, and commercial infrastructures to shape the entire region. In this era before activist government or urban planning, Eads demonstrated how a private-sector entrepreneur could build wealth for himself by offering a blueprint for his city as a whole. Executing those innovations mostly fell to others, before and after his death in 1887. It speaks to the power of Eads's mind that later men would take the paths he laid down.

An Integrated Whole

The chief engineer's appreciation for St. Louis as a functioning amalgam of people, geography, the river, work, housing, and commerce, all deserving improvement, recurs throughout his 1868 report. His bridge should arrive at Washington Avenue, where its centrality would benefit all residents, north and south, while that location would readily disperse the traffic equitably into the city's neighborhoods. The tunnel under downtown would sidestep Lucius Boomer's "preposterous" notion of running steam trains down street-laid tracks to cross the city's "great thoroughfares."[1] The new horse-drawn street railway on Eads's upper deck would provide mass transit options to the working classes on both sides of the river. Those connections would make East St. Louis "a large and populous manufacturing place" where laboring families would be able to afford housing. He planned the dual-lane roadways and two sidewalks on that deck to accommodate the growth of those two cities for decades to come.[2] All this came true. His report also forecast a system of freight terminals and movement on both sides of the river. Given its cheaper land costs, East St. Louis was the logical place for "a great union freight depot" serving all carriers, east and west.[3] As matters unfolded, by 1886 St. Louis Bridge owned and operated a "ranging yard" in Illinois, where upwards of a thousand freight cars from innumerable railroads were sorted, then forwarded in all directions.

By that year, St. Louis Bridge had created an extensive rail infrastructure on the growing west side of St. Louis. Back in 1868, Eads had not foreseen these operational roles for his company. It built these expensive facilities in town because the eastern railroads still refused to do it. Its Union Transit subsidiary provided direct services into its three freight warehouses and six yards for freight and passenger cars. Equally important, St. Louis Bridge allied with new manufacturers, providing the industrial sites and rail connections that sustained Missouri Glass, Goodwin Soap, and a dozen other factory employers. All these facilities boosted the city.[4] And they transformed St. Louis Bridge into a full-fledged railroad, employing 860 men and 22 locomotives in 1887.[5] That railroad hauled across the river all passenger trains and most freight cars destined for local businesses in the city.

In proposing joint facilities for passengers and freight, the chief engineer encountered opposition from politicians, hoteliers, draymen, and plain citizens. They saw profit in making goods and people stop in the city.[6] By contrast, Eads wanted to secure for St. Louis a growing portion of "the trade of the West . . . by expediting it in every way in our power. . . . The marvelous rapidity with which it will then multiply on our hands will give our people greatly increased profit and employment."[7] Again his

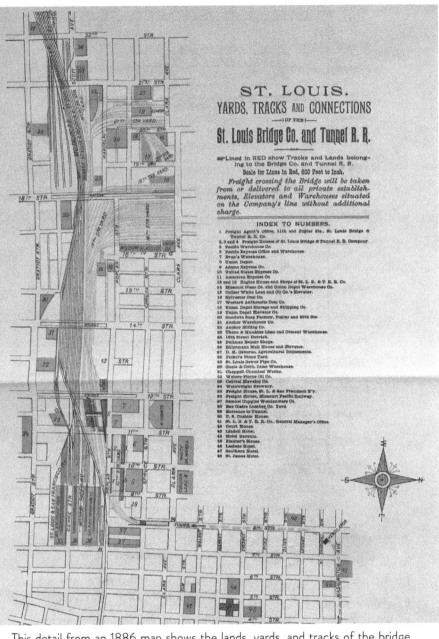

This detail from an 1886 map shows the lands, yards, and tracks of the bridge company in St. Louis (*west is up in this view*). It owned nearly all the shaded properties, except Union Depot, which is #8 on the map, located east of (*below*) 12th Street. The Missouri Pacific owned about half the yard trackage shown here; the bridge company accounted for the rest. The tunnel crossed near the bottom. Map by St. Louis Bridge, box 15, folder 6, JSM.

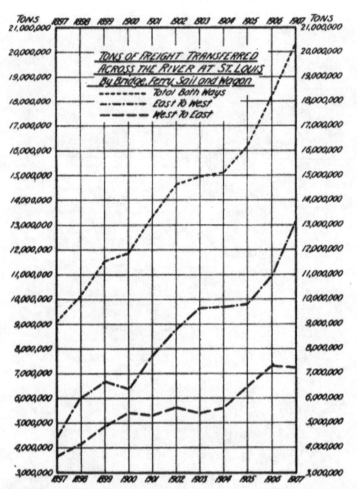

This graph from the trade press shows the growing crush of freight tonnage that crossed the Mississippi at St. Louis from 1897 to 1907 over all routes and by all methods—"by bridge, ferry, [r]ail, and wagon." Image from "Solving the Terminal Problem at St. Louis, Part 1," *Railway Age* 45, no. 7 (Feb. 14, 1908) : 214.

predictions proved accurate. In 1880 St. Louis ranked sixth in population among all US cities. A decade later it was fifth, and in 1900 it had moved up to the fourth position.* By the lights of the era, its growth and wealth signaled smoky, gritty success for the company and the region. Many

---

* Across the river, East St. Louis also grew dramatically thanks to the bridge, its population increasing by 100 percent during the 1890s, then doubling again by

factors drove these developments; all depended on the foundations Eads had planned and provided.

The pace of the city's growth accelerated around 1900. Between 1897 and 1907 the total weight of cargos traveling over the river crossings at St. Louis doubled.[8] By then, the city's striving merchant and dry goods houses supplied retailers across a hinterland stretching from Kentucky and Tennessee to Louisiana, Texas, New Mexico, and Colorado. Its location at the junction of the American East, West, and South remained the city's leading asset. In 1910, the Merchants Exchange published a pie chart to illustrate St. Louis as the terminus of twenty-seven carriers. The bridge knitted them together. The rail lines became a steel circulatory system, connecting the city, the region, and the nation.

Cooperative Ownership and Monopoly Control

In his 1868 report, Eads forecast the benefits of union freight and passenger services—joint facilities to serve multiple railroads—but he said nothing about how those businesses would be financed or structured. William Taussig provided the answer. In June 1886 he sketched a proposal to reinvent St. Louis Bridge. His central concept: "all the lines East and West which lead into St-Louis should be invited to join in the Bridge Lease." From that fresh start he foresaw a host of blessings. "It would virtually change the terminus of the Eastern lines from East St-Louis to St-Louis." The new association would enlarge the freight terminals serving the region. With those facilities, he predicted, freight tonnage over the bridge would double. The associated lines "would, after taking over the Union Depot property, solve the problem of a new Passenger Station, and perhaps run their Passenger trains—in and out—with thier [sic] own road Engines." The problem at the station was a crush of business, a fine problem to have.[9]

Taussig's proposal represented a radical break for the political economy of his day. To be sure, union stations were no longer rare by 1886.[10] At the

---

1910 (reaching 58,540, most in railroad and manufacturing jobs). The town remained a rough place, so in 1901 a daily average of 8,000 commuters preferred to live in St. Louis, crossing the Eads Bridge by electric streetcars to their Illinois jobs. A fine history is Theising, *Made in USA*.

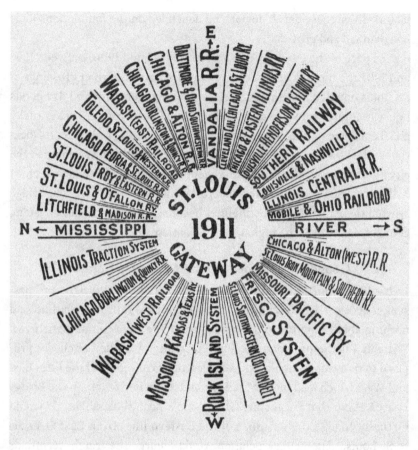

The St. Louis Merchants Exchange illustrated the city's prime commercial position this way. The Terminal Railroad connected each with every. Image from St. Louis Merchants Exchange, *Annual Statement* (1910), 88.

larger ones, joint ownership of the building and tracks often led to the creation of a small terminal railroad to handle switching duties for all tenants. Taussig was planning on a vastly larger scale. He proposed to give a single company the responsibility for handling the movement of freight cars and passenger traffic for a large metropolitan region spread across two states. Few Americans of the era could have imagined any corporate entity or governmental agency in that role. Eads had suggested it in 1868.

Jay Gould signed off on Taussig's plans, forming the Terminal Railroad Association of St. Louis (TRRA) in 1889. Junius Morgan's approval was even more important.[11] At the start, six carriers constituted the association's "proprietary members," sharing joint ownership of the property. Their new association pledged to serve equally all carriers entering the city from east and west.[12] The TRRA immediately began to deliver on Eads's vision and Taussig's memo. It combined the bridge and the tunnel, freight yards on both sides of the Mississippi, the Union Depot and nearby passenger yards, the operating company (Union Railway & Transit), and the landholding company (Terminal Railroad of St. Louis).[13] The new firm was immediately profitable, earning $331,700 in 1890 ($11.3 million in 2023) after paying all its fixed charges.[14]

Taussig's predictions of 1886 proved misguided on one key point. He believed that this association "would put a stop to the cry of Monopoly and to the unjust odium attached to it." In his view, "a property owned and operated in common by many and diverging interests will command the sympathy of the public."[15] He could not have been more wrong. By its own actions in the next few years, the TRRA aroused passionate fear about its growing control over the markets and people of the region.

In 1889, a group of leading St. Louis financiers and industrialists, all members of the Merchants Exchange, began construction of the city's second rail bridge. They vowed that their Merchants Bridge would compete in services and rates with the St. Louis Bridge. Built three miles upstream, the new venture did not violate the Illinois charter restrictions that protected the first bridge.[16] The new route had backing from the Burlington and other lines. Working through its Vandalia connection, the Pennsylvania promised to put much business over the Merchants Bridge.[17] These investments suggest three insights: St. Louis was underserved; it had a bright future; and powerful carriers were happy to work with the Merchants company to improve their access to the St. Louis market and gateway.[18] Accurately or not, the local press trumpeted the new bridge as deliverance from Gould. Now that the city had its second bridge, the first needed a new name. The St. Louis Bridge became the Eads Bridge.[19] This fitting honor, a bridge named for its designer, remains nearly unique.

The Merchants company did good business from its opening in June 1890. In the contest to move freight cars across the river, the new

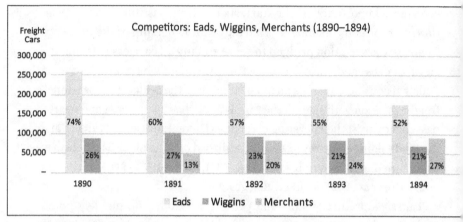

The Merchants Bridge quickly upset the comfortable collusion of the Eads Bridge and Wiggins Ferry. Just four years after it opened, it took more business than the long-established ferry company (bars include market shares as percentages). The left axis shows total freight cars, both ways over the river. The Merchants Bridge opened in March 1890, and it had some traffic that year; apparently, the Merchants Exchange lacked a source to acquire the data. Chart by author from data compiled from St. Louis Merchants Exchange, *Annual Statement* (1890–94).

bridge proved to be a serious competitor with the TRRA's Eads Bridge and the Wiggins Ferry. As demonstrated in a chart showing traffic for the years 1890 to 1894, it had the connections and alliances required to take business away from long-established routes.

While the new crossing took a growing share of the burgeoning freight volume, it did not compete on prices with the Eads/TRRA crossing or the Wiggins Ferry. From 1891, the three companies had an "amicable arrangement as to maintaining a fairly remunerative and uniform [rate] schedule," as William Taussig wrote to the Morgan bank in London.[20] They were setting rates collaboratively. Price-fixing. That behavior showed no regard for the Sherman Antitrust Act, an 1890 statute that outlawed *any* combination in restraint of trade.

Even that hidden collusion, however, proved insufficient for the men running the TRRA. In 1893, a consortium of railroads nearly acquired control of the Merchants Bridge and its terminal railroad. Their plans included construction of a new belt line to serve industries and carriers on the west side of St. Louis. They also began "purchasing real estate for a [new] up-town Passenger Depot," whose tracks would connect with the

Merchants Bridge. Writing to the Morgan bank in New York, William Taussig cast these developments as a dark threat to the Terminal Association.[21] To be sure, these long-hidden maneuvers could have created a powerful competitor to the TRRA for river-crossing traffic and rates in the St. Louis region (freight and passenger).[22]

To prevent that calamity, the TRRA outbid the consortium, buying a controlling interest in the Merchants Bridge Terminal Railway in August 1893. It paid a premium price, with financing provided by J. P. Morgan & Company. A similar situation unfolded in 1902 when the Rock Island nearly acquired control of the Wiggins Ferry Company. Again, the TRRA blocked this incipient competition by buying outright its longtime partner in price-fixing.[23] Again it had to pay a high price for the new acquisition, again financed by the Morgan bank in New York. These maneuvers demonstrated the TRRA's resolve to protect its profit margins. Its customers and the entire region would continue paying tribute to the monopoly, just as James Eads had foreseen.[24] But all was not dark here. Once the Merchants and Wiggins companies became components of the TRRA, their capacity and connections could aid the systemic goals of the association while bolstering its services and profits.

The Largest Station in the World

In 1868, Eads advocated the construction of a union passenger depot near the western mouth of the tunnel. He predicted that the passenger trains of every railroad, east and west, would start or end their runs at this common point. This was a radical notion in his day. As matters played out, the 1875 Union Depot served the city well for fifteen years, a credible record considering its difficult origins. In its first year, the station hosted sixty daily arrivals and departures by nine railroads. In 1891, the same facility accommodated a daily average of 230 passenger trains for nineteen lines.[25] Success begat success, just as Captain Eads had predicted. The station demonstrated network effects that benefitted the carriers, their passengers, and the city at large.

After its creation in 1889, the TRRA had the incentive, the membership, and the resources to build a much-enlarged station to serve the growing metropolis. Alongside these practicalities, symbolism provided another motive for a grand new station. By then, many North American railroads

Union Station in St. Louis was the largest in the world for many years. Seldom photographed from this angle, the station (*left*) had an attached hotel that opened in 1895 (*entrance at right*). A modern steel and glass train shed stretched out behind this monumental facade. Photograph by Russell Froelich, MHS, N 34529.

serving larger metropolises had new monumental terminals. These impressive edifices—such as Windsor Station in Montreal (1889) or Grand Central in Chicago (1890)—have been described as secular cathedrals by architectural historians. To impart grandeur on the surrounding city, they boasted features far exceeding the mundane needs of travelers on the 8:32 departure to anywhere. The imposing structures presented railroad corporations as public benefactors. This savvy marketing strategy aimed to counter the widespread conviction that the carriers were no more than grasping monopolies. Ironically, railroads poured capital into these grand stations even though passenger services provided comparatively little profit.[26]

The new terminal for St. Louis opened on September 1, 1894.[27] The front facade, 600 feet long, was modeled on the medieval city walls of Carcassonne, France. A stone clock tower soared 230 feet above the city. Those features reflected the American embrace of an architectural style

The 1894 train shed was so large that it taxed the skills of a professional photographer to convey its vast capacity. The signs (*at right*) directed patrons to their tracks and trains. A Louisville & Nashville limited with connections to Atlanta and New Orleans stands ready to depart on track 30. Photograph by Emil Boehl, MHS, N 11334.

known as Richardson Romanesque. The head house and tower gave no hint of the practical business of railroading. Behind that ornate front, tracks and trains took center stage under a massive train shed. Its steel trusses supported a roof of tin and glass pierced by skylights and smoke vents. Covering 424,000 square feet, it was the largest train shed in the world. Its thirty tracks served twenty-two carriers, both records.[28] All this pleased a city happy to boast, but the structure wasn't an overstatement. It quickly filled to capacity, requiring substantial enlargements just eight years later. In all, the station had eleven acres under the roof and cost nearly $6.5 million ($234.5 million in 2023).[29] In stone, glass, and steel the Terminal Railroad Association demonstrated what cooperation could achieve.

The cyclone of May 1896 blew a section of the roadway deck into the river while toppling the stone arcade that had supported it. More than two hundred feet long and sixteen feet tall, the arcade's stone pillars were thirty inches thick. Here they lie tumbled across the tracks. Photograph from author's collection.

Striking Again

On Wednesday, May 27, 1896, the cataclysm that James Eads had foreseen and Jay Gould had feared fell upon St. Louis with awesome ferocity. Morning sun gave way to fluffy clouds, and the barometer fell all day. At 5:00 p.m. a boiling, black cyclone roared out of the southwest. Crossing the city, it tore through Captain Eads's old neighborhood of Compton Hill, leveling scores of houses. It missed the Union Station, blasted the City Hospital, obliterated warehouses, then crossed the river. The Wiggins fleet was destroyed, save one vessel. Moving ashore, it devastated East St. Louis, leveling half the structures in that city of twenty-five thousand. On the ground for thirty minutes, the twister killed more than two hundred people, injured thousands, and inflicted $10 million in property losses ($370 million in 2023).[30] The cyclone shattered all telegraph, telephone,

Alton train number 7 was crossing the bridge eastbound when the engineer, William Swoncutt, saw the approaching fury (sometime earlier a TRRA towerman had routed his train onto the track typically reserved for westbound running). Fearing that the cyclone would blow his train into the river—or overwhelm the bridge entirely—Swoncutt increased his speed to reach the Illinois shore. Thanks to his quick decision, his train passed over the tracks shown in the preceding image seconds before the winds inflicted the damage shown there. Here on the rarely photographed eastern approaches, the roadways descend at a steeper grade than the railway deck (the winds have blown off most of the roadway railings). While many passengers were injured when the cyclone derailed these coaches, none were killed. Photograph from author's collection.

electricity, and streetcar services. Retrospective analysis rated the twister as an F-4 event, with winds up to 260 miles per hour. The Fujita scale has only one higher rating for winds and damage. Measured by its death toll, this cyclone ranked as the third worst in US history up to the date of this writing (2023).

The Eads Bridge took a direct hit. On the east span, the roadway and sidewalks were torn loose. Further east, the cyclone toppled the stone

Shown here is the battered east section of the bridge after temporary repairs put the roadway deck back in operation. The railway deck is also open. In the foreground, the sole Wiggins ferry to survive the storm has a wagon and team aboard. To the north, a Wiggins car-transfer barge has broken its keel. Further north, a wrecked Wiggins ferry lies amid the bridge debris. The preceding image depicts a spot just to the east (*right*) of the destruction shown here. *Temporary Repairs on Eads Bridge*, Keystone stereoview no. 2214 (1896).

arcade supporting the roadway deck, blowing that deck away. Despite all this destructive force, the steel arches came through unscathed.[31] If St. Louis had ended up with Lucius Boomer's bridge, the twister would surely have obliterated that crossing, dropping half its spans or more in the river. By contrast, Eads had ordered his engineers to redesign the upper deck after the tornado that hit the city in 1871. Twenty-five years later, that decision probably saved the bridge. After two days of repairs, a succession of trains over the Eads and Merchants Bridges brought relief workers and supplies into the battered city.*

Making a Gateway

For James Eads, the St. Louis bridge, tunnel, union depot, and joint freight facilities would have simple purposes: to draw freight and passengers to his city and to speed their outbound passage. But the eastern railroads refused to cross the river, causing the venture to evolve in ways he had not foreseen. Instead of a railway tollgate, the company became a delivery service. From 1875, its new locomotives hauled passenger trains over the three miles of track separating the Union Depot downtown from the Relay Depot in East St. Louis, Illinois. When it came to freight, the St. Louis Bridge and Tunnel Railroad struggled initially to build its business. To deliver cars from the east, it had to build yards, freight houses, and industrial spurs in downtown St. Louis. With Morgan's money and Gould's support, the company largely achieved those goals during the 1880s (see the image on p. 251). When operations and control passed to the new Terminal Railroad Association in 1889, the work of deliveries grew. Crucially, the business entered a third phase across the 1880s, expanding its interchange services.

Transforming individual railways into regional or national networks required, above all, that freight cars would flow without hindrance from one carrier to the next. For many decades, however, car interchange appeared to be an impossible dream or a foolish notion. The California senator James McDougall had ample cause to believe in 1863 that "proper business interests" would not allow "strange cars" to run over their railroads (ch. 2). Eventually, railroad managers realized that this traffic between home and "foreign" rails was essential and profitable.[32]

Lifting the barriers blocking car interchange required three decades of innovation in technologies, interfirm agreements, and managerial mind-sets.[33] As matters played out, the technical problems took time and money to solve, but at least those solutions were comparatively clear. Adopt the standard track gauge, agree to a common design for couplers and brakes, build actual track connections. The bridge became one of

---

* The cyclone's path bypassed the Merchants Bridge entirely.

those links. Eads had seen the value of the standard gauge for its tracks from the start.*

Construction of the bridge motivated carriers serving St. Louis to adopt the standard gauge. That standard grew in importance for any single railroad only after interconnections like the bridge allowed freight cars to flow from one line to another. The North Missouri adopted the standard gauge in August 1867, the Missouri Pacific on July 18, 1869, the Ohio & Mississippi on July 23, 1871, and the Iron Mountain on June 28, 1879.[34] Railroads typically converted their tracks in a single day because all traffic necessarily stopped during the transition. Leading up to that day, however, were months of planning, huge investments in new equipment, and the coordinated effort of hundreds of laboring men. For those reasons, many southern railroads delayed the conversion. Finally, on May 31–June 1, 1886, carriers with more than thirteen thousand route miles of track across the old Confederacy adopted the standard.[35]

Organizational hurdles also stymied the interchange of cars for many years. On some railroads, managers loathed sending their own cars offline but delighted in holding, using, and profiting from foreign cars. In 1873 the *Railroad Gazette* noted: "Cars have stayed away from their owners, frequently for months and have returned almost worn out." Even so, "the managers of well-equipped roads have allowed the miserable system of interchange now in vogue to exist so long."[36] Connections like the bridge finally forced these managers to act. During the 1880s, railway officers developed comprehensive agreements to govern car movements, payments for the use of foreign cars, apportionment of freight revenues, and rules about defects and repairs of interchanging cars.[37] By 1890, 17 to 45 percent of freight cars owned by major carriers were off-line.[38]

---

* A standard of 4 feet, 8.5 inches, for track gauge (the distance between the rails) was eventually adopted by most American railroads (Taylor and Neu, *Network*, 81–82). In his 1868 report, Eads sidestepped the question of gauge despite its centrality to all he was proposing, probably because he saw no value in alienating lines that had not yet adopted the standard. But his plans to link union freight and passenger depots by means of the bridge and tunnel nearly required that common choice. By contrast, Lucius Boomer's 1867 proposal envisioned four rails on his bridge so that its single-track right-of-way could accommodate trains in three different gauges. Brown, "Not the Eads," 533.

Table 2   Growth of Interchanging Traffic on the Eads Bridge, 1881 and 1902

Year	Local	Interchange	Total	Interchange (%)
1881	115,992	24,136	140,128	17
1902	85,031	202,547	288,578	70

Source: Data from St. Louis Bridge, *Report*, 1881, 6; and Terminal Railroad Association of St. Louis, *Annual Report*, 1902, 9.

Note: These numbers, showing an eightfold increase in interchange traffic, reveal a transformation shaping St. Louis and the commerce of the nation. These data include all loaded freight cars moving over the Eads Bridge (east- and westbound). The final column shows interchanging traffic as a percentage of all loaded cars. The absolute decline in local business appears surprising given the city's prosperity in this period, but rail traffic over the river had a new route available by 1902, the Merchants Bridge.

This national phenomenon resulted in dramatic changes at St. Louis Bridge. In 1881, loaded freight cars passing through the city accounted for only 17 percent of its total freight traffic (the rest were local deliveries). By 1902, through or interchange cars made up 70 percent of a much larger total, as shown in table 2.

The Terminal Railroad Association began operations at the right juncture to grow, speed, and benefit from this traffic. Car interchange became its raison d'être. Every railroad serving the city had its own yards and freight houses to handle shipments that began or ended on its own rails. Each carrier also hauled interchange cars destined for those railroads to which it had direct track connections.[39] For shipments beyond home rails and direct connections, for switching to and from industrial sidings, and for nearly everything that crossed the river the TRRA did the work. By 1900 the volume of interchanging traffic through St. Louis had grown to staggering proportions, ratifying Eads's insight that by expediting the trade of the West "it will then multiply on our hands."

In his 1868 report, Eads did not explicitly focus on St. Louis as an interchange point, although he presented the city as a gateway. The joint facilities that he envisioned gave his company and its successors considerable advantages in building interchange traffic. No other American city had anything like the TRRA.[40] Chicago became a bigger gateway, the country's largest. But size alone proved a disadvantage. According to a 1910 profile, the TRRA gave incalculable advantages to St. Louis, operating its freight and passenger terminals "with an efficiency and economy" unknown elsewhere in the country. The same reporter noted that Chicago's rail network periodically became paralyzed, its yards congested with cars, the carriers unable to move them to interchange points. Disastrous for any rail hub, those delays cost Chicago railroads upwards of $3 million ($102.3 million in 2023) during the winter of 1909–10 alone.[41]

Proving Its Worth

In 1904 St. Louis took a star turn in front of the country, hosting the Louisiana Purchase Exposition, known popularly as the St. Louis World's Fair. With the exposition as an attractive drawing card, more than three hundred fifty organizations convened their annual conventions in the city during the seven-month fair. They included the Anti-Horse Thief Association and the Democratic National Convention, the American Federation of Labor and the National Association of Manufacturers, the Mexican War Veterans and the United Daughters of the Confederacy.[42] Nearly everyone arrived by train. Significant in the history of the city and the country, the world's fair resulted directly from larger transformations then making America modern. For its part, the Terminal Railroad Association—its bridges, stations, and networks—was both a cause and a product of those changes.

Forest Park, a vast swath of the city's west side, hosted pavilions, exhibit halls, and midway attractions—all built for the fair. St. Louisans proudly boasted that their great event spread across twice as much land as Chicago had devoted to the World Columbian Exposition in 1893. Grand buildings in a white neoclassical style were set in artfully designed landscapes of sweeping roadways, luxuriant gardens, lagoons, and vistas near and far. Although they resembled the marble temples of ancient Rome, the columned halls were actually framed in wood, their ornate facades made of a plaster and straw mix called "staff."

Besides all that it displayed, the fair mattered for what it *conveyed* about the larger transformations of the country. In contrast to the characteristic squalor of American cities during the Gilded Age, this magnificent display of collective design by experts created a landscape of beauty and harmony. Its elements rewarded the eye and ennobled the mind. James Eads would

The US Government Building at the Louisiana Purchase Exhibition. The distance from the boat in the foreground to the domed Government Building exceeded 1,500 feet. To the left, the Palace of Liberal Arts; to the right, the Palace of Mines and Metallurgy. Evoking imperial Rome, these scenes impressed and pleased visitors. That cities could be beautiful was a novel idea for many Americans. *Neoclassical Architecture in the Government Building*, photograph from Francis, *Universal*, 91.

have approved. Crucially, the exhibition demonstrated both the feasibility and the benefits of reaching for a common good. In 1868 it had been inconceivable that governments at the local, state, and national levels would fund projects like the St. Louis Bridge directly. For that reason, men like James Eads had stepped up with plans that advanced their own self-interest alongside the general good, more or less. By 1904 a country with far more wealth, much larger cities, and new leadership had adopted new policies to advance the common wealth. To put on the fair, the city of St. Louis contributed $5 million, and the federal government matched that sum.[43] That munificent funding did not stumble over the temporary

In August 1905, the *Post-Dispatch* ran this cartoon on page 1, beneath a banner headline that read, "Bumper Crops Bring Biggest Trade St. Louis Ever Saw: Record Sales of World's Fair Year Far Surpassed." The cartoon shows the TRRA head, William McChesney, bent over sheaves of invoices and manifests. Outside his window are loaded freight trains and the Eads Bridge. Cartoon by D. Widman, *St. Louis Post-Dispatch*, Aug. 27, 1905, 1.

nature of the exposition, open for just seven months. Then its grand buildings were dismantled, removed from Forest Park, and burned. Thirty years earlier government funds for *permanent* public improvements such as bridges or sewers could garner minimal public support. American society and governance had traveled far from those days and notions.

The Terminal Association delivered on its promises to provide rail services equal to the demands of the fair. In contrast to its normal daily average of 300 passenger trains (in and out), in September and October the TRRA handled 520 daily trains at Union Station.[44] That impressive score on an exceptional test appeared to prove the benefits of coordination and investment, although not everyone agreed. The exposition closed on De-

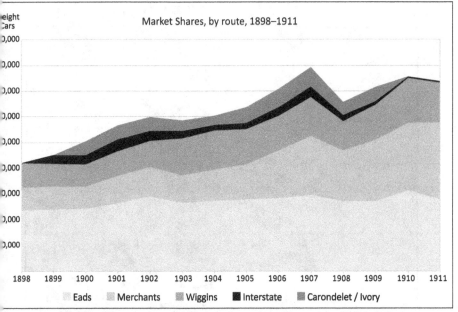

The TRRA monopoly at work. The Missouri Pacific (a TRRA proprietor, hardly a competitor) owned and operated the Carondelet ferry. After 1902, all other routes over the river at St. Louis were controlled by the Terminal Association. The totals shown combine all east- and westbound freight cars crossing the Mississippi at the city. Chart by author, derived from data in St. Louis Merchants Exchange, *Annual Statement* (1898–1911).

cember 1, 1904, its grand structures razed soon thereafter. The Terminal Association's enlarged physical plant was permanent. The passenger crush eased in 1905, a good thing as freight traffic that year again broke records.[45] The gateway city boomed ahead.

These developments around 1900 may appear far removed from Captain Eads and his bridge. Not so. That structure was simply the most visible component of the integrated urban system that sprang from his mind and plans then grew across decades. The visionary innovator of caissons and the steel bridge was the great rationalizer of transportation to serve the city, region, and nation. This ordered system, however, came with its own costs. After 1902, the Terminal Railroad Association monopolized the business of moving railcars over the river at St. Louis.

PARDON ME, I'LL ATTEND TO THIS!

The *Post-Dispatch* ran this cartoon after news broke that the US Department of Justice would prosecute the TRRA for violations of the Sherman Antitrust Act. The stylized knight represents Saint Louis. The city took its name from that one-time king of France, Louis IX, who died in 1270 while on his second crusade to the Holy Land. He was later canonized. The smoking dragon stepping from the river brilliantly combines the Eads Bridge with the Terminal Association (chest), Wiggins (jaw), and the Merchants Terminal Railway (locomotive head). President Roosevelt advances to battle the monster. Cartoon by Harry B. Martin, *St. Louis Post-Dispatch*, Oct. 8, 1905, 29.

Split Verdicts on a Monopoly

Even as the TRRA demonstrated its value, it was fighting for its very existence. In their utter dependence on this single company, many St. Louisans saw the Terminal Association as a sinister monopoly, pure and simple. People across Missouri and southern Illinois shared that dark view, as did political leaders in those states and further afield. Whether malevolent or public spirited, its market power had become impossible to dispute after its takeovers of the Merchants and Wiggins companies. In August 1903, the Missouri attorney general sued to revoke the

In 1908 the Terminal Association proposed to build this massive freight house on riverfront land in St. Louis adjacent to the Eads Bridge. Found in every US city and town, freight houses processed incoming and outbound shipments by rail. Its scale suggests that TRRA managers believed St. Louis needed more and better freight-handling services. Its size also shows why St. Louisans saw the association as a devouring monster. City leaders rejected this proposal, refusing to condemn the land for this purpose. Drawing from "Solving the Terminal Problem at St. Louis, Part 2," *Railway Age* 45, no. 8 (Feb. 21, 1908): 242.

TRRA's charter and dissolve the association. Joseph Crow lost the case in the state supreme court by a vote of 4 to 3.[46] Soon thereafter, President Theodore Roosevelt launched a federal antitrust case against the TRRA.

Public opinion in St. Louis backed the president enthusiastically. It is easy to understand why. This commercial hub for the Midwest and the Southwest had been captive to a succession of monopolies in river commerce for a century. With the railway age, St. Louis Bridge had wielded its market power, created by its Illinois charter, to exploit the people, although James Eads and Junius Morgan would surely have disputed that verdict. The TRRA's control over rates, services, and workers prompted a stream of crusading headlines in Pulitzer's *Post-Dispatch* and other papers.[47] The journalists told people what they already knew: the entire region depended utterly on this single company to move the products of their labor and deliver all of life's necessities and luxuries. Even when the TRRA sought to improve its services—for example, by building additional facilities—politicians and citizens condemned the company as a devouring monster.

Federal prosecutors based their case against the TRRA on the Sherman Antitrust Act, the 1890 statute that banned every "contract, combination . . . or conspiracy in restraint of trade among the several States."[48] Their evidence was compelling.[49] No less an authority than J. P. Morgan & Company had marketed Terminal Association bonds in 1902 by claiming that the company "controls practically the only available entrance to St. Louis."[50]

But the case against the TRRA stumbled. A special master took nearly three years to gather evidence; then the US Circuit Court for the Eighth Circuit (sitting in St. Louis) heard two days of oral arguments in early April 1909.[51] Seven weeks later, that four-judge panel delivered a hung verdict. Two judges voted to break up the Terminal Association, reestablishing Wiggins and Merchants as independent properties, and two favored acquittal.[52] The deadlock meant that the court issued no ruling in the case. Again, the TRRA avoided its own dismemberment by the narrowest margin. Six months later, US Attorney General George Wickersham appealed the case to the US Supreme Court.[53] In earlier antitrust cases against other companies, the federal courts had delivered conflicting interpretations of the Sherman Act. The circuit deadlock in *United States v. Terminal Railroad Association of St. Louis* arose from those muddled readings of the statute, whose sweeping language ensured that judges would craft its meanings for real-world conditions.[54]

In 1911 a new chief justice of the US Supreme Court advanced a new doctrine in antitrust, the "rule of reason." Chief Justice Edwin Douglass White laid down tests that federal courts would apply to distinguish between corporations that had violated the statute to exploit their markets in contrast to a business that benefitted consumers—even if such a monopoly had used illegal methods to achieve market control. As a practical matter, White declared that the justices had the knowledge and insight required to distinguish "bad trusts" from beneficial monopolies. For better or worse, the justices would recast the Sherman Act as a statute on economics rather than the law.

In May 1911, the chief justice launched his new rule in two closely watched cases: *Standard Oil* and *American Tobacco*. Both companies had trampled on the antitrust statute, their competitors, and customers. The court ordered their dissolution. That outcome, however, scarcely concealed the pro-business bias of White's new rule. Responding to *American Tobacco*, the *New York Times* wrote that "the decision was all that the big corporations could ask." The editor of the *Wall Street Journal* admitted that "the Court had read into the Sherman law an amendment that never could have passed the Congress."[55]

Terminal Railroad was the next antitrust case on the docket. By a vote of 7 to 0, the justices wielded the rule of reason to reject the prosecution's arguments and preserve the company.[56] It was a major pivot in the law. For the first time, a monopoly received the sanction of America's highest court.[57] The justices' decision tinkered around the margins of the TRRA, with directives on its rates, membership rules, and operations. But the verdict affirmed the company's claim to be an efficient and fair provider of networked services to all carriers. The court declared that the TRRA was a natural monopoly.[58] According to this view, the high cost of bridging the river and the narrow constraints for westward-reaching tracks in the Mill Creek Valley meant that all carriers (east and west) should share access to the city via the facilities of the Terminal Association. The logic, lifted from the TRRA's defense in the case, proved sufficiently compelling to achieve a unanimous decision of the justices.[59] But they had to ignore a lot to achieve the outcome they sought. The monopoly had originated in the Illinois legislature with its statutory gifts to the ferry and bridge companies. Decades of illegal collusion had sustained it. Then the TRRA buyout of the Merchants company (1893) had preserved it. There was nothing natural in this history.

Even as the circuit court was considering the legality of the TRRA under the Sherman Act—before the case reached the Supreme Court—the Illinois Transit System (ITS) found itself compelled to build the third rail bridge over the river and into St. Louis.[60] This interurban railroad was forced to construct its McKinley Bridge (completed in 1910) after the TRRA refused to grant access to its network of bridges and tracks. The new crossing constituted 30 million pounds of evidence that the justices' natural-monopoly logic was fallacious. Notwithstanding the court's claims, it was TRRA policy—not geographic conditions—that had created the situation in which no railroad could "even enter St. Louis . . . without using the facilities entirely controlled by the Terminal Company."[61] By its adamant refusal to accommodate the ITS, even as the federal antitrust case unfolded, the Terminal Association behaved like a powerful overlord, not the all-access benefactor it claimed to be.

Speculation on the logic behind a fiction is a chancy endeavor at best. Still, the court's decision demands some explanation. At just this moment, companies with strong claims to actually being natural monopolies were beginning to dominate electricity generation, telecommunications, and other industries. By its finding in *Terminal Railroad*, the court established protections under law for these new corporate giants. While the justices achieved their desired outcome for antitrust law at large, their logic in this case suggests that they were poor students of economics, disinterested in accurate history, and little concerned with the facts in the dispute.[62] The lead defense counsel for the TRRA, Harry Priest, stated that "the decision will affect the Terminal on no vital point, and for that reason I consider it a great victory."[63] As it surely was. The court also crafted a regulatory framework to oversee the rates and operation of the company, placing the monopoly under the rule of law. For many observers, that too was worth celebrating.

Terminal Railroad illustrates how the meaning and significance of law shifts with time. In 1912 the justices chose to ignore the company's past, its illegal acts to control its markets. At that time they were looking ahead, using the case to legalize natural monopolies, notwithstanding the Sherman Antitrust Act. In a 1945 decision, the court looked back to *Terminal Railroad*, seeing there a precedent for a new doctrine known as "essential facilities."[64] That principle has grown in importance in recent decades. The body of antitrust law that today governs telecommunication companies, internet service providers, and other networks takes its origins from *Terminal Railroad*. Like the TRRA's tracks and yards, the central technologies provided by those firms may dominate a market serving many other businesses. But if they *are* deemed essential, those networks must remain equally accessible to all users.[65]

Turning from what the Terminal Association did to what it was, a hypothetical question begs for some consideration here. Given the same facts, would antitrust prosecutors at the Department of Justice today want to dissolve the TRRA? Probably. After all, the company could not prove that it delivered economies of scale, its local rates were high by all accounts, and it had abused its control over market access. Furthermore, in four trials or appeals from 1905 to 1912, state and federal prosecutors argued that dissolving the association into its operational components (namely, the once independent firms that in the past had each provided switching and interchange services—the Terminal, Merchants, and Wiggins companies) would serve the country better than the monopoly had done. That result could have finally created competition in St. Louis terminal services, to the benefit of shippers, railroads, and consumers.

But the choice to file any significant antitrust case depends on decisions by political actors: the attorney general and the president. And the resolution of those cases rests upon evolving interpretations originating at the US Supreme Court. Since the 1980s, the justices have returned to approving of corporate size and market power if defendants could plausibly argue that they resulted in value to consumers: lower prices or better products. That logic traces back ultimately to Chief Justice White and his pivot from legal interpretation to economic assertions in *Standard Oil* and *Terminal Railroad*.[66]

In his 1868 report, James Eads finessed the core question raised by any monopoly: its power over price. He promised that St. Louis Bridge would undercut the rates charged by Wiggins Ferry and its partner, the St. Louis Transfer Company. He predicted that "in 1871 when [the bridge] will be finished, the saving [from rate cuts] to our citizens will amount to millions." He anticipated that the bridge would soon drive the ferry and transfer companies out of business.[67] If those companies *had* disappeared, a bridge monopoly would simply have replaced the ferry monopoly, a point Eads chose to overlook. But those alternatives survived, thanks to the long-lived pooling and price-fixing agreements.[68] That fact in turn meant that every shipper and resident of St. Louis paid tribute (or monopoly rent) until the federal courts and the Interstate Commerce Commission began to exercise their oversight over rates for St. Louis freight around 1912.

Nonetheless, St. Louis drew immeasurable benefits from the connections fostered by the Eads Bridge. Compared with the city's commerce before 1874, the volume of regional trade grew exponentially, while freight rates and passenger fares fell. Furthermore, the TRRA had to offer competitive pricing on its interchanging traffic, given that those shippers could choose alternative gateways such as Kansas City, Omaha, or Chicago. The through segment grew to dominate its freight business after 1880, just as Eads had predicted.

The TRRA's market power had an additional purpose: its coordination of freight and passenger services made each of its connecting railroads work better.[69] For example, the Terminal Railroad ensured the priorities for the train and car movements of every carrier passing over its tracks to, from, or through the region. For all lines, scheduled passenger services went through first, followed, in descending priority, by livestock trains, cars with perishable products such as fruit or beer, scheduled or time freights, extra (nonscheduled) freight trains, interchange cars, and industrial deliveries.[70] No single corporation or agency in Chicago or any other American city exerted this kind of direction over such a large and vital crossroads.

Furthermore, the key challenge for the railroads, shippers, and industries of the region was not monopoly pricing, despite all the heat and smoke of that issue. Well into the 1920s, the pace of economic growth in greater St. Louis outstripped all the efforts and facilities of individual railroads to carry the raw materials, finished goods, and agricultural products of the growing region. With its federal sanction, the TRRA became a well-funded, nonprofit public utility, growing to bear the loads of the city. By the 1920s its three belt lines connected the tracks of twenty-eight railroads. Its six thousand employees and 185 locomotives handled 4.5 million freight cars annually.[71] No one could have foreseen these developments. But they all flowed from the blueprint that James Eads drew in 1868 for his fellow citizens and the entire region.

Epilogue

THE PROMOTER AND CHIEF ENGINEER of the St. Louis Bridge died on March 8, 1887. The obituaries and tributes mostly lionized the man, and justly so. Upon completing the bridge, James Eads immediately plunged into his next great work, creating a navigable channel through the Mississippi River Delta so that ocean freighters could reach the docks at New Orleans. Again, he crafted an audacious financing plan, this time by act of Congress. The statute specified generous payments, but only if he succeeded in deepening the channel. Again, he challenged the Army Corps of Engineers. True to form, General Humphreys had advocated a canal.[1] By constructing jetties to concentrate the river's natural currents, Eads avoided the high cost of a canal, while he deepened the South Pass channel from fourteen feet (1874) to thirty-one in 1879.[2] Success earned him another fortune, $5,950,000, paid by the US Treasury ($185.1 million in 2023).* After ocean freighters could dock at New Orleans, that city rose from eleventh to second place among American export hubs.[3]

---

* That munificent renumeration originated in the savvy pitch that Captain Eads made to the US Congress. To secure its authorization (and to block the canal alternative advocated by the Corps of Engineers), Eads proposed taking his fee only after his jetties had succeeded in producing the channel depths specified by statute. Until then, he bore all the costs and risks. Again, his blend of audacity, finance, and engineering won the day. To secure working capital to fund jetty construction, he enlisted investors, further proof of his charisma. Reuss, "Humphreys," 18; Barry, *Rising*, ch. 6.

SOUTH PASS JETTIES.

Although somewhat stylized, this engraving gives a sense of the extensive, wet work of creating the South Pass jetties. Most of their structures lay beneath the surface. Berms of woven willow mats were weighted with rock and submerged to build the jetties. They channeled the current that carried river sediment out into the Gulf of Mexico. The image includes depths achieved in the channel. Engraving from Dacus and Buel, *Tour*, 28.

Eads's creativity and drive never left him. In September 1879 he embarked on his seventh voyage to Europe and beyond. The trip included inspections of the Danube River and the Suez Canal and consultations in Constantinople about a bridge over the Bosporus.[4] As early as 1874 he had begun talks with the grand vizier "to unite Asia and Europe."[5] He even prepared drawings and estimates for the 3,500-foot structure.[6] In this he was a century ahead of his time.

At this point in his life, the Captain's primary focus was the Tehuantepec Ship Railway to connect the Atlantic and Pacific oceans. The venture had all the Eads hallmarks. Rejecting a traditional canal, then under consideration across the Isthmus of Panama, he proposed instead a massive railroad over the Isthmus of Tehuantepec in Mexico, a ship-hauling rail-road capable of carrying iron freighters weighing eight thousand tons with their cargos at speeds up to ten miles per hour over the sixty-mile route. To persuade skeptics that his notions were sound, he rounded up support-ing letters from old friends and colleagues, including Octave Chanute and Henry Flad. *Scientific American* ran a full description.[7] Perhaps it was a fanciful idea, yet not unknown. A few years later, Canadian and British promoters secured $4 million from investors to build the Chignecto Ship Railway connecting the Bay of Fundy to the Gulf of St. Lawrence.* In the end, the Tehuantepec project died with Eads. With three-quarters of their work done, the Chignecto promoters gave up in 1890 after their financing collapsed.[8] It is impossible to say whether a ship railway was visionary or looney. Many serious people took the concept seriously at the time. The Bosporus project and the ship railway are reminders that the Captain's ego had always towered alongside his abilities as an engineer and a promoter.

During the last decade of his life, Eads earned international regard for his expertise in the hydraulics of rivers and harbors. In 1884 he served as an expert witness in a controversy racking the civil-engineering commu-nity in Britain, over the benefits and environmental costs of the proposed Manchester Ship Canal. Opponents feared that the new canal would dis-rupt the currents and depths of the River Mersey, a key shipping artery. To advance their case, they hired Eads to review detailed studies and report his findings to a parliamentary committee.[9] He received £4,000 in consulting fees, reportedly a record. The journal *Engineering* found that the expenditure "was fully justified by the results" since "it was largely on Captain Eads' evidence that the Bill was thrown out."[10]

Also in 1884, the Fellows of the Royal Society of Arts selected Eads to receive the Albert Medal, its annual award for "distinguished merit in pro-moting the arts, manufactures, and commerce." The Prince of Wales personally bestowed the honor, named for his father, at a London cere-mony that revealed how far the Captain had traveled from the bottom of the Mississippi River.[11] He was the first American to receive the distinction,

---

* The railway's seventeen-mile route aimed to save shipowners and their crews from the cost and difficulties of a 500-mile voyage in often stormy seas around Nova Scotia.

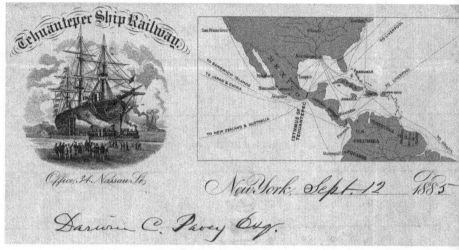

Detail of the letterhead Eads used in correspondence about the ship railway. Contemporary facts bolstered a concept that appears far-fetched today. The Panama Canal was shrouded in an unknowable future (it would open in 1914). In Captain Eads's day, most oceanic merchant shipping was still powered by sail or sail-assisted steam, as at left. Eads's concept would have allowed shipowners and masters to avoid long and expensive voyages against adverse winds or through the doldrums. The map shows Tehuantepec's locational advantage over Panama; dotted lines illustrate major trade routes. Detail from a letter from Eads to Darwin C. Pavey, box 1, folder 12, MHS, DO 3715.

joining a group that included Michael Faraday, Ferdinand de Lesseps, Henry Bessemer, and Louis Pasteur. In 1892 Thomas Edison would become the second American so honored.[12]

Eads's accomplishments elicited envy and scorn as well as laurels. The designer of the famous ironclad USS *Monitor*, John Ericsson, denounced the Captain as "a huge sham sustained by hired brains."[13] Their wartime conflicts over ironclad design had spiked Ericsson's vitriol. By contrast, a friend and professional colleague offered warm praise. William Sellers stood at the top rank of mechanical engineers for a half century. In a memorial tribute, Sellers credited Eads with nearly fifty patents issued by the American and British governments for innovations in ironclads, bridge design and construction, and a broad range of other fields. Equally important in Sellers's view, Eads had a "special characteristic of true greatness," never fearing that "praise for his associates might detract from his own fame."[14]

Some people saw the Captain's ship railway as the chimerical scheme of a char-latan. Others took him seriously. His plans detailed the route, methods for load-ing and launching ships at each end, speed and power requirements, costs and tariffs—and potential profits. The concept received authorizing legislation from the Mexican government and the US Senate. Eads died before he could lobby for passage of the statute in the US House of Representatives. Engraving by J. David-son, *Mining and Scientific Press*, Mar. 28, 1885, 206.

Sellers had come to know the Captain while making steel for the bridge, a demanding, difficult time for both men. His praise was hard-earned.

James Eads died of pneumonia when he was just shy of age 67. That ending harkened back to his bouts of respiratory illness and his years of breathing in and under the Mississippi. His death evoked praise through-out the English-speaking world. The editors of *Engineering* called him "one of America's most gifted engineers." Those close friends genuinely grieved for the loss of "a great man, admired for his genius and beloved for his simple and winning nature."[15]

Echoing similar testimonies, the *Railroad Gazette* offered a detailed evaluation of his character and work: "He combined courage, enthusiasm, persistence, insight and judgement in such measure as to amount to genius." The *Gazette* reviewed the five chapters of his remarkable pro-fessional life: river salvage and riches, the Civil War gunboats, his great bridge, triumph against powerful opposition at the South Pass jetties, and "the last great scheme" at Tehuantepec. From that resume, the memorial offered three insights. The work of millions had secured victory in the Civil War. Even so, Eads's Mississippi ironclads stood out for their signifi-cance to the course of the war and for being the work of one man, reflect-ing his "daring, energy and resource." The *Gazette* predicted that his South Pass jetties would prove his leading contribution to "the world's wealth." As for the bridge, it was "an instructive story of new problems and new solutions." His pioneering method for deep foundations, the pneumatic caissons, had already proven central to modern civil engineer-ing. But the bridge "has its lessons of what to avoid, and it stands as a monumental work which will never be repeated."[16] Harsh words for an obituary. Later verdicts about the bridge would evolve alongside develop-ments in engineering, technology, and culture. Later verdicts about Eads crowned him with laurels, praising him as a brilliant innovator and a fault-less hero.

The Railroad Men

Tom Scott ascended to the presidency of the Pennsylvania Railroad in 1874. As the depression deepened, it was a difficult time to run any rail-road, let alone the world's largest. His own frothy speculations across the West had pumped up the bubble that had burst all over the country. While running the PRR, Scott worked desperately to salvage his investments in the Texas & Pacific. As usual, he needed more capital to keep his kites aloft. Given the depression, private financing was out of the question, whether in New York or London, so he tried to entice politicians, a reliable source of aid in the past. In three attempts, he failed to move a subsidy bill through the US Congress.[17] The blows continued: bitter strikes on his PRR in 1877, a stroke and partial paralysis in October 1878, final capitulation and sale of the bankrupt Texas & Pacific to Gould in 1879, resignation from the Pennsylvania Railroad in June 1880, and death from a heart attack a year later at 57.[18]

When remembered at all, Scott typically appears near the top of the roster of Gilded Age robber barons. He certainly bought and bribed his way onto that list. Like Eads and Carnegie, however, he was a maker as well as a taker—creating the railroads, manufacturing companies, and infrastructures of his time. Their methods strike us today as often unsavory and sometimes fraudulent. Their careers arose from the dynamic, inchoate qualities of capitalism, money markets, government, corporations, and engineering in an era of rapid change, often of their own making.[19]

Scott's life and legacy fell into obscurity, while Jay Gould's reputation remained resolutely big and black for decades after his death. Tuberculosis carried him off in 1892 at age 56. Gould deserves a nuanced assessment.[20] He merged an assortment of weak or bankrupt lines into an often profitable system across the Midwest and Southwest. In that effort, he salvaged many pieces that Scott had planned and then lost in the financial storm after 1873. And he reinvented himself, evolving from a market manipulator into a system-building promoter. In approving creation of the Terminal Railroad Association, Gould forecast an organizational type that would become common after 1900: the privately owned public-service corporation.

The Money Men

Junius Morgan is universally ranked among the top financiers in London and the world during the Gilded Age. True, investors in St. Louis Bridge endured some troubled years. In the end, Morgan's bank and his bondholders profited from their bridge investments.[21] Even before those accounts shifted to black ink, J. S. Morgan & Company made cash loans to finance construction of the Captain's jetties.[22] That credit derived from the banker's faith in the charismatic and visionary Eads. The loans again underscore Morgan's ongoing penchant for courting risk. From 1879, the London bank participated in the syndicate financing Britain's first steel bridge.[23] These investments in technological potential show that Junius Morgan was a venturesome financier of the revolutions that were transforming his times—as well as a hardheaded, principled banker. In April 1890, he died in Monte Carlo after suffering fatal injuries in a carriage accident. His son Pierpont took over as senior partner of the London house, although he remained in the United States.[24] That change ratified the ascendence of New York City over London as the center of world finance.

Andrew Carnegie left St. Louis in April 1874, embittered by Eads and his bridge. By nature upbeat, he nursed those grievances for years. From his insider's seat, Carnegie knew that Eads and other principals had pocketed funds that were arguably his due. Worse, the bonus of $250,000 ($6.8 million in 2023) owed to Carnegie & Associates became a liability. With nothing else in the till, the bridge company had paid that bonus with its third mortgage bonds. After foreclosure, the bankruptcy court determined that those bonds had been fraudulently distributed. Now the judge wanted the money back to pay legitimate creditors.[25] An 1879 lawsuit revealed more shady dealings. The US Circuit Court for the Southern District of New York found that the stock-for-bonds swap (described in ch. 9) had the effect of defrauding the creditors of St. Louis Bridge. As a result of that judgment, counsel for the plaintiff went looking for $265,000. Among the forty-five stockholders in St. Louis Bridge who were dunned to pay up voluntary, these names stand out: Solon Humphreys and Edwin D. Morgan, Morris K. Jesup and Robert Lenox Kennedy, James Eads and William Taussig, Junius and Pierpont Morgan. And Andrew Carnegie.[26]

His troubles with St. Louis Bridge may have played a role in Carnegie's decision to make a fresh start in his career. In September 1872, he entered the steel business with a series of cautious hedges. The new steelworks was jointly owned and managed with capable operational and financial partners. It joined the cartel that managed steel patents and prices for the entire country. The partners hired a brilliant engineer, Alexander Holley, to design their new plant. And they built that factory on land near Pittsburgh with rail connections to the PRR and the B&O. Those railroads would have to compete for its business. Naming the works after Edgar Thomson didn't hurt its prospects.* This time, Carnegie became the indisputable boss after buying out his partners during the 1870s depression.[27] If St. Louis Bridge was a risky venture, Carnegie Steel was anything but.

The Engineers

In his memorial tribute, William Sellers described James Eads's "peculiar capacity . . . for selecting the best men to aid him in the various departments of his great enterprises."[28] The careers of his subordinates confirm this insight. Until the 1890s, most young men learned their engineering on the job, not in colleges. Their work on St. Louis Bridge taught them a lot. Eads's principal assistant engineer, Henry Flad, had designed the cantilevering system to erect the arches. In later years, he patented railway brake systems and designs for rapid transit and cable-car lines.[29] In 1886, he was elected president of the American Society of Civil Engineers. At 19, Onward Bates supervised the construction of Eads's iron caissons. He went on to run the Pittsburg Bridge Company, then became superintendent of bridges for the Chicago, Milwaukee & St. Paul.[30] He too became ASCE president (1909).

After his work in St. Louis (and his tumble into the Mississippi), Theodore Cooper had a distinguished career as a consulting engineer designing iron railroad bridges across the country. His research and publishing laid the theoretical and practical foundations required for steel to replace iron in bridging after 1880. He also established a standard measure for live

---

* Alexander Holley took the Bessemer process out of Britain and adapted it to American conditions, designing the first eleven successful steel plants in the United States. The Edgar Thomson works was his tenth. A patent cartel, the Bessemer Association, controlled access to needed patents, thus restricting entry into the US steel industry. It also "largely determined" the price of steel rails from 1877 to 1915. Misa, *Nation*, 21–22. Every company in the Bessemer Association enjoyed another perk: for three decades, starting in 1870, the US government imposed a tariff on imported British steel, upwards of $28 per ton. By 1877, American-produced rails sold for $42 per ton, ensuring that British producers would never get past the American tariff wall. Nasaw, *Carnegie*, 141, 174. With all these props and the protections from the state, it is difficult to see how any American steel plant could have failed to turn a profit.

loads, known as Cooper units, used in bridge design to the present. Despite all the contributions of his distinguished career, he is chiefly remembered as the designer of a disaster. His Quebec Bridge, a steel railway crossing over the St. Lawrence River, collapsed while under construction in August 1907. Seventy-five steelworkers died.[31] This grievous failure shocked Canada, the Empire, and the United States. The collapse revealed that despite all its progress in design, methods, and materials, engineering remained an art bounded by uncertainty.[32] As it still is. That fact surprised the public and confounded more than a few engineers.

Lucius Boomer had exactly the career that any fair observer could have expected when he capitulated to Eads in St. Louis in March 1868. Well into the 1870s, his American Bridge Company remained the largest and most successful bridge-building firm in North America.[33] By 1876, American had erected seven railroad bridges over the Mississippi and three more over the Missouri. All used conventional iron trusses.[34] Boomer, however, suffered his own reversal. In 1879, American Bridge contracted to erect the second steel bridge in the country, at Glasgow, Missouri. Carnegie's Edgar Thomson works produced the components for its pin-connected trusses.[35] One of those 314-foot spans fell during construction after a sudden rise in the Missouri River overwhelmed its supporting falsework. The Chicago & Alton Railway would complete this bridge in May 1879.[36] By then the collapse had thrown Boomer's company into bankruptcy.

Engineering Legacies

Eads's bridge reflected his qualities, for better and worse. It originated in his singular imagination. His driven nature saw it to completion against innumerable obstacles. But Jacob Linville was right: the steel of 1867 was flawed and expensive, while Eads's arches were problematic at best, wracked by temperature-induced stresses.[37] To manage those strains, the ribs had to be substantially overbuilt, the main reason the bridge can remain in use today. In 1874, that fact raised costs considerably, contributing to the foreclosure. Linville and Carnegie were correct in believing that Keystone could have built a perfectly adequate, pin-connected truss bridge at St. Louis with wrought iron and entirely compliant with the Omnibus Bridge Act. Keystone's bridge over the Ohio River at Cincinnati revealed its impressive competencies at that time.* Such an alternative would have cost less, opened two years earlier, and probably avoided receivership. It would not have dodged the cyclones of 1871 and 1896. It would not have affected the rivalry with Chicago, and it would not have become the graceful icon of St. Louis.

During the twentieth century, the Terminal Railroad Association made extensive improvements and repairs to the Eads Bridge. It underwent thorough structural inspections in 1940 and 1950, receiving high marks overall. Its strength ratings increased from a Cooper E-36 rating in 1921 to Cooper's E-45 in 1970. The shift from steam to diesel locomotives eased the stresses on the arches.[38] And its extraordinary strength has proven essential on those occasions when strong river currents overpowered towboats or their barges, causing them to strike the bridge piers, even the superstructure.†

The strength derived partly from Eads's decision to use steel for the main structural members. That decision cut two ways. On one hand, it imposed two years of delay, nearly ensuring foreclosure for St. Louis Bridge. Given the challenges overcome in that learning curve, however, this inaugural success for structural steel was a noteworthy achievement. If William Butcher failed, his successor largely succeeded. At Midvale Steel Company, William Sellers put that hard-won knowledge to good use. In 1881, his company supplied thousands of tons of rolled and forged steel to construct the Brooklyn Bridge.[39] Midvale remained a leading manufacturer of specialty steels well into the twentieth century.

Thanks mostly to the troubles with steel, Carnegie developed a caustic view of Eads and the bridge, describing him decades later as "an

---

* This Newport & Cincinnati Bridge had a 415-foot channel span, with eight more Linville truss spans in wrought iron, for a total length of 1,821 feet. It opened in March 1872. *Railroad Gazette*, Mar. 30, 1872, 141–42.

† On October 14, 1969, with the river in flood thirty feet above normal stage, the towboat *Elaine Jones* was overpowered by the current, striking a pier of the Veterans Bridge, then continuing downstream to hit the superstructure of the Eads Bridge. The accident broke two tube segments of the upstream rib of the center arch, tore off the boat's steel pilot house, and killed the pilot. The bridge was subsequently repaired.

original genius *minus* scientific knowledge to guide his erratic ideas of things mechanical."[40] By contrast, Sellers, who worked closely with the Captain to produce suitable steel, had only praise for the self-taught Westerner. Upon Eads's death, Sellers wrote a lengthy memorial for the National Academy of Science, detailing Eads's life. He closed his portrait saying: "Perhaps few men are heroes to those who know them intimately, but those who were nearest to Mr. Eads honored and loved him most."[41]

Eads's innovations in the methods of design endured and changed civil engineering in America. His team of Roberts, Flad, and Pfeifer gave St. Louis Bridge the ability to mandate design specifications to suppliers. Their pioneering use of calculus to chart stresses, select materials, and size components became standard practice. Their insistence on materials testing became commonplace. Their concern for the modulus of elasticity entered the designer's standard toolkit. In these ways, civil engineers gained new powers to specify their design needs and impose their will on contractors like Midvale, Keystone, and American Bridge. The self-taught amateur played essential roles in making a profession out of civil engineering.[42]

Although Eads encountered a harsh reception at the 1874 convention of the American Society of Civil Engineers, with the passage of time engineers' ideals and verdicts shifted. By the 1880s, a new generation of designers aspired to more than low-cost utilitarian structures. They sought to marry engineering and architecture, function and beauty. The steel arches of the Washington Bridge over the Harlem River in New York City (1888) and those of the Whirlpool Rapids Bridge over the Niagara River (1897) owed a measure of inspiration to James Eads. Aesthetic qualities seemed especially worthy for great structures in great cities, where design could please the eye and ennoble the minds of millions. Still, the curves of Eads's arches remained unusual for North America. Their shallow quality suggests a kind of coiled energy, springing across the river in three leaps.

To the Present

As the twentieth century dawned, the bridge bore heavy and contradictory symbolic freight. For many citizens and politicians, it tangibly rep-

Linking northern Manhattan to the Bronx, the Washington Bridge over the Harlem River was completed in 1888. Its 510-foot steel arches nearly equaled the center span at St. Louis. This crossing exemplified new methods for creating urban infrastructures, methods that replaced Eads's private-sector model and became dominant in twentieth-century cities. A governmental body, the Harlem River Bridge Commission, spearheaded a design competition, selected this winning plan, and oversaw construction. Built at public expense as a roadway crossing, it carries six lanes of traffic without tolls. This 1905 image also shows the Harlem River Speedway beneath the bridge, along the Manhattan shore, a favorite run for afficionados of fast horses. The equestrians are long gone; the Washington Bridge remains a key artery. Photograph titled *The Harlem from High Bridge*, Detroit Publishing Company, 1905, Library of Congress, LC-DIG-det=4a12688.

resented the TRRA's monopoly. In those years, advocates for a city-owned Free Bridge secured its charter and funding. Located a mile south of the Eads Bridge, this double-decked road and rail bridge finally opened to car traffic in 1917, its automobile patrons happy to avoid the tolls of the older crossing. Long before that deliverance, however, many St. Louisans had adopted the Captain's bridge as a proud symbol of their city. For them, the structure embodied a triumph of will and the boldness of innovation.

The Free, or Municipal, Bridge in 1928, looking east. Three Pennsylvania through trusses carried a roadway deck above a level with dual railroad tracks. Trains needed approaches with much easier (shallower) grades than required for cars and trucks, so the roadway deck had an S curve to clear the main line railway approach (obscured by the road, so not visible here; an incomplete spur line appears at right). For many decades, too many drivers failed to negotiate the curves, often with fatal results. Compared with this hodgepodge, the Eads Bridge incorporated elegant design solutions. Renamed the MacArthur Bridge in 1942, its rail deck remains in use. Photograph by W. C. Persons, MHS, P0079.

The Eads Bridge became an important influence in modern architecture. Louis Sullivan's precept that "form follows function" derived from its example. At age 16, Sullivan avidly followed the construction progress of the bridge in the pages of the *Railroad Gazette,* watching it grow and growing with it.[43] That example set him on the path "to make an architecture that fitted its functions—a realistic architecture based on well-defined utilitarian needs."[44] From that genesis, he drew his first skyscraper, all of ten stories, the Wainwright Building in St. Louis (1892).

This guest ticket to the 1896 Republican National Convention offered two icons to please the delegates. The log cabin, Ulysses S. Grant's Missouri home in the 1850s, harkened back to the GOP commitment to preserving the nation. The steel bridge was a badge of greatness for St. Louis. The masonry arcade shown here (*left*) appears as the tumbled stonework strewn across the track in the figure on p. 260. "Guest Ticket," MHS, A1229.

Sullivan's aesthetic inspiration depended in turn upon the concealed strength of its framing in structural steel, the material pioneered by Captain Eads.

With the post-1945 rise of modernist architecture, critics rediscovered in Eads and his bridge an original American triumph of pure form. By then the members of the American Society of Civil Engineers had overcome their initial doubts. In 1971 the organization dedicated the bridge as a National Historic Civil Engineering Landmark. In its centennial year, the influential architecture critic Ada Louise Huxtable declared that it ranked "among the most beautiful works of man."[45] Walt Whitman had given a similar assessment in 1879, writing in *Specimen Days*, "I have haunted the river every night lately, where I could get a look at the bridge by moonlight. It is indeed a structure of perfection and beauty unsurpassable, and I never tire of it."[46]

Shown here under construction, the terminal building at St. Louis Lambert International Airport opened in 1956, hailed as a modernist triumph. The design reflected an earlier achievement in another St. Louis transport link. Photograph by Henry Mizuki, MHS, P0374.

His bridge remains the Captain's chief legacy. In 1989 the city of St. Louis swapped its MacArthur Bridge (originally known as the Municipal, or Free, Bridge) for the Terminal Railroad Association's Eads crossing.[47] City planners wanted the bridge and tunnel to create a route for electric-powered light-rail trains to serve downtown. With its rededication in 2003, this oldest bridge on the Mississippi was back in business. The restored roadway deck carries auto traffic in four lanes without tolls. Broad sidewalks again offer unobstructed prospects of the river and the city. On the rail deck beneath, trains of the MetroLink light-rail system connect St. Louis's eastern and western suburbs to the urban core and the airport. Each weekday, MetroLink trains cross the Mississippi three hundred times. Passengers enjoy fleeting views of the river and the bridge's iron and steel sinews before plunging into the darkness of the 1874 tunnel with its two new downtown stations. The system also serves Union Station, which now houses shops, tourists, and convention visitors rather than main line railroad travelers. In 2016, the bridge and its approaches returned to like-new condition thanks to a $48 million restoration. With its original strengths intact, the restored bridge should have a service life reaching to 2091 or beyond, according to engineering projections.[48]

James Eads would greet that news with his characteristic self-assurance. In his remarks on opening day, he predicted that "this bridge will endure as long as it is useful to man."[49]

ACKNOWLEDGMENTS

It is a great pleasure to thank the individuals whose contributions made this book possible. Beyond a reckoning of my considerable debts, these acknowledgments amount to a history of the book itself. It began before I was born.

In 1948, John Atlee Kouwenhoven (1909–1990) began to research the Eads Bridge. An accomplished scholar in American Studies, he was then a professor of English at Barnard College. His interest in the bridge followed publication of his *Made in America: The Arts in Modern Civilization* (1948). In that book, Kouwenhoven rejected distinctions between art and design, and he celebrated vernacular, or home-grown, design for its uniquely American values. *Made in America* ignited his fascination with the Eads Bridge. In his view, "the honesty and majesty of the finished structure" marked it as a milestone in engineering and a turning point in design (p. 71). In 1958, Kouwenhoven joined a diverse group of scholars to create a new academic field, the history of technology. Over the following thirty years, he published on many topics, including two articles about the bridge, but he never wrote the big book. After his death, his son and literary executor gave Kouwenhoven's research materials to me. Gerrit made that gift with the injunction that his father's work should serve as a point of departure for my own inquiries, thoughts, and book.

John Kouwenhoven did his research the old-fashioned way: reading widely, tracking down original sources, and taking notes in perfectly formed longhand with scrupulously complete citations. Although I never met him, I am deeply in his debt. Fulfilling his written instructions, I will donate all his Eads material to the Missouri Historical Society, a fine institution in St. Louis.

When this project came to me, I was fresh out of graduate school with a PhD in the history of American technology. Like anyone working that turf, I was familiar with James Eads and his remarkable bridge. But the bridge project could not take my own center stage until I had received tenure in the Department of Science, Technology and Society at the University of Virginia. A sabbatical in 2003 provided time to review the trove that had come my way. My own research efforts began in 2004 amid other commitments. Retirement from teaching in 2015 freed up time for further research. I began to write the manuscript in 2019, completing it in 2023.

Over the decades of its gestation, this account has been improved by a teeming multitude of smart and generous people. John Kouwenhoven would surely have thanked Charles van Ravenswaay and Frances H. Stadler at the Missouri Historical Society. George P. Mueller opened doors and files at the Terminal Railroad; George Ketchum did the same at United States Steel. Ernst A. Stadler translated contemporary articles about the bridge from a lively German-language St. Louis newspaper, the *Westliche Post*. Historians who freely shared ideas and sources with Kouwenhoven included Ari Hoogenboom, David McCullough, and Richard C. Overton.

My own debts to others began in their books and articles that framed facts and interpretations about the bridge. David Billington, Robert Jackson, John Kouwenhoven, Howard Miller, Henry Petroski, and Calvin Woodward all did foundational work. Studies on diverse topics by John Barry, Vincent Carosso, Albert Churella, Rebecca Edwards, Sharon Grimberg, Richard John, Morton Keller, Maury Klein, David Nasaw, James Neal Primm, Heather Cox Richardson, Robert Vogel, James Ward, and Richard White were immensely helpful in placing Eads into the larger contexts of Gilded Age America.

Archivists and rare-book librarians make scholarly work possible. My debts here are too numerous to list by name, but I thank the professional and helpful staffs at the repositories listed before the Notes. Lauren Sallwasser at the Missouri Historical Society graciously tracked down many of the illustrations that enrich the text. An online resource managed by James Baugh, Bridgehunter.com, has proven invaluable. The maestros of the University of Virginia's Interlibrary Loan office, Lew Purifoy and Holly Shifflett, brought the riches of other libraries to Charlottesville, tracking down obscure engineering reports and articles from long-defunct technical journals.

For many years, friends, family, and colleagues have heard about my bridge and sharpened my thoughts and prose. Those generous people include Mark Aldrich, David Aynardi, Hewson Baltzell, Locke Brown, Wendy Brown, Sally Clarke, Robert Friedel, Meg Graham, Naomi Lamoreaux, Mike Matejka, Betsy Mendelsohn, William Middleton, Marty Reuss, Keith Revell, Ed Russell, Larry Thomas, and John Willis. Three former students in the Engineering School at UVA—Judy Butler, Kaela Mattson, and Chris Wenderoth—educated me about structural analysis.

I first developed many ideas presented here in conference papers at meetings of the Society for the History of Technology, the Business History Conference, and the Society for Industrial Archaeology. I gave invited talks at Johns Hopkins and at the University of Minnesota, where hosts and audiences replied with valuable feedback. At the Economic History Seminar at the University of Pennsylvania, Walter Licht hosted a memorable meeting in which probing comments from Maury Klein sharpened my insights into Andrew Carnegie. At Penn's Wharton School, Dan Raff and Phil Scranton sponsored a conference on the historical development and uses of knowledge inside and among firms, ideas that improved chapter 7.

Academic papers often grow up to become published articles or chapters. My contribution at Wharton became a chapter in *The Emergence of Routines,* published in 2017. In that transformation, editors Raff and Scranton, fellow contributors, and the anonymous referees at the Oxford University Press all improved the original paper. In 2014, *Technology and Culture* published my article "Not the Eads Bridge." That exploration of counterfactual history underscored the quirky, unpredictable unfolding of events that we then retrospectively present as history. Editor Suzanne Moon and anonymous referees at *T&C* sharpened my understanding of contingency in history, an awareness that became central to this book. At a 2022 conference honoring the career of Olivier Zunz, I explored the important, unheralded place of the Terminal Railroad Association in American antitrust law. That work evolved into another article, "The 'Rule of Reason' and an Unnatural Monopoly," much improved by editors Nicolas Barreyre and Derek Hoff. It informs chapter 10 here.

Along with ideas, money and time make books. Among the community of historians at UVA, Olivier Zunz and Brian Balogh created opportunities for me to focus on this work. My colleagues in the Department of Science, Technology, and Society generously picked up my teaching during two sabbaticals. One proved especially rewarding and productive, thanks to François Weil and Pap Ndaie at the Center for North American Studies at the University of Paris VI.

As draft chapters emerged from my computer, more talented friends stepped up to share expertise, provide leads, and offer critical feedback. Lee Buck personified the target that many authors aim to hit: the educated general reader. With his experience in corporate finance, John Reid offered valuable insights. Mike Henke and Daniel Crane sharpened my understanding of antitrust law. A leading authority on the railroads of St. Louis, Ron Goldfeder, read the manuscript and saved me from many factual derailments. Two freelance editors, Charlie Feigenoff and Peter Behrens, read and improved the entire draft. Near the end of a project that often appeared interminable, good friends Dorothy Massey and Olivier Zunz gave valuable feedback on the manuscript. Another friend, James Gelly provided time, insight, and support beyond measure.

Reflecting its long gestation, this book had a somewhat difficult birth. Across that process, the series editor, Richard John at Columbia University, shared his mastery of the issues and period covered here, providing invaluable critical feedback. The editor Matt McAdam at Johns Hopkins University Press guided the manuscript through acceptance and into production. Two anonymous referees did their best to put my wandering drafts onto a straight path. All these guides gave generously of their patience, wisdom, and experience. The book is immeasurably better thanks to their efforts.

During their research and writing, historians often appear to work in isolation. As these acknowledgments testify, that impression is wildly incorrect. These smart people, however, could not possibly save me from all misjudgments, errors of fact, or dubious conclusions. They remain mine alone.

NOTES

This account draws from twenty-four collections of business or personal papers, listed immediately below. Citations to these sources appear in the notes, which often include references such as K9/7 or B35/5. These refer to notes or photocopies of original sources in files amassed by John A. Kouwenhoven or by the author. For example, K9/7 refers to the seventh document in Kouwenhoven's ninth folder. To aid further scholarship, this material will be donated to the Missouri Historical Society (MHS) along with the manuscript of a detailed daily chronology cited here as Kouwenhoven, "Chronology."

Archival Collections

AC   Papers of Andrew Carnegie, Heinz History Center, Pittsburgh

ACE   Records of the Army Corps of Engineers, Record Group 77, War Records Division, National Archives

BB   Barton Bates Papers, MHS

CP   Carl Pfeifer Papers, MHS

EDM   Edwin D. Morgan Papers, New York State Library

GB   George Bliss Papers, New-York Historical Society

GMD   Grenville Mellen Dodge Papers, State Historical Society of Iowa

GVF   Gustavus Vasa Fox Papers, New-York Historical Society

JA   James Andrews Papers, MHS

JBE   James Buchanan Eads Papers, MHS

J&M   Letters of James B. Eads and Martha Dillon Eads, Churchill Memorial and Library, Westminster College, Fulton, MO

JSM   J. S. Morgan & Company Papers, Morgan Grenfell Archive, London Metropolitan Archives

JSW   Julius S. Walsh Papers, MHS

JWR   Papers of John A. and Washington A. Roebling, Rensselaer Polytechnic Institute, Troy, NY

PC   Peugnet Collection, MHS

PF   Pliny Fisk Collection, Barringer Library, St. Louis Mercantile Library
RGD   R. G. Dun credit reports, Baker Library, Harvard Business School
TRRA   Terminal Railroad Association Papers, MHS
TRRA/B   Terminal Railroad Association Collection, Barringer Library, St. Louis
          Mercantile Library
USS   Carnegie Papers, held by United States Steel Corporation, Pittsburgh, in
          1972, now at Heinz History Center, Pittsburgh
VPC   Vincent P. Carosso Papers, The Morgan Library, New York City
WJP   William Jackson Palmer Papers, Colorado Historical Society, Denver
WMR   William Milnor Roberts Papers, Montana State University, Bozeman
WT   William Taussig Papers, MHS

Preface

1. Woodward, *History*, 56–57.
2. Andrew Carnegie to William Taussig (confidential), Dec. 30, 1870, USS,
    K10/73.
3. The Eads Bridge provided the focus for four earlier books and innumerable
    articles. Without exception, they describe little of its history following its
    triumphant opening in 1874. Its foreclosure a year later largely explains the
    truncation. See Dorsey, *Road*; Jackson, *Rails*; Miller and Scott, *Eads*; and
    Woodward, *History*.
4. The best overview of these bridges is Warren, *Bridging*, an 1878 report by the
    Army Corps of Engineers. A comparatively small number were owned outright
    by railroads. For more on this topic, see chapter 3.
5. In the largest cities, such as Philadelphia and New York, a municipally owned
    water monopoly serving wealthy districts was common before the Civil War, but
    a study from 1860 found that 79 of 136 American waterworks were privately
    owned. In 1888, most American towns still relied on the private sector for water
    service. Hunter, *Steam Power*, 519–21.
6. Historians have largely overlooked the American city gas industry. By 1859 the
    sector had its own trade publication, the *American Gas-Light Journal*, with a
    biweekly press run of ten thousand copies. Its first issue (July 1, 1859) listed 183
    private-sector gas companies in the United States (pp. 2–3).
7. Since the 1970s, historians have examined the effects of street railways on
    urban development, but the business history of that sector remains unwritten.
    A fine biography of one traction magnate is Franch, *Yerkes*. This private-sector
    model for public infrastructure also built the first subway lines in New York
    City (1904). By then, a new field, administrative law, had grown in force and
    reach with a mandate to ensure that these entities balanced private profit with
    public good.
8. McCullough's history of the Brooklyn Bridge, *The Great Bridge*, offers narrative
    riches but no analysis of that structure as an exemplar of the larger forces
    shaping the country.

9. American urban growth in this period was staggering. In 1850, the country had 62 towns with populations exceeding 10,000 people. By 1890 that total had risen to 354. Over the same decades, the nation's urban population grew by 600 percent, to over 22 million. US Department of Commerce, Bureau of the Census, *Historical Statistics*, 1:11–12.

Prologue. The Celebration

1. Most parade details are from Kouwenhoven, "Celebration." Also see "Tramping Thousands," *St. Louis Dispatch*, July 2, 1874, 4. A dose of hyperbole shaded all the newspaper accounts, booster fervor then being as essential to journalism as ink.
2. John A. Kouwenhoven (hereafter JAK), notes to Reavis, "Bridge," K20/43.
3. The mileage for 1874 is derived from Ringwalt, *Development*, 211.
4. "Magna Pons," *St. Louis Dispatch*, July 4, 1874, 1.
5. "Magna Pons," *St. Louis Dispatch*, July 4, 1874, 2.
6. Woodward, *History*, 197.
7. Jackson, *Rails*, 192.
8. Hetherington "Union," 5.
9. Cronon, *Chicago*, 372–73.
10. Grant, *Papers*, xxiii, 406. On June 21, 1874, Eads (then in Washington, DC) invited Grant to speak at the grand opening. A day later the president's secretary replied that no formal invitation had yet been received. The bridge men's cavalier attitude toward the president's schedule may have reflected nothing more than disorganization. Perhaps as likely, the late-arriving request may have been a wily stratagem to make Grant look bad: invited to attend, but a no-show at the city's big event.
11. Peterson, *Freedom*.
12. Slap, *Doom*, ch. 7.
13. B. Gratz Brown, quoted in Gilbert and Billington, "River Politics," 88.
14. Zerah Colburn, "The Mississippi Bridge," *Engineering* (London) 6 (Sept. 25, 1868): 285.
15. The original subscriber roll of February 1869 included fifty-nine individuals or firms. Woodward, *History*, 56–57. By July 1874, St. Louis Bridge had more than one hundred shareholders; see JAK table, "Stockholders of Illinois & St. Louis Bridge Co.," K19/1.
16. Minutes of the Executive Committee, June 2, 1874, vol. 2, p. 49, TRRA, K17/10.
17. Hyde and Conard, "William Taussig," *Encyclopedia*, 4:2218–20.
18. For Eads's complete address, remarkably short for the era, see Eads, *Addresses*, 42–45.
19. Woodward, *History*, 196; Jackson, *Rails*, 187.
20. J. S. Morgan & Company (hereafter JSM Co.) to Dabney, Morgan & Company (DM Co.), Mar. 24, 1870, in "Private Copy Out-Letter Book, Nov. 1867–Mar. 1875," 366–68, MS21795/001, JSM.

21. "Magna Pons," *St. Louis Dispatch*, July 4, 1874, 4.
22. Kouwenhoven, "Celebration," 179.
23. Strouse, *Morgan*, 152.
24. JAK, notes to Taussig, "Origin," 6, K7/2.

Chapter 1. Captain Eads

1. Dacus and Buel, *Tour*, 140.
2. Adler, *Yankee*, ch. 6.
3. Woodward, *History*, 10.
4. Woodward, *History*, 1–2.
5. JAK, notes to DeVoto, *Empire*, 170–71, K1/33.
6. Robert Fulton, quoted in Sale, *Fire*, 14.
7. Gould, *Fifty*, 99–100.
8. Brown, *Limbs*, 7.
9. Gould, *Fifty*, 118–22; Hunter, *Steamboats*, 101–2.
10. How, *Eads*, 1. Although his full name was James Buchanan Eads, no reliable evidence supports an off-repeated claim that he was named for the man who would serve as the fifteenth president. The biography written by Eads's grandson, Louis How, makes no such claim. For more on this point, see JAK notes, K1/28.
11. Sellers, "Memoir," 61.
12. JAK, notes (p. 1) to Snyder, "Eads of Argyle," K1/17.
13. Beitz, "Nantucket," K1/14.
14. Gould, *Fifty*, 485.
15. Miller and Scott, *Eads*, 68.
16. Eads's collected writings of a lifetime fill 650 pages, set in eight-point type. Eads, *Addresses*.
17. Gould, *Fifty*, 485.
18. Gould, *Fifty*, 738–39.
19. Dickens, *American*, 163.
20. "Hiram Hill," 213, B34/11.
21. Dorsey, *Road*, 19.
22. James Eads to Martha Eads, Oct. 1, 1848, and Dec. 9, 1849, letters 122 and 146, respectively, J&M.
23. James to Martha, [1844], letter 2, J&M.
24. Martha to James, [1844], letter 1, J&M.
25. Eads included these direct quotations in James to Martha, June 24, 1845, letter 8, J&M.
26. James to Martha, Aug. 6, 1845, letter 15, J&M.
27. James to Martha, Aug. 12, 1845, letter 18, J&M.
28. James to Martha, Aug. 24, 1845, letter 22, J&M.
29. James to Martha, Aug. 9, 1845, letter 16, J&M.
30. James to Martha, Sept. 6, 1845, letter 25, J&M.
31. James to Martha, Sept. 8, 1845. letter 26, J&M.

32. James to Martha, Sept. 1, 1848, letter 116, J&M.

33. Dorsey, *Road*, 26–28.

34. James to Martha, May 6, 1847, and Apr. 2, 1847, letters 41 and 36, respectively, J&M.

35. James to Martha, Aug. 5, 1847, letter 83, J&M.

36. James to Martha, July 9, 1847, letter 69, J&M.

37. Dorsey, *Road*, 29.

38. Dorsey, *Road*, 29. On Eads's debts, see How, *Eads*, 16.

39. James to Martha, May 23, 1847, letter 49, J&M.

40. James to Martha, July 30, 1848, letter 106, J&M.

41. James to Martha, Aug. 27, 1848, letter 115, J&M.

42. James to Martha, Sept. 1, 1848, letter 116, J&M.

43. Martha to James, Sept. 10, 1848, letter 117, J&M.

44. James to Martha, Mar. 8, 1849, letter 132, J&M.

45. Taylor and Crooks, *Sketch*, 115.

46. Taylor and Crooks, *Sketch*, 116.

47. Martha to James, Aug. 2, 1848, and Aug. 13, 1848, letters 107 and 111, respectively, J&M. James and Martha had two surviving children, Eliza and Martha.

48. Dorsey, *Road*, 30–32.

49. James to Martha, Aug. 14, 1852, letter 211, J&M.

50. Martha to James, Oct. 6, 1851, letter 213, J&M.

51. Dorsey, *Road*, 37–38.

52. Martha to James, Apr. 23, 1847, letter 35, J&M.

53. Gordon, *Fire*, 103, 106, 117, 118.

54. James to Martha, Dec. 23, 1849, letter 150, J&M.

55. Elliott, *Notes*, 249.

56. Dacus and Buel, *Tour*, 24.

57. Parton, "City," 657, B34/17.

58. Elliott, *Notes*, 287–88.

59. Hodes, *Divided*, 11, 20.

60. Dacus and Buel, *Tour*, 140.

61. Dacus and Buel, *Tour*, 140. The railroad was originally (optimistically) named the Pacific Railroad.

62. Elliott, *Notes*, 275–76.

63. Taylor and Crooks, *Sketch*, 116.

64. Taylor and Crooks, *Sketch*, 117.

65. Taylor and Crooks, *Sketch*, 120–25.

66. Gould, *Fifty*, 486.

67. Credit report of May 14, 1861, Eads & Nelson, a.k.a. Western River Improvement and Wrecking Co., Missouri, vol. 38, p. 212, RGD.

68. By 1861 Eads claimed to only have a $2,500 investment in the wrecking company at a time when its capital stock totaled $235,000. US Congress, House, *House Report #2*, 958, K25.1/32.

69. How, *Eads*, 20.
70. Entries in the 1860 manuscript census schedules for St. Louis, Second Ward, p. 342, National Archives, in entry for James Eads, Ancestry.com.
71. Jackson, *Rails*, 8.
72. Scharf, *History*, 2:1073.
73. Wikipedia, s.v. "Bogy, Lewis V.," accessed Jan. 26, 2020, https://en.wikipedia .org/wiki/Lewis_V._Bogy.
74. JAK, notes to Wiggins Ferry Co., Stock Certificates, Book 1, TRRA, K19/16.
75. Goldfeder, "Railroads," 20–22.
76. JAK, notes (pp. 6–7) to Polinsky, "Construction," K16/1.
77. Parton, "City," 655, B34/17.
78. JAK, notes (p. 13) to Taussig, "Development," K7/1.
79. "Illinois and St. Louis Bridge."
80. [Boomer], *Board*, 96.
81. JAK, notes (p. 6) to Polinsky, "Construction," K16/1.
82. Woodward, *History*, 7–9.
83. Mark Twain, quoted in McCullough, *Great*, 79.
84. Timmerman, "Bridging," 22.
85. JAK, notes to Emerson, *Journals*, 525, K16/5.
86. Bain, *Empire*, ch. 5.
87. Arenson, *Heart*.
88. James to Martha, Feb. 2, 1851; Martha to James, July 1, 1852; and James to Martha, Aug. 14, 1852, letters 193, 208, and 211, respectively, J&M.
89. James to Martha, Aug. 14, 1852, letter 211, J&M.
90. Primm, *Lion*, 246.
91. Bearss, *Hardluck*, 11; Barton Bates to Eads, Apr. 17, 1861, box 1, folder 1, JBE.

Chapter 2. Advances from War

1. Hodes, *Divided*, 371.
2. Parton, "City," 662, B34/17.
3. Kouwenhoven, "Chronology," Apr. 23, 29, 1861; JAK, notes to Scharf, *History*, 1:536; Canney, *Ironclads*, 41.
4. Bearss, *Hardluck*, ch. 2.
5. Canney, *Ironclads*, ch. 5.
6. Boynton, *Navy*, 544–49.
7. Hodes, *Divided*, 359–60. Primus Emerson built the Confederate ironclad CSS *Arkansas*.
8. James to Martha, Dec. 9, 1849 (emphasis in original), letter 146, J&M.
9. US Congress, House, *House Report #2*, 954, K25.1/32.
10. Eads to President Lincoln, Sept. 19, 1863 (emphasis in original), found in an entry for James Buchanan Eads in Ancestry.com, accessed Jan. 30, 2020 (no archival citation given there). The eighth ship referred to in Eads's letter to Lincoln was the USS *Benton*, an ironclad converted from Eads's old *Submarine*

No. 7. Before their engagements with the forts, the Union ironclads had fought a Confederate flotilla on the Mississippi.

11. Canney, *Ironclads*, 116.
12. See, e.g., James B. Eads, "Improvement in Operating Guns and Gun Towers," #1811, reissued Nov. 8, 1864.
13. "Income List for the Year 1864," *Missouri Democrat*, Aug. 19, 1865, 3, B35/21.
14. Letter of appointment, Gideon Welles to Eads, June 21, 1864, JBE, K5/26.
15. Eads to Gustavus Fox, Aug. 17, 1864, box 8, folder 13, GVF, K4/20.
16. Churella, *Pennsylvania*, 213–14.
17. Nasaw, *Carnegie*, 62.
18. Organized as a partnership, Piper & Shiffler, the firm incorporated in 1865 (capitalized at $300,000) as Keystone Bridge, the name used here throughout. The original partnership included those two named principals as well as Linville, Carnegie, and Scott. Bridge, *Inside*, 39.
19. For the "Carnegie company," see Nasaw, *Carnegie*, 102. "Scott's Andy" is quoted from Krass, *Carnegie*, 50.
20. Bridge discusses the extensive PRR parentage and influence over Keystone Bridge in *Inside*, 50. Nasaw also emphasizes Keystone's sales to the PRR.
21. The overall proportion of Carnegie/Thomson's holdings in Piper & Shiffler ca. 1862 is unclear. Evidence from 1868 suggests that Carnegie then held a one-eighth interest in Keystone Bridge, but as Thomson's proxy. In this arrangement, he kept the dividends paid on those shares. See Carnegie to Jacob Linville, Oct. 1, 1868, USS, K9/29.
22. Churella, *Pennsylvania*, 269–70. Thomson planned to use an Ohio railroad to operate this link, as his PRR lacked charter authority to operate in Ohio.
23. *Congressional Globe*, 37th Cong., 2nd Sess., p. 3115 (1862), K25.1/49.
24. Scott frequently bribed legislators to advance bills of interest to the PRR. The politicians, however, may have demanded the payoffs as a precondition to advancing his legislation, so historians cannot know who was corrupting whom.
25. *Appendix to the Congressional Globe* (Washington, DC, 1862), 37th Cong., 2nd Sess., p. 406, K25.1/51.
26. JAK, notes to Doyle, *Steubenville*, 242, K29/15; Condit, *Building*, 141–42.
27. DeLony, "Golden," 16.
28. JAK, notes to "Kansas Question," K15/4; Adler, *Yankee*, 145.
29. MacGill, *History*, 512.
30. Elliott, *Notes*, 286.
31. Owned by an independent corporation, this first bridge at Rock Island connected the Mississippi and Missouri Railroad (the first carrier to operate in Iowa) with the Chicago and Rock Island in Illinois (hereafter the Rock Island).
32. Pfeiffer, "Bridging"; McGinty, *Lincoln's*.
33. Warren, *Bridging*, 142

34. Known in 1856 as Stone & Boomer, the business incorporated in 1870 as the American Bridge Company, the name used hereafter.

35. JAK, notes to Parker, *Iowa*, 96, K16/6.

36. McGinty, *Lincoln's*, 62–63.

37. McGinty, *Lincoln's*, 74.

38. "Uncover New Evidence in 'Effie Afton Case,'" *Sunday Times Democrat* (Davenport-Bettendorf, IA), Jan. 20, 1963, 4d, K1/63. Also see letter from Eads & Nelson to Charles Case (agent), Sept. 3, 1856, Davenport Public Museum, accession #1962-196, K4/3.

39. McGinty, *Lincoln's*, ch. 5.

40. Immediately after its completion, the Chicago & Northwestern bought control of this bridge. Greenhill, *Witness*, 94.

41. Warren, *Bridging*, 94–96.

42. Ringwalt, *Development*, 203. The work of an unknown builder, the original Clinton Bridge mostly used composite (wood and iron) Howe trusses, which deteriorated quickly. They were replaced in 1868–69 with iron spans fabricated by four factory producers (American, Detroit, Phoenix, and Keystone), evidence of the rapid development of this new industry. Warren, *Bridging*, 94–96.

43. Ringwalt, *Development*, 142, 178.

44. JAK, notes to Turner, *Victory*, 98–99, K25.1/28. More on this connection appears in testimony in US Congress, House, *House Report #2*, 924–35, K25.1/32.

45. Hodes, *Divided*, 378.

46. JAK, notes to Taylor and Neu, *Network*, 53, K25.5/6.

47. *Congressional Globe*, 37th Cong., 3rd Sess., pt. 2, p. 1048 (1862).

48. JAK, notes (p. 1) to First Pacific Railroad Act, July 1, 1862, *Appendix to the Congressional Globe*, 37th Cong., 2nd Sess., pp. 381–84 (1862), K25.1/46.

49. Klein, *Union*, 14; JAK, notes to Riegel, *America*, 469–70, K25.1/47.

50. Originally the Leavenworth, Pawnee & Western, in 1863 this line became the Union Pacific, Eastern Division, a name chosen to attract investors. It became the Kansas Pacific in 1869, the name used hereafter.

51. Klein, *Union*, 65.

52. Klein, *Union*, 28–29.

53. Pacific Railroad Act of 1864, *Appendix to the Congressional Globe*, 38th Cong., 1st Sess., pp. 250–53 (1864), K25.2/7.

54. Borneman, *Iron*, 47.

55. Ward, *Thomson*, 191.

56. Borneman, *Iron*, 47.

57. Ward, *Thomson*, 161–62.

58. Storey, "Palmer," 146.

59. Kamm, "Scott," 35–40, 47, 101.

60. Eads to Fox, July 19, 1867, box 11, folder 10, GVF.

61. Ward, *Thomson*, 192–94. By 1866 Congress "had removed the stipulations" that had motivated this race in the first place. Thomson's personal secretary tallied the costs of "fixing" new legislation to boost a KP line to the Pacific.

62. JAK, notes to An Act authorizing the Atlantic & Pacific Railroad, *Appendix to the Congressional Globe*, 39th Cong., 1st Sess., pp. 406–9 (1866), K25.2/33.

63. Eads to Fox, June 8, 1867, box 11, folder 10, GVF.

Chapter 3. Conventional or Radical

1. Jackson, *Rails*, 41.

2. Gilbert and Billington, "River Politics," 107n8.

3. Details from JAK, notes on Boomer, K12.1/10–11.

4. Linville, quoted in Woodward, *History*, 16.

5. The first postwar proposal by Truman Homer is detailed below. A partnership of experienced engineers, Shaler Smith and Benjamin Latrobe, promoted a lift bridge just upstream of the city. General William Jackson Palmer (a railroad man before and after the war, linked to Edgar Thomson) argued for a tunnel under the river. For various reasons, these three proposals faded quickly, leaving Boomer and Eads as the serious contenders.

6. For the political power and taxing priorities of elites in another Midwest city of the era, see Einhorn, *Property*.

7. Woodward, *History*, 10.

8. Dated Feb. 11, 1865, the plans are "Perspective View of Proposed Bridge . . . by Th. Schrader," f11390, Department of Prints and Photographs, MHS, TRRA, B34/26.

9. Homer, *Reports*, 14–15, K16/7.

10. Ward, *Thomson*, 179.

11. The Prairie du Chien bridge authorized in the 1868 statute was not built for some time. Warren, *Bridging*, 87. A pontoon bridge alternative (used seasonally) existed by 1878.

12. Among the many studies on the Chicago–St. Louis rivalry, Wyatt Belcher (*Rivalry*) suggested that St. Louis merchants became complacent during the 1850s, committed to their steamboat and river trades, while their Chicago counterparts embraced railroads; Jeffrey Adler (*Yankee*) emphasized growing national discord over slavery during that decade in retarding investment in St. Louis's businesses and railroads; and William Cronon (*Chicago*) stressed Chicago's advantageous ties via railroads to New York City's markets and financiers.

13. To clarify the point, the qualities of the Mississippi River and the capabilities of contemporary civil engineering were foundational to Chicago's growing dominance as a rail hub for the West. For their part, railroad managers had little interest in boosting any particular city over another. As a contemporary writer described: "Skillful railroad men . . . do not work for the development of . . . this or that city, but for the development of the traffic of their road. They afford all the facilities possible for transportation to any place—St. Louis or Chicago, New York

or Nauvoo—when their road can do it profitably." See "The St. Louis, Kansas City & Northern Railway," *Railroad Gazette*, Jan. 27, 1872, 40.

14. The independents were at Dubuque, Burlington, Keokuk, Quincy, Hannibal, and St. Louis. The Chicago and Northwestern Railway built the Winona crossing. See Warren, *Bridging*, 180, for the capital-saving rationale for private-sector toll bridges on the western rivers.

15. A legal issue argued against direct ownership by the railroads. Their charters were granted by states, so the companies lacked clear legal authority to operate interstate crossings.

16. JAK, notes to the Omnibus Bridge Act, K19/12.

17. John K. Brown, "Heuristics, Specifications, and Routines in Building Long-Span Railway Bridges on the Western Rivers, 1865–1880," in Raff and Scranton, *Emergence*, 176. The only suspension bridge in railway service, John Roebling's Niagara Falls Bridge (1855–97), had an 825-foot center span.

18. Kouwenhoven, "Designing," 543.

19. Goldfeder, "Railroads"; Taylor and Neu, *Network*, map 2.

20. Its full name was the St. Louis, Alton & Terre Haute. Wallis, *At Bay*, 17.

21. Grant, *"Follow the Flag,"* 39–40.

22. Quotation from Eads to Fox, June 30, 1867, box 11, folder 10, GVF. For the takeover method, see "St. Louis and the North Missouri Railroad," *Railroad Gazette*, Nov. 11, 1871, 330.

23. PRR bond prospectus, May 31, 1852, enclosed in J. Edgar Thomson to Joseph Shipley, June 20, 1852, K25.1/12.

24. Wallis, *At Bay*, 15–21. With the link provided by the Terre Haute & Indianapolis, the promoters aimed to build a route from St. Louis to Cleveland, then onward via the Lake Shore line and the New York Central to New York City.

25. Boomer's partner was Andros B. Stone. For more on the Stone and Boomer families, see Stone, *Memoir*; and Taliaferro, *All*.

26. Wallis, *At Bay*, 21. Its full name, the St. Louis, Vandalia & Terre Haute, is shortened hereafter as the Vandalia.

27. Dorsey, *Road*, 101. In his semi-official history of the bridge, Calvin Woodward wrote that "after a very full and impartial examination of all the facts" he was convinced that Boomer was "entirely sincere" in his plans "to build a bridge at St. Louis." *History*, 18–19.

28. See National Park Service, "Amazing Bridge"; and Brown, "Not the Eads," 529–30.

29. Details from JAK, notes on Boomer, K12.1/10–11.

30. Woodward, *History*, 14. Cutter's original 1864 corporation was chartered as the St. Louis and Illinois Bridge Company, while Boomer's 1867 venture was the Illinois and St. Louis Bridge Company. To sidestep confusion, this account designates the companies by the promoters' names.

31. Woodward, *History*, 14. In its *Charles River Bridge* decision (1837), the US Supreme Court approved a state-granted monopoly to a private bridge

company if its terms and duration were clearly fixed by its charter. Hoven-kamp, *Enterprise*, 110–14.

32. Boomer to Eads, May 31, 1868, 2, TRRA, B35/13. The consultants were Adna Anderson and George Parker. Parker had just completed a long and difficult rail bridge over the Susquehanna River at Havre de Grace, Maryland.

33. ASCE, *Biographical*, 98–99.

34. Brown, "Not the Eads," 533. Boomer's plans were detailed in "Bridge over the Mississippi River at St. Louis," *Engineering* (London) 6 (July 24, 1868): 75, 82; and "Proposed Bridge across the Mississippi at St. Louis," *Engineering* (London) 6 (Aug. 7, 1868): 116–17.

35. Gilbert and Billington, "River Politics," 108n11.

36. [Boomer], *Board*.

37. Kouwenhoven, "Designing," 543.

38. Woodward, *History*, 15. John Kouwenhoven found an error in Woodward of significance to the Boomer-Eads rivalry. Eads first joined a meeting of the Cutter company on February 26 (after the new Illinois charter came through), not February 17. The timing is significant because most writers since Dorsey have cast Boomer as a late-arriving spoiler, determined to undermine the great Eads, bolster Chicago, and/or prevent construction of a rail bridge for St. Louis.

39. Woodward, *History*, 15.

40. R. L. Lamborn (Philadelphia) to William Jackson Palmer (St. Louis), Mar. 22, 1867 (confidential), WJP, K13/6.

41. JAK, notes to *Chicago Tribune*, July 23, 1867, K19/25.

42. As built, the side spans are 502 feet, and the center arch is 520 feet.

43. Eads to Linville, June 3, 1867, box 14, folder 2, TRRA, K23.1/19.

44. Linville (Philadelphia) to Eads, June 13, 1867, box 14, folder 2, TRRA, K23.1/20.

45. Linville, quoted in Woodward, *History*, 16.

46. Eads, "Report," 1868, 501–4; Brown, "Not the Eads," 536.

47. Eads, "Report," 1868, 507.

48. Around 1870, new iron bridges in railway service commonly had a margin of safety five times the anticipated design loading, according to Alfred Boller, "Papers on Bridge Construction, part 5," *Railroad Gazette*, Mar. 2, 1872, 91. As built, the St. Louis Bridge would have a safety margin of six times its design loads, far less than that of this hyperbolic structure crawling with people and covered with locomotives. Miller and Scott, *Eads*, 95.

49. Misa, *Nation*, 17.

50. Carnegie to Eads, Nov. 9, 1867, USS, K9/17.

51. Delony, "Golden."

52. That any firm's core function lies in aggregating knowledge was the key insight in Coase, "Nature." That the firm wields knowledge to create markets and influence their behavior is elaborated in Lazonick, *Myth*.

53. In the early 1870s the Phoenix Bridge Company commonly put up a 160-foot span in fewer than nine hours. Howland, "Bridges," 22.

54. Medley, *Tour*, 146.

55. Woodward, *History*, 16, 33.

56. [Boomer], *Board*, 15.

57. [Boomer], *Board*, 48.

58. [Boomer], *Board*, 78.

59. Zerah Colburn, "The Mississippi Bridge," *Engineering* (London) 6 (Sept. 25, 1868): 285–86. Like Eads, Colburn was a visionary engineer who had mastered the state of the art in many fields, while refusing to be shackled by conventional wisdom. See Mortimer, *Colburn*.

60. Kouwenhoven, "Designing," 559n68.

61. Miller and Scott, *Eads*, 90.

62. *St. Louis Republican*, quoted in Woodward, *History*, 22.

63. Eads to Col. John Knapp, May 31, 1883, MHS, K4/62.

64. Kouwenhoven, "Knew It," 182–84.

65. Henry Blow, quoted in Woodward, *History*, 19.

66. Woodward, *History*, 23.

67. "List of Stockholders," Dec. 18, 1867, TRRA, K19/28.

68. Carnegie to Eades [*sic*], Oct. 29, 1867, marked in Carnegie's hand "Confidential" (emphasis in original), USS, K9/16.

69. The Terre Haute & Indianapolis threw its lot in with the PRR in late December 1867, signaling defeat for Stone's coalition to control that carrier. Wallis, *At Bay*, 21.

70. Klein, *Union*, ch. 13.

71. Grant, *"Follow the Flag,"* 40

72. *Missouri Democrat*, Jan. 11, 1868, K19/32.

73. JAK, notes (p. 3) to Taussig, "Origin," K7/2; "Suggestions by Harrison and Eads," n.d., box 14, folder 2, TRRA, B35/11. After the merger, the combined corporation took the name of Boomer's venture: the Illinois and St. Louis Bridge Company. Eads and his associates probably chose to use the corporate name of his rival to ensure that its rights, granted by the Illinois charter, indisputably continued in the consolidated company.

74. Edwin D. Morgan to Eads, Mar. 13, 1868, Letter Copy Book, vol. 27 (reel 86), EDM.

75. Woodward, *History*, 30–31.

76. The bill's sponsor was Representative William Pile of St. Louis. *Congressional Globe*, 40th Cong., 2nd Sess., p. 2974 (1868). Also see Woodward, *History*, 30.

77. Roebling, *Long*, 39. If a new company succeeded in financing and building Roebling's design for a St. Louis bridge, its legality under the Omnibus Bridge Act likely would require resolution by the federal courts or Congress.

78. The leading historian of American bridges, Henry Petroski, makes this point. Petroski, *Dreams*, 327.

79. The accident happened near the beginning of that demanding project, while John Roebling and his engineers were fixing the locations of the two great

towers. Standing on a ferry dock, he placed a foot wrong just as the ferry came into its slip. Shoved by the boat's hard landing, a timber balk in the dock crushed his boot and the toes of his right foot. A less headstrong man would likely have recovered quickly with conventional medicine. Insisting on his own treatment, Roebling died on July 22, 1869, in terrible agony after two weeks of seizures, lockjaw, and coma. McCullough, *Great*, 90–93.

Chapter 4. The Art of a Promoter

1. Eads, "Report," 1868, 537.
2. Barton Bates to Edward Bates, Oct. 10, 1861, BB. The investors—Albert Pierce, Oliver B. Filly, and Barton (wrongly named Martin) Bates—are given in US Congress, House, *House Report #2*, 954, K25.1/32.
3. Gould, *Fifty*, 486 (emphasis in original).
4. Kane, *Banking*, 91–93.
5. Bensel, *Economy*, ch. 2.
6. Unger, *Greenback*.
7. Unger, *Greenback*. A monetary policy based on the gold standard assured lenders and investors that repayment would not be eroded by inflation over the term of their loans. Indeed, they might hope for additional return as loan payments—made in gold—increased in value relative to paper currency or other instruments.
8. Woodward, *History*, 15.
9. White, "Corruption," 22.
10. Eads, "Report," 1868, 537.
11. Woodward, *History*, 30.
12. Taussig to W. Milnor Roberts, Oct. 6, 1868, USS, K7/19.
13. JAK table, "Stockholders of Illinois & St. Louis Bridge Co.," K19/1. Share ownership evolved over time; these men had all subscribed to the Cutter/Eads company as of Dec. 18, 1867.
14. Eads, "Report," 1868, 535–37.
15. ASCE, *Biographical*, 102.
16. W. Milnor Roberts, "Autobiography written on *Scotia*," 25, box 1, folder 4, WMR.
17. Months later, Roberts wrote to his friend John Roebling that Eads's plans "were practicable and . . . his estimate was fair. In round numbers, I put the cost at $5,000,000." Roberts to Roebling, Oct. 1, 1868, K23.2/4.
18. Roberts to Roebling, Oct. 1, 1868, K23.2/4.
19. Roberts, "Autobiography written on *Scotia*," 25.
20. Kouwenhoven, "Chronology"; Woodward, *History*, 55. The timing suggests that Eads embarked on the SS *Cuba* given the entry for the *Cuba* in Hubbard, *North*, 57.
21. Fox to Eads, July 10, 1868, box 4, folder 7, GVF, K5/34.
22. Woodward, *History*, 56.

23. Henry Flad, quoted in Woodward, *History*, 55.

24. Taussig to Thomson, Sept. 17, 1868, WT, K7/13.

25. Taussig to Roberts, Sept. 18, 1868, WT, K7/14.

26. Taussig to Roberts, Sept. 30, 1868, WT, K7/18.

27. Benson and Rossman, "Scott"; Ward, *Thomson*, ch. 9; Wall, *Carnegie*, ch. 8; Nasaw, *Carnegie*, chs. 4–6.

28. JAK, notes (p. 17) to *Railroad Gazette*, Aug. 12, 1871, quoting George Alfred Townsend, K8/3.

29. Quoted in Miner, *St. Louis–San Francisco*, 87.

30. Benson and Rossman, "Scott," 2.

31. Barton Bates (president of the North Missouri and Eads's associate) floated that idea in Bates to Eads (New York City), Dec. 23, 1868, JBE, K5/39.

32. *New York Times*, Dec. 15, 1868, 10.

33. JAK, notes (p. 2) to board meeting minutes for Dec. 19, 1868, and Jan. 12, 1869, TRRA, K17/6.

34. William M. McPherson to Eads, Jan. 12, 1869 (emphasis in original), JBE, K5/41.

35. McPherson to Eads, Jan. 12, 1869, JBE, K5/41.

36. Woodward credits McPherson and Taussig as the plan's authors (*History*, 56). Inferences suggest to me that Eads and McPherson were its protagonists.

37. Prospectus, Feb. 1869, box 14, folder 2, TRRA, B35/10.

38. The bankers would take their customary markup when reselling to investors. Eads and McPherson drafted this plan after consulting with financiers, so the notion that the bonds could yield 90 percent to the bridge company was more than simply an assumption.

39. John Franch details the frothy mix of municipal franchises, politicians, and financiers that created street railway companies in this period. Franch, *Yerkes*, 85–87, 121, 129, 213.

40. Stockholders Minute Book (1868–73), 6:10–12, TRRA, B35/14.

41. Although the stock offering was fully subscribed, execution varied slightly from the plan. See JAK, list of subscribers and amounts (which for some reason fall just shy of $3 million, at $2,980,000), K19/1.

42. Carosso, *Morgans*, and Strouse, *Morgan*, have little on the Morgans' private-account transactions, saying nothing about their equity investments in St. Louis Bridge.

43. In her insightful history of one such bank, Dolores Greenberg notes that "the 'legitimate' image that financiers deliberately conveyed [as capital intermediaries between investors and corporations] has been too readily perpetuated" by business historians. *Financiers*, 47–54. Notwithstanding her insight, Greenberg also overlooks this Eads transaction, even though Morton, Bliss, took an equity stake of $50,000, and Levi Morton subscribed for $15,000 in bridge company stock.

44. I wanted very much to understand how these experienced investors viewed Eads and his deal. The few sources available said nearly nothing. The most

generous archive, that of Edwin Denison Morgan (EDM), has thousands of letters. A thorough review showed Morgan's high regard for Eads, but the letters had nothing on this transaction.

45. Prospectus, Feb. 1869, 2, box 14, folder 2, TRRA, B35/10.

46. The Ohio & Mississippi, the St. Louis & Alton, and the Chicago, Alton & St. Louis.

47. In his revenue projections of June 1868, Eads had assumed that all the ferry traffic over the river would switch over to the bridge after its completion in 1871. Eads, "Report," 1868, 531–38.

48. The *St. Louis Republican* of Feb. 21, 1869, greeted McPherson's announcement with the headline "The Bridge Problem Solved." Clipping in box 9, folder 11, WMR.

49. JAK, notes to *East St. Louis Gazette*, Feb 20, 1869, K19/44. This announcement was a general invitation to the party that followed a day later.

50. Carosso, *Morgans*, 45. The rule on speculations appeared in the partnership articles governing George Peabody & Company, the London bank in which Junius served as junior partner (1854–64).

51. According to an economist of the period, the term *speculation* described short-term plays to harvest a profit from fluctuations in the value of an asset. Hadley, *Transportation*, 48.

52. J. S. Morgan to J. P. Morgan, Mar. 25, 1858, quoted in Carosso, *Morgans*, 85.

53. Carosso, *Morgans*, 80.

54. Quoted in Carosso, *Morgans*, 109.

55. Carnegie was credited with eight hundred shares in an 1870 ledger, although his holdings fluctuated over time. See "List of Stockholders," USS, K19/43. Also see JAK table, "Stockholders of Illinois & St. Louis Bridge Co.," K19/1.

56. C. W. Quin, "France—Mining and Metallurgical Products," *Laboratory: A Weekly Record of Scientific Research* 1 (Aug. 3, 1867): 313–14.

57. Woodward, *History*, 57–58.

58. Woodward, *History*, 60.

Chapter 5. To Bedrock

1. Eads, "Report," 1868, 500.

2. Woodward, *History*, 59.

3. Eads, "Report," 1868, 496–97.

4. Woodward, *History*, 242–45, gives a detailed survey of these and other innovators. Cochran's career inspired the exploits of the fictional Jack Aubrey in the novels of Patrick O'Brian.

5. M. Malézieux, "Fondations à l'air comprimé," in *Annales*, 7:336–37.

6. Woodward, *History*, 244.

7. Woodward, *History*, 59–60.

8. For Moreaux's career, see Wikipedia, s.v. "Moreaux, Félix," accessed Apr. 30, 2020, https://fr.wikipedia.org/wiki/F%C3%A9lix_Moreaux.

9. "Extracts from Annual Report of Col. Simpson," *Engineering News*, Feb. 28, 1878, 67.

10. Chanute, "Pneumatic," K20/48–50.

11. Ignoring Chanute's articles, Woodward presented Eads's trip to Europe and discovery of French methods as a fortunate "accident." *History*, 59. In June 1868, after Eads was stricken by the illness that would send him to recuperate in Europe, he had tried to hire Chanute to replace him as chief engineer. After Chanute declined, Eads had turned to Milnor Roberts. Miller and Scott, *Eads*, 97.

12. Chanute, "Pneumatic," 17–18, K20/49.

13. Woodward, *History*, 379.

14. Eads, "Report," 1869, 540.

15. W. Milnor Roberts, "Ms. account of constructing the St. Louis Bridge," 5, box 6, folder 7, WMR.

16. Woodward, *History*, 60.

17. Taussig to Roberts, Oct. 6, 1868, Letterbook, WT, K7/19. The Andrews contract was to build the masonry towers and abutments. The bridge company, however, contracted to provide the stone to Andrews. That fact gave Eads the ability to maintain quality, done by extensive testing. It also meant that Eads picked the vendors, and he could skim a markup off the stone. Eads claimed in his 1869 report (p. 548) that Andrews contracted for the masonry, but another source suggests otherwise. See James Laurie, Report to Stockholders, Apr. 10, 1872, 3–4, TRRA, K23.2/12.

18. Eads, "Report," 1869, 548–49.

19. Progress report, Flad to Charles K. Dickson, forwarded to JSM Co., July 10, 1870, HC/03/001/001(006)/box 1, JSM.

20. Granite contracts, box 1, folder 4, TRRA, B34/27 and B34/29; Clough, *Head*, 31.

21. Eads, "Report," 1869, 551.

22. Eads, "Report," 1869, 551.

23. Woodward, *History*, 211.

24. Woodward, *History*, 239–42.

25. "Local Weather History," WLFI.com, accessed May 8, 2020, https://www.wlfi.com/content/news/Local-Weather-History-Two-Weekly-Top-10-Lists-498051421.html.

26. "The St. Louis Bridge," *New York Times*, Oct. 30, 1869, 8.

27. Eads to Fox, Nov. 15, 1869, JBE, K4/37.

28. Woodward, *History*, 205, 210.

29. Woodward, *History*, 208–9, 213, 215.

30. JAK, notes to Reavis, "Bridge," 11, K20/43.

31. Woodward, *History*, 216–17. For a detailed report by a trained engineer who visited the caisson, see Louis Nickerson, "The Illinois and St. Louis Bridge," *Railroad Gazette*, Nov. 4, 1871, 316–18, B35/27.

32. Jaminet, *Effects*, 39–40.

33. Roberts, "Ms. account of constructing the St. Louis Bridge," 23.
34. Woodward, *History*, 213–14, 218.
35. Jaminet, *Effects*, 8, 32–39.
36. JAK, notes to Reavis, "Bridge," K20/43. Also see Roberts, "Ms. account of constructing the St. Louis Bridge," 16.
37. Julius Walsh, quoted in undated clipping (ca. June 1921) from *St. Louis Globe Democrat*, K12.2/40.
38. Eads to Fox, Dec. 22, 1869, JBE, K4/39.
39. Woodward, *History*, 216.
40. Woodward, *History*, 247.
41. JAK, notes to *Missouri Democrat*, Apr. 1, 1870, K20/52.
42. Eads, "Report," 1870, 561.
43. Woodward, *History*, 62, 218, 248.
44. Woodward, *History*, 219, 249.
45. "Wages."
46. Woodward, *History*, 220.
47. "Death from Compressed Air," *Missouri Democrat*, Mar. 25, 1870, K20/53.
48. Woodward, *History*, 253.
49. Woodward, *History*, 221.
50. Roberts, "Autobiography written on *Scotia*," 28.
51. Woodward, *History*, ch. 19.
52. Woodward, *History*, ch. 20.
53. Woodward, *History*, 231–34.
54. Jaminet, *Effects*, 123.
55. Wikipedia, s.v. "Decompression Practice," accessed May 15, 2020, https://en.wikipedia.org/wiki/Decompression_practice#Continuous_decompression.
56. Jaminet, *Effects*, 131.
57. Philips, *Bends*, ch. 6. Also see Butler, "Caisson."
58. Woodward, *History*, 253–55.
59. Alfred H. Smith, quoted in McCullough, *Great*, 306.
60. Woodward, *History*, 232, 239–42.
61. Eads, quoted in Woodward, *History*, 64. A single pontoon boat served the east abutment caisson, reflecting its location at the river's edge.
62. J. S. Walsh to Josephine Walsh, ca. Mar. 30, 1871, JSW, K5/58.
63. "The St. Louis Bridge," *Railroad Gazette*, July 15, 1871, 177–78, K20/39.
64. Woodward, *History*, 64–65.

Chapter 6. London and Real Money

1. JAK, transcription of McPherson to Carnegie (Philadelphia), Dec. 2, 1869, USS, K9/72.
2. Morgan's work brokering the bonds of St. Louis Bridge receives scant consideration in histories of the bank, and most of those brief accounts draw from Andrew Carnegie's *Autobiography*, which presents self-serving and

erroneous descriptions of Eads, Morgan, and the bond deal. See Carosso, *Morgans*, 242–43; Morris, *Tycoons*, 93; and Strouse, *Morgan*, 138–39; of the three, only Carosso sidestepped the errors introduced by Carnegie's book.

3. Carnegie, *Autobiography*, 155.
4. Carnegie, "St. Louis Bridge Memoranda," Dec. 1869, USS, K9/74.
5. Keystone Bridge Company, "Statement of Stock Ledger," Aug 14, 1871, USS, K10/118.
6. Undated document, ca. Dec. 15, 1869, in Carnegie's hand, USS, K9/74.
7. Gow, "Dynasty."
8. Eads, "Report," 1868, 529–30.
9. Eads, "Memorandum of Agreement," Feb. 3, 1870, USS, K9/88.
10. Kouwenhoven, "Designing," 557. Carnegie's willingness to accept those dated estimates is clearly stated in his "St. Louis Bridge Memoranda," Dec. 1869, USS, K9/74.
11. Carnegie, "Memorandum of an Agreement," n.d. [ca. Feb. 2, 1870], USS, K9/87.
12. Stockholders Minute Book (1868–73), 6:48, minutes of Feb. 5, 1870, TRRA, B35/15.
13. By April 6, 1870, Keystone had signed its own contract with St. Louis Bridge to supply the superstructure, a contract that also made Keystone the prime contractor to supply steel. David McCandless to Linville, Apr. 6, 1870, USS, K10/24.
14. Linville (Philadelphia) to Eads, June 13, 1867, box 14, folder 3, TRRA, K23.1/20.
15. White, "Corruption," 20.
16. The parties *not* bound by this deal included the PRR, Keystone Bridge, a steel supplier, and a banker.
17. Carnegie to McPherson, Feb. 16, 1870, USS, K9/90.
18. Carnegie to Eads, Feb. 25, 1870, USS, K9/91.
19. Carnegie, *Autobiography*, 154–57.
20. DM Co. to JSM Co., Mar. 7, 1870, HC/03/001/001(006)/box 1, JSM.
21. Thomson to "Gentlemen" (JSM Co.), Mar. 4, 1870, HC/03/001/001(006)/box 1, JSM.
22. Woodward, *History*, 249.
23. Telegrams, Thomson to Carnegie, Mar. 27, 1870, and Carnegie to Thomson, Mar. 28, 1870, USS, K10/16.
24. See Geo. McCandless to T. M. Carnegie, Mar. 25, 1870, USS, K10/15.
25. JSM Co. to DM Co., Mar. 24, 1870 (emphasis in original), in "Private Copy Out-Letter Book, Nov. 1867–Mar. 1875," 366–68, MS21795/001, JSM.
26. JSM Co. to DM Co., Mar. 26, 1870, in "Private Copy Out-Letter Book, Nov. 1867–Mar. 1875," 366–68, MS21795/001, JSM.
27. Atlantic cable message, McPherson to Carnegie (London), Mar. 28, 1870, HC/03/001/001(006)/box 1, JSM.
28. JSM Co. to Carnegie & Associates, Mar. 31, 1870, USS, K10/19.

29. No evidence supports Nasaw's claim (*Carnegie*, 115) that Carnegie "had priced the issue" at any specific amount, let alone at 85. Jackson (*Rails*), Krass (*Carnegie*), and Nasaw repeat some picturesque hokum from Carnegie's *Autobiography* that he saved three weeks of time (and perhaps the deal itself) by suggesting to Junius Morgan that they use the Atlantic cable instead of the mails to secure approvals from St. Louis (*Autobiography*, 155–56). The Morgans *always* used telegrams for time-sensitive matters.

30. JAK, notes to Carnegie to Thomson, Apr. 2, 1870, USS, K10/21.

31. JSM Co. to DM Co., Apr. 2, 1870, in "Private Copy Out-Letter Book, Nov 1867–Mar. 1875," 372–74, MS21795/001, JSM.

32. JAK, notes to Carnegie to Thomson, Apr. 2, 1870, USS, K10/21.

33. JAK, notes to Carnegie to Thomson, Apr. 6, 1870, USS, K10/25. A half century later, Carnegie repeated these lies in his *Autobiography* (156).

34. The Smith option is in James Smith to JSM Co., Mar. 31, 1870, HC/03/001/001(006)/box 1, JSM.

35. Jameson, Smith & Cotting to JSM Co., May 25, 1870, HC/03/001/001(006)/box 1, JSM.

36. Taussig to Louis B. Fisher, Apr. 16, 1870, trans. Ernst Stadler, in Letterbook, WT, K7/42.

37. "Bankers' Price Current," *Economist* (London), Apr. 15, 1871, 452, and Apr. 27, 1872, 525.

38. Daunton, *Wealth*, ch. 7. From 1865 to 1914, American ventures accounted for 21 percent of all British overseas investment (US railroads and public utilities took two-thirds of that total). The Dominion of Canada ranked second (10 percent).

39. "List of Holders," Feb. 20, 1882, TRRA, K19/72.

40. The bonds of many western railway ventures in the period paid 7 percent, a comparatively high rate that reflected the general thirst for investment capital at that time and place and "high risks." Grodinsky, *Strategy*, 10–12.

41. Stockholders Minute Book (1868–73), 6:47–55, TRRA, B35/15.

42. Nasaw, *Carnegie*, 106–8.

43. Besides the returns to insiders and equity investors discussed in the text, Carnegie aimed to put the trio into plans to finance the station and tunnel. For all eleven gambits, see Brown, "Eleven Avenues to Gains for Andrew Carnegie & Associates (ACA) Via St. Louis Bridge," B40/10.

44. Nasaw, *Carnegie*, 110.

45. Hendrick, *Life*, 1:146; Wall, *Carnegie*, 222–23; Nasaw, *Carnegie*, 113–14.

46. JAK, notes to Carnegie to J. N. McCullough, Jan. 11, 1870, USS, K9/83.

47. Carnegie to Scott, Jan. 24, 1870, quoted in Nasaw, *Carnegie*, 112.

48. JAK, notes to *Railroad Gazette*, May 21, 1870, 176, quoting *St. Louis Journal of Commerce* (emphasis added), K25.2/62.

49. *Description and Plans*, 4, 7, 10.

50. Taussig to Carnegie, Oct. 6, 1870, typescript, box 51, folder 2, AC, B40/3.

51. JAK, notes (pp. 12–13)1 to Taussig, "Development," K 7/1.

52. "An Act Authorizing the Formation of Union Depots and Stations for Railroads in Cities of this State," Mar. 21, 1871, box 5, folder 19, TRRA, B38/2; Taussig to Carnegie, Mar. 18, 1871, box 51, folder 3, AC, B40/4.

53. Scott signed for the Pennsylvania Company and for the Pittsburgh, Cincinnati & St. Louis. Other subscribers included the Ohio & Mississippi (allied to the Baltimore & Ohio), the Alton, and the Wabash.

54. JAK, notes (pp. 13–14) to Taussig, "Development," K7.1.

55. Carnegie to Taussig, Mar. 22, 1871, USS, K10/98.

56. Taussig to Carnegie, Mar. 4, 1871, USS, K10/91.

57. Glendenning, *Chicago & Alton*, 67. One new car ferry forwarder was the East St. Louis Transfer Company; Wiggins was the other.

58. JAK, notes to Carnegie to Thomson, Apr. 2, 1870, USS, K10/21.

59. JAK, notes to Carnegie to James Britton, Sept. 21, 1870, USS, K10/46.

Chapter 7. Troubles with Steel

1. Carnegie to McPherson, Oct. 5, 1870, USS, K10/54.

2. Carnegie to Eads, Oct. 18, 1870, USS, K10/63.

3. Carnegie to Taussig, June 5, 1871 (emphasis in original), USS, K10/109.

4. JAK, design chronology, 11, K23.1/5.

5. Carnegie to Eads, Oct. 18, 1870, USS, K10/63. On steel and Butcher, see Kanigel, *Taylor*, 155–60.

6. Kanigel, *Taylor*, 156.

7. Miller and Scott, *Eads*, 94.

8. Ward, *Thomson*, 176.

9. Eads, "Report," 1868, 517.

10. Miller and Scott, *Eads*, 95.

11. Eads, "Report," 1868, 521.

12. Eads, "Report," 1868, 528.

13. Miller and Scott, *Eads*, 95.

14. Haupt, *Theory*, 62.

15. Eads, "Report," 1868, 529; Woodward, *History*, 379.

16. The engineers computed the maximum variation in the height of the crown for the center arch between winter and summer at 8.7 inches. The extensive bracing of the chords transmits temperature-induced stresses throughout each arch. Woodward, *History*, 347.

17. Woodward's semi-official account devotes seventeen pages to explaining the mathematics. *History*, 331–47; also see John K. Brown, "Heuristics, Specifications, and Routines in Building Long-Span Railway Bridges on the Western Rivers, 1865–1880," in Raff and Scranton, *Emergence*, 192–97. The calculating burden for iron truss bridges was far lower, and their makers had already mastered the issues involved.

18. Woodward, *History*, 293–95.

19. Woodward, *History*, 68.

20. Miller and Scott, *Eads*, 111–12.

21. Miller and Scott, *Eads*, 120.

22. Woodward, *History*, 379.

23. Woodward, *History*, 90–91. The anchor rods varied in length, depending on their placement.

24. Eads, "Report," 1871, 591.

25. Miller and Scott, *Eads*, 115–16; Eads, "Report," 1871, 592–93.

26. Eads, "Report," 1871, 593.

27. Vitiello, *Sellers*, ch. 5; Brown, "Machines." As of February 1870, Sellers controlled negotiations, contracts, and pricing at the steelworks. Carnegie to Eads, Feb. 25, 1870, USS, K9/91. Sellers had peerless credentials in engineering, so his roles at the Butcher works meant that Keystone and St. Louis Bridge were not simply at the mercy of an alchemist.

28. Sinclair, "Thread."

29. Cox and Foley, "Pioneering," 36.

30. Miller and Scott, *Eads*, 117.

31. Woodward, *History*, plate 29.

32. Woodward, *History*, 89. To make rails, I-beams, or the Captain's staves, a white-hot billet of steel passed through a steam-powered rolling mill with dies (in pairs, top and bottom) to form the desired cross section. The billet needed many passes through the rolling stand, reheating, and a succession of dies to acquire its final shape.

33. Miller and Scott, *Eads*, 116.

34. Woodward, *History*, 158.

35. Bridge, *Inside*, 31.

36. Miller and Scott, *Eads*, 118; Woodward, *History*, 101.

37. Eads, quoted in Woodward, *History*, 103.

38. Ferguson, "American-ness," 12–13.

39. JAK, transcription of Carnegie to Taussig (confidential), Dec. 30, 1870, USS, K10/73.

40. JAK, transcription of Carnegie to Taussig (confidential), Dec. 30, 1870 (emphasis in original), USS, K10/73.

41. Carnegie to Thomson, Nov. 27, 1871, USS, K11/13.

42. Carnegie to Taussig, Dec. 8, 1871, USS, K11/14.

43. Jackson, *Rails*, 128.

44. American Society of Civil Engineers. "James Laurie."

45. James Laurie to McPherson, Feb. 28, 1872, TRRA, K23.2/10.

46. Eads to President & Directors, St. Louis Bridge, Mar. 18, 1872, TRRA, K23.2/11.

47. Laurie to McPherson, Apr. 10, 1872, TRRA, K23.2/12. Also see Woodward, *History*, 113.

48. Eads to President & Directors, St. Louis Bridge, June 1, 1872, TRRA, K23.2/14.

49. Keystone had signed its contract with Butcher on October 24, 1870. Reliable deliveries of acceptable steel began in March 1873.
50. Glaab, *Kansas*, ch. 7.
51. Chanute and Morison, *Bridge*.
52. The Chicago, Burlington & Quincy will be called simply the Burlington hereafter.
53. Miner, *St. Louis–San Francisco*, 87. By 1870, Scott and Thomson had invested in four lines with transcontinental aspirations. Churella, *Pennsylvania*, 406.
54. "The St. Charles Bridge," *Railroad Gazette*, June 3, 1871, 114.
55. Grant, *"Follow the Flag,"* 40, 47.
56. "North Missouri," *Railroad Gazette*, Aug. 5, 1871, 217; "The End of the North Missouri," *Railroad Gazette*, Jan. 20, 1872, 27; "St. Louis and the North Missouri Railroad," *Railroad Gazette*, Nov. 11, 1871, 330.
57. "The End of the North Missouri," *Railroad Gazette*, Jan. 20, 1872, 27. The North Missouri became the St. Louis, Kansas City & Northern after reorganization; this account will continue to use its original name.
58. JAK, notes (p. 16) to *Railroad Gazette*, Sept. 18, 1875, K25.3/14.
59. Taussig to Carnegie, Nov. 6, 1871, USS, K11/7. The Great Chicago Fire had given that city a chance to consolidate its passenger terminals, an opportunity that went unrealized.
60. "Cincinnati & Newport Bridge," *Railroad Gazette*, Mar. 30, 1872, 141–42.
61. Carnegie to Taussig, Jan. 24, 1871, USS, K10/83.
62. Taussig to Carnegie, Mar. 29, 1871, USS, K10/100.
63. Taussig to Carnegie (confidential), June 20, 1871, USS, K10/111.
64. Carnegie to Scott, June 7, 1871 (emphasis in original), USS, K10/110.
65. JAK, transcription of Taussig to Carnegie, Sept. 6, 1871, WT, K7/70.
66. Taussig to Carnegie, Dec. 8, 1871, WT, K7/77.
67. Carnegie to Taussig, Dec. 8, 1871, USS, K11/14.
68. Eads, "Report," 1868, 531.
69. Carnegie wrote to Taussig on March 22, 1871: "The tunnel is an inseparable part of the Depot Scheme + should be so treated. Total cost will be as we thought 2 and ¼ millions. This as % isn't a very heavy tax for Passenger accommodation + Freight line through the City." USS, K10/98.
70. Taussig to Solon Humphreys, May 25, 1871 (emphasis in original), WT, K7/63.
71. Taussig to Eads, May 18, 1871, WT, K7/62.
72. JAK, notes (p. 13) to Stockholders and Directors Minutes, July 9, 1868–Mar. 12, 1873, TRRA, K17/6.
73. Eads to JSM, Aug. 10, 1872, HC/03/001/001(006)/box 2, JSM. In eight typeset pages, Eads spun tales of progress (real and fictive), need, and profit. Morgan circulated Eads's letter among other bankers and potential investors.
74. Memorandum of agreement between Eads and JSM Co., Oct. 10, 1872, HC/03/001/001(006)/box 2, JSM.
75. JAK, notes to *East St. Louis Gazette*, June 10, 1871, K21/63.

76. Taussig to Senator Carl Schurz, Feb. 19, 1873, WT, K7/117.

77. According to Taussig, the amendments forced into the bill called for two 500-foot clear spans carrying a double-track railroad at 100 vertical feet above the river. Taussig to JSM Co., Apr. 5, 1873, HC/03/001/001(006)/box 1, JSM.

Chapter 8. Arches over the River

1. In a typical case, in 1871 a team from Detroit Bridge and Iron assembled a 300-foot span on the Missouri River in thirty-six working hours. Detroit Bridge and Iron Works, *Memoir*, 45.

2. Woodward, *History*, ch. 24.

3. Woodward, *History*, 66, 143, 172.

4. JAK, notes to *East St. Louis Gazette*, Sept. 7, 1872, K24/6.

5. JAK, notes to Taussig, "Origin," 4, K7/2.

6. Taussig, "My Father's," 181.

7. Allen, "President's Report," May 7, 1873, 10, TRRA, K17/7.

8. Eads to President and Directors [of SLB], Mar. 31, 1873, 3, 7, TRRA, K21/37.

9. Agreement, Midvale Steel and Keystone Bridge, Mar. 5, 1873, USS, K22/20.

10. E. E. Thrum, "Alloy Bridge Steel," *Iron Age* 124, no. 828 (1929): 734.

11. Miller and Scott, *Eads*, 120.

12. Woodward, *History*, 161.

13. Gerard Allen to Carnegie, June 3, 1873, USS, K11/40.

14. Woodward, *History*, 166.

15. Woodward, *History*, 164–67.

16. Eads (Pittsburgh) to Carnegie, Aug. 11, 1873, USS, K11/42.

17. Woodward, *History*, 166–67.

18. Trask, "History," 26.

19. JAK notes (pp. 21–22) to Minutes of Executive Committee, Sept. 6, 1873, vol. 1, TRRA, K17/10.

20. Woodward, *History*, 176–77, 187. The loan terms are in JSM Co. to Eads, Sept. 9 and Sept. 13, 1873, HC/03/001/001(006)/box 2, JSM.

21. Woodward, *History*, 163.

22. Woodward, *History*, 162.

23. Woodward, *History*, 173.

24. Keystone claimed a substantial design role in these arrangements. Keystone Bridge Company, *Catalogue*, 16.

25. Theodore Cooper, quoted in Woodward, *History*, 171.

26. Woodward, *History*, 174.

27. Woodward, *History*, 172.

28. Woodward, *History*, 188. Thanks to David Aynardi for this point.

29. Woodward, *History*, 168.

30. Woodward, *History*, 175.

31. Woodward, *History*, 175.

32. "The St. Louis Bridge," *New York Times*, Sept. 20, 1873, 8.

33. White, *Railroaded*, 81–84.

34. Unger, *Greenback*, 220–21.

35. The letter that ignited this incident survives in the National Archives. See JAK, notes to John S. McCune (president, Keokuk Company) to Major Charles Suter (St. Louis), July 19, 1873, in "Letters Received," no. 1631, ACE, K24/23. By August, captains of the Wiggins and Keokuk lines were circulating petitions among steamer owners "requesting the Secretary of War to arrest the construction of the Bridge." Many recipients "indignantly refused," wanting to see the bridge completed. Taussig, "Report," 33, B36/6.

36. Gilbert and Billington, "River Politics," 94.

37. Many sources detail the enmity between Eads and the Corps, including Barry, *Rising*, 62–65; Gilbert and Billington, "River Politics"; Reuss, "Humphreys"; and Shallatt, *Structures*, 195–99.

38. U.S. Congress, House, "Reports on the Construction," 2.

39. Reuss, "Humphreys," 11–12.

40. Taussig, "Report," 36, B36/6.

41. In its final form, this Cincinnati rail, wagon, and foot bridge stood 100 feet above the low watermark. "Cincinnati & Newport Bridge," *Railroad Gazette*, Mar. 30, 1872, 141–42. For the cost of the Corps-mandated alterations, see "Newport & Cincinnati Bridge," *Railroad Gazette*, June 6, 1874, 217.

42. "The Great Bridge," *St. Louis Democrat*, Oct. 19, 1873, details the Corps report. See clippings in box 14, folder 1, TRRA, B36/1.

43. Gilbert and Billington, "River Politics," 98–99.

44. Shallatt, *Structures*, 196.

45. Henry Flad, quoted in "The Great Bridge," *St. Louis Democrat*, Oct. 19, 1873, B36/1.

46. Taussig, "Report," 37, B36/6.

47. With its direct quotations, this account draws from JAK notes to Taussig, "Personal," 10–13, K7/10.

48. Taussig, "Personal," 12–13, WT, K7/10.

49. A professional photographer, Robert Benecke, thoroughly documented each stage in the construction of the St. Louis Bridge. As he later recounted, "The bridge was built with English money, and the photographs were to show the syndicate the progress that was being made." See "Reminiscences of Robert Benecke, Oldest St. Louis Photographer," *St. Louis Republic*, Sept. 27, 1903, 2.

50. Woodward, *History*, 184–85.

51. JAK, notes to Cooper, "Notes," 248, K24/34.

52. "The St. Louis Bridge," *New York Times*, Dec. 19, 1873, 4.

53. The bank priced the issue in pounds sterling at 80 percent of par and sold out. JAK, notes (p. 1) to *Railroad Gazette*, Jan. 10, 1874, K25.3/14.

54. JAK, notes (pp. 24–26) to Minutes of the Executive Committee, vol. 2, Oct.–Dec. 1872, K17/10, TRRA.

55. Woodward, *History*, 188.

56. John Piper to Carnegie, Mar. 4, 1874, USS, K11/55.

57. JAK, notes (p. 5) to *Railroad Gazette*, Apr. 25, 1874, K25.3/14.

58. Carnegie to Morgan, Feb. 9, 1874 (emphasis in original), USS, K11/52. Carnegie concluded this side arrangement with St. Louis Bridge on Feb. 7, 1874. Woodward, *History*, 174. Given the forward state of the work on that date, it is unclear why the company believed itself compelled to strike this deal.

59. "The Bridge," *St. Louis Dispatch*, Apr. 20, 1874, 4.

60. Woodward, *History*, 195–96.

61. Linville to Carnegie, Oct. 24, 1873, USS, K11/43.

62. Piper to Linville, Jan. 28, 1874, USS, K11/46.

63. Woodward *History*, 195.

64. Charles Edward Tracy, arbitration statement, June 23, 1874, USS, K11/65.

65. J. W. Vandevort to Carnegie, Dec. 16, 1874, USS, K11/66.

66. Total from bridge company's October 1873 estimates of assets and liabilities, box 7, folder 56, TRRA, B35/9.

67. Explanation of Capital Stock and Bonus (statement by St. Louis Bridge), July 9, 1878, HC/03/001/001(006)/box 6, JSM. These gifts of bonus stock came from the $1 million block of its own equities that St. Louis Bridge had reserved in the original Eads-McPherson prospectus of February 1869.

68. Taussig (confidential) to Carnegie, Feb. 21, 1874, USS, K11/53. "Penn. Interests" referred to Carnegie, Scott, and Thomson.

69. JAK, notes (p. 23) to Minutes of the Executive Committee, vol. 2, Oct. 1873–Apr. 1876, TRRA, K17/10.

70. Woodward, *History*, 196.

71. Taussig to JSM Co., July 8, 1874, HC/03/001/001(006)/box 2, JSM.

72. "The St. Louis Tunnel," *New York Times*, Oct. 14, 1873, 1.

73. Eads to JSM, July 12, 1874 (emphasis in original), HC/03/001/001(006)/box 2, JSM.

Chapter 9. Foreclosure and a Pool

1. Kouwenhoven, "Bridge," 13.

2. JAK, notes to *East St. Louis Gazette*, Aug. 22, 1874, K26/3. The first regular westbound freight, nine cars of building stone, crossed on July 16. "Illinois and St. Louis Bridge," *Railroad Gazette*, July 18, 1874, 278.

3. Theising, *Made*, 101–5.

4. "Traffic on Bridge and in Tunnel," trans. Ernst Stadler, *Westliche Post* (St. Louis), Sept. 7, 1874, K21/81.

5. "The St. Louis Bridge," *New York Times*, May 17, 1873, 4.

6. Taussig to JSM Co., July 8, 1874, HC/03/001/001(006)/box 2, JSM.

7. JAK, notes (pp. 1–2) to Minutes of Stockholders and Directors Meetings, vol. 3, May 1875–Mar. 1879, TRRA, K17/8.

8. That meeting included representatives from the Wabash, the Alton, two lines affiliated with the PRR (the Vandalia and the Indianapolis & St. Louis), and the

Pennsylvania Company itself. JAK, notes (p. 11) to *Railroad Gazette*, July 18, 1874, 278, K25.3/14. Also see Jackson, *Rails*, 209.

9. Working through Keystone, Scott likely could have forced St. Louis Bridge into foreclosure at any point (given the monies it owed to that prime contractor), but that step would have alienated the Morgans, enmity the PRR could ill afford. If Scott wanted the bridge, a better strategy would be to wait until the Morgans triggered a foreclosure themselves, with an understanding that Scott's PRR would become the new owner-operator of the bridge.

10. St. Louis Bridge, *Report*, 1875, 8.

11. JAK, notes to Taussig, "Origin," 6, K7/2. As noted in chapter 6, the bridge-friendly directors of the MoPac had sold their interest in the carrier back in March 1871.

12. Use contract, n.d., HC/03/001/001(006)/box 1, JSM.

13. JAK, notes to Taussig, "Origin," 7, K7/2.

14. "Annual Statement W. F. Co., Ending April 30, 1874," folder 1874, PC, B35/26. "Money Then and Now," above in the front matter, discusses price comparators and observes that the historical consumer price index is an imperfect tool for charting changing money values for different kinds of assets. Real estate illustrates the point. A comparator devised to measure the relative worth of such tangible assets places the modern value of the Wiggins properties at $1.01 billion in 2023.

15. Wallace, "Wiggins," 13; Eads, "Report," 1869, 550.

16. *Wiggins Ferry Co. v. City of East St. Louis*, 107 U.S. 365.

17. JAK, notes to Taussig, "Development," 11, K7/1.

18. Wallace, "Wiggins," 13.

19. JAK, notes to Taussig, "Origin," 5, K7/2.

20. Goldfeder, "Railroads," 36. An article in the *St. Louis Daily Democrat* on December 29, 1869, indicated that the Wiggins car ferry had begun operating by then. JAK, notes (p. 1), K7/1.

21. A similar story played out after the Union Pacific opened its bridge over the Missouri River from Omaha to Council Bluffs in April 1872. Carriers terminating in Iowa refused to use the new bridge, preferring the ferry "as their officers are part owners of the transfer." Klein, *Union*, 278.

22. St. Louis Bridge, *Report*, 1884, 10.

23. Kouwenhoven, "Celebration," 177.

24. Kouwenhoven, "Celebration," 177–78, with quotation from the *St. Louis Republican*, July 5, 1874.

25. Eads to J. S. Morgan, July 12, 1874, HC/03/001/001/(006)/box 2, JSM.

26. JAK, notes (p. 11) to *Railroad Gazette*, Aug. 8, 1874, K25.3/14. The fourteen locomotives used on July 2 in testing the bridge's strength were all borrowed.

27. Jackson, *Rails*, 213.

28. JAK, notes to *Mines, Metals and Arts*, Sept. 24, 1874, 229, K24/32.

29. *Commercial and Financial Chronicle* (New York), Dec. 19, 1874, 630.

30. Letters from folder titled "JSM's Letters from America, August–December 1874," HC/01/006/004, JSM.

31. "Floating debt" meant those obligations that St. Louis Bridge had incurred outside of its bonds, chiefly its accounts payable to vendors and contractors plus short-term loans from banks. Some evidence survives to show that Eads used the National Bank of the State of Missouri (discussed in ch. 4) as a ready cash box. On July 1, 1873, he took out a demand loan of $132,000 ($3.4 million in 2023) from the bank, knocking the interest rate down from its customary 10 percent to 8 percent and providing as collateral 1,577 shares of bridge company stock. (The transaction thus fixed the value of that security at that time at $83.70.) It is unclear whether or how this personal loan related to the bridge company's finances, but it reveals the nifty convenience Eads derived from having his own bank. He failed to repay the loan in full, settling a portion of the debt in 1879. National Bank of the State of Missouri, standard form for loans, July 1, 1873, box 1, folder 1, JBE, B35/26.

32. Letters from folder titled "JSM's Letters from America, August–December 1874," HC/01/006/004, JSM.

33. Telegram, DM Co. to JSM Co., Feb. 27, 1875, Private Telegram Book, 1:67, MS21802/001, JSM.

34. St. Louis Bridge, *Report*, 1875, 5. The fourth mortgage was dated Jan. 1, 1875. Anticipating the likelihood of foreclosure, it formalized the protection in default of the loans that J. S. Morgan & Company was then making to sustain St. Louis Bridge.

35. JAK, notes (p. 18) to *Railroad Gazette*, June 12, 1875, 248, K25.3/14.

36. The original receipt for the office contents is K19/67. Also see St. Louis Bridge, *Report*, 1875, 5.

37. Credit report of Oct. 30, 1875, for Illinois & St. Louis Bridge Co., Missouri, vol. 39, p. 346, RGD.

38. St. Louis Bridge, *Report*, 1875, 14.

39. Julius Walsh to JSM Co., July 1875, 2, TRRA, K26/7.

40. "The Terminal Railroad Association of St. Louis," *Railway and Engineering Review*, Sept. 8, 1900, 513. Goods from the east flowed over the bridge in high volumes, justifying regularly scheduled freight trains. Eastbound freights out of St. Louis were more likely to move as extras (not formally scheduled trains). Earlier studies contain an oft-repeated error that the depot initially served just fourteen trains a day.

41. JSM Co., "Circular to Bondholders," June 29, 1875, HC/03/001/001(006)/box 3, JSM.

42. JAK, notes to Taussig, "Origin," 7, K7/2.

43. Vernon, "Report," 40 (emphasis in original).

44. Vernon, "Report," 40 (emphasis in original).

45. Eads, "Report," 1868, 537. Fudging the comparison, Eads included the costs of acquiring the land and building the approaches in the 1868 estimate but

excluded those items from the 1875 reckoning. The 1868 total also included the tunnel, which disappeared from the tally seven years later.

46. JAK, notes to "Discussion on Upright Arched Bridges," reprinted in Eads, *Addresses*, 107–21, K19/75.

47. A source from 1889 puts the total cost at $14,245,000 (including the tunnel, salaries, and interest) up to the date of completion, the most accurate accounting. Switzer, *Report*, 356.

48. Taussig to Milnor Roberts, Oct. 6, 1868, WT, K7/19.

49. The judge in an 1878 suit filed by a creditor was appalled by the stock frauds, writing with dry reserve that "it would be a dangerous precedent to sanction this mode of disposing of the capital stock of a corporation." See "Judgement of Mr. Justice Thayer in *Skrainka & Others v. Allen & Others*," July 1, 1878, HC/03/001/001(006)/box 4, JSM. This document, a reprint by J. S. Morgan & Company, errs in its title for Judge Thayer.

50. Eads to Taussig, July 28, 1874 (emphasis in original), TRRA, K4/50. In this letter, Eads volunteered to pay the debt with a note, a credit IOU. That meager concession could do little to help the cash-starved company given the general depression.

51. White, "Corruption," 20.

52. The story ran in the *St. Louis Globe Democrat* on June 20, 1877, as detailed in William Glasgow Jr. to Sudy Glasgow, June 21, 1877, BB.

53. Kane, *Banking*, 91–96.

54. Shaler Smith to Onward Bates, Feb. 7, 1879, BB.

55. Kane, *Banking*, 91.

56. By modern lights, fraud is an anomaly when capitalists go bad. A provocative inquiry into nineteenth-century British financial history argues that fraud was endemic to and inseparable from that era's capitalism. Klaus, *Forging*.

57. J. S. Morgan (New York City) to JSM Co., Sept. 23, 1874, in folder titled "JSM's Letters from America, August–December 1874," HC/01/006/005, JSM.

58. Nearly all the stockholders were wealthy bankers and merchants in New York and St. Louis. As Carnegie later wrote, "When the remaining 25% due on St. Louis Bridge Stock was called in, shareholders were dissatisfied as they had been led to believe that not more than 40 percent of their subscription would be required. The Bridge Co. therefore resolved to *give* each stockholder, who would pay up his stock in full, 2nd Mtge Bonds equal to the 25% due upon the Stock. This was done + subsequently these bonds were sold in London and the shareholders received back their respective proportions of the proceeds." Carnegie to Wm. Spackman, Feb. 23, 1877 (emphasis added), USS, K11/78.

59. By this transaction, the Morgans committed no fraud, since the bank stood behind the Seconds. Nonetheless, the episode leaves a rotten smell. Here a narrow group of stockholders benefitted because they had ties to the Morgans. The British buyers of those bonds lacked the insiders' knowledge of the peril in the books of St. Louis Bridge. This trick shifted liability away from

those who had taken it avidly in good times and pushed it onto unwitting bondholders.

60. Eads, "Upright," 3:319.

61. Eads, "Upright," 3:222, 216; 4:172.

62. For a thorough description of the financial aspects of this reorganization, see Carosso, *Morgans*, 242–45. No account details the Morgans' work in managing distressed properties.

63. Measured by par value, 65 percent of all US railroad bonded debt held in Great Britain and Europe had defaulted by 1876. According to the *Banker's Magazine and Statistical Register* (Boston), the default rate for railroad debt held in the United States was 14 percent. White, "Corruption," 28.

64. St. Louis Bridge, *Report*, 1875. For the year ending April 30, 1875, net earnings were $188,000 without reckoning the cost of debt servicing. The report says nothing about that year's interest charges, probably because its London bankers extended cash loans to pay the bondholders. With minimal rail traffic, the tunnel company was in worse shape.

65. JAK, notes to *East St. Louis Press*, Oct. 30, 1875, K26/10. Presumably, St. Louis Bridge could remain competitive while charging that higher rate because it moved cars more readily than did Wiggins.

66. Untitled typeset agreement, Apr. 23, 1877, HC/03/001/001(006)/box 3, JSM.

67. Morganization receives extensive coverage in Corey, *House*.

68. For just six months (to Nov 30, 1877), the net earnings (profits) of the bridge and tunnel exceeded $175,000; the ferry's profit was $37,500. However, the bridge and tunnel were not yet paying any bond interest. Taussig to J. P. Morgan and Solon Humphreys, Receivers, Dec. 26, 1877, HC/03/001/001(006)/box 3, JSM.

69. Thomas, "Connecting," 67.

70. A 1906 study counted a daily average of 2,683 pedestrians crossing on the bridge, compared with 4,301 on the ferries. Offering convenience to many residents, the water route owed its survival partly to collusive pricing. Considered broadly, this outcome was socially desirable, if economically inefficient. Perkins, "Conditions," 18.

71. St. Louis Bridge, *Report*, 1877. Data for the fiscal year ending May 1, 1877; percentage growth compared with a year earlier.

72. Chandler, *Visible Hand*, 138.

73. Churella, *Pennsylvania*, 476–80.

74. White, *Republic*, 346–47.

75. Laurie, *Artisans*, 144.

76. Historians have detailed the Morgans' frequent interventions in railroad policy decisions, their exhortations to avoid self-destructive rate wars and territorial incursions that bled profits and beggared returns on invested capital. That the bankers sided actively with management during national labor disputes is both unsurprising and unnoted in histories of the Morgan banks. See Carosso, *Morgans*, 226–30; and Strouse, *Morgan*, 173, 255–56.

77. Telegrams, JSM Co. to J. W. Garrett, July 19, 1877, and Garrett to JSM Co., July 20, 1877, Private Telegram Book, 2:74–75, MS 21802/002, JSM.
78. Bellesiles, *1877*, 150.
79. Foner, *Workingmen's Party*, 81–88.
80. "Trouble Brewing at St. Louis," *New York Times*, July 22, 1877, 7.
81. "The Situation at St. Louis," *New York Times*, July 23, 1877, 2.
82. "The Strikes Extending: Complete Blockade at St. Louis," *New York Times*, July 24, 1877, 5.
83. "The Strikes Extending: Complete Blockade at St. Louis," *New York Times*, July 24, 1877, 5.
84. Foner, *Workingmen's Party*, 81.
85. Telegram, DM Co. to JSM Co., July 25, 1877, Private Telegraph Book, 2:80, MS 21802/002, JSM. As the recipients of this telegram knew, the PRR ran trains to Pittsburgh, the New York Central went to Albany, the Erie served Hornellsville, New York, and the B&O main line ran to Martinsburg, West Virginia.
86. Foner, *Workingmen's Party*, 88.
87. Roediger, "Overcome."
88. J. S. Morgan (Hartford) to JSM Co., Oct. 2, 1877, HC/01/006/005, JSM.
89. J. S. Morgan (New York City) to JSM Co., Oct. 23, 1877, HC/01/006/005, JSM.
90. JSM Co. to J. S. Morgan (New York City), Sept. 26, 1877, HC/03/001/001(006), JSM.
91. These loans apparently began with the coupons due on April 1, 1875. Trask, "Corporate Data," 51.
92. William Taussig contributed to reorganization planning by detailing the convoluted finances that had built the bridge and lined many pockets. His memo ran to fifteen pages, despite omitting "many matters . . . of policy and motives" because they had originated in "personal and corporate relations which had better not be put in writing." Behind that veiled language lurked the self-dealing and frauds that tarnished the principal players. See "Historical Statement," [ca. July 1878], HC/03/001/001(006), JSM.
93. Charles Branch to J. S. Morgan, Dec. 20, 1878, HC/03/001/001(006), JSM.
94. Branch to J. S. Morgan, Dec. 20, 1878, HC/03/001/001(006), JSM.
95. "Finis: The Great Bridge Finally Turned Over to the Purchaser," *St. Louis Republican*, Mar. 30, 1879, clipping from Minute Book of the Bondholder's Committee, box 5, folder 1, JSM. Readers are reminded that the legal name for the Eads-led firm (after consolidation with Boomer's venture) had been the Illinois & St. Louis Bridge Company. This account will continue to refer to the company as St. Louis Bridge.
96. For three years, the new bond issue paid only half of its regular 7 percent interest, buying time to boost profits.
97. These details given in Trask, "Corporate Data," 33–38; and Carosso, *Morgans*, 243–44.

98. In 1887 (a good year), the reorganized company paid nearly $600,000 in dividends and interest on securities. This broke down to the mandated $350,000 (or 7 percent) on its bonds, $149,400 in dividends (6 percent) on its first preferred stock, and $90,000 (3 percent) on the second preferred. By this time, Jay Gould's Missouri Pacific and his Wabash lines had leased the bridge and tunnel, lifting traffic substantially.

99. "The St. Louis Bridge," *Railroad Gazette*, July 8, 1881, 375.

100. Baughman, *Charles Morgan*, 73–75. Also see Renehan, *Commodore*, 197–201.

101. St. Louis Bridge, *Report*, 1876, 9. Also see Taussig to J. P. Morgan and Solon Humphreys, Receivers, Sept. 19, 1877, 5, HC/03/001/001(006), JSM.

102. Klein, *Gould*, 233–34; Grant, *"Follow the Flag,"* 50–52; Hansen, Hofsommer, and Schwantes, *Crossroads*, 138–39.

103. "The Opinions of Jay Gould," *Railroad Gazette*, Sept. 7, 1883, 587.

104. Joseph Pulitzer, quoted in Klein, *Gould*, 3.

105. Klein, *Gould*, chs. 20–21.

106. Telegram, J. S. Morgan (New York) to Walter H. Burns, Dec. 15, 1879, Private Telegraph Book, 3:64–65, MS 21802/003, JSM.

107. Telegram, J. S. Morgan to J. P. Morgan, June 8, 1881, Private Telegraph Book, 3:165, MS 21802/003, JSM.

108. Telegrams, J. S. Morgan to J. P. Morgan, July 2, 1881, and J. P. Morgan to J. S. Morgan, July 9, 1881, Private Telegram Book, 3:172, 175–76, MS 21802/003, JSM.

109. St. Louis Bridge, *Report*, 1882, 3.

110. JSM Co. to DM Co., Mar. 24, 1870 (emphasis in original), in "Private Copy Out-Letter Book, Nov. 1867–Mar. 1875," 366–68, MS21795/001, JSM. Earlier accounts of Eads and his bridge stumble over the foreclosure, mistakenly concluding that "in economic terms, it never really worked very well." Jackson, *Rails*, 223.

Chapter 10. Successes across Time

1. Eads, "Report," 1868, 483.

2. Eads, "Report," 1868, 536.

3. Eads, "Report," 1868, 484.

4. See maps included with copy of new tariff, Aug. 1886, in folder titled "Various re St. Louis Bridge," box 15, JSM.

5. St. Louis Bridge, *Report*, 1887, 5.

6. This parochial view continued for years. See *Railroad Gazette*, Feb. 8, 1889, 99–100.

7. Eads, "Report," 1868, 485.

8. "Solving the Terminal Problem at St. Louis, Part 1," *Railway Age* 45, no. 7 (Feb. 14, 1908): 215.

9. William Taussig, "Scheme for the transfer of the Bridge and Tunnel Lease," [June 1886], HC/03/001/001/(006), box 15, JSM.

10. While somewhat unusual in the 1880s, union stations became a growing phenomenon after 1900. See "Editorial Notes," *Railroad Gazette*, Nov. 1, 1912, 821.

11. With the 1881 lease to Gould, the Morgans stepped back from issues of management and strategy at St. Louis Bridge, but they remained keenly interested in the company for two reasons. Gould was only a lessee, so the bankers worked directly with William Taussig to safeguard the interests of the bond- and stockholders under the 1879 reorganization. The Morgans also planned to underwrite its future capital needs, such as a $7 million bond issue to finance the new (1894) union station. Carosso, *Morgans*, 357.

12. The original six carriers were the MoPac, the Iron Mountain, and the Wabash from the west (all Gould) and the Ohio & Mississippi, the Louisville & Nashville, and the Big Four route from the east.

13. Taussig to JSM Co., Nov. 6, 1889, HC/03/001/001(006), box 15, JSM.

14. Terminal Railroad Association of St. Louis, *Annual Report* (1891), 5.

15. Taussig, "Scheme for the transfer of the Bridge and Tunnel Lease," n.d. (June 1886), HC/03/001/001(006), box 15, JSM.

16. That location placed the Merchants Bridge's east abutment in Madison County, Illinois, just north of the exclusive privilege held by St. Louis Bridge in St. Clair County.

17. St. Louis Merchants Exchange, *Annual Statement*, 1891, 80–82. For simplicity, this account treats the Merchants Bridge and its associated terminal railroad, the St. Louis Merchants Bridge Terminal Railway, as a single venture, although they were separate corporations. See Taber, "Celebrating," 89–95.

18. Taber, "Celebrating," 93.

19. The name Eads Bridge grew in popular usage after Eads's death in 1887. In the annual reports of the TRRA, the name shift happened gradually across the 1890s. See Terminal Railroad Association of St. Louis, *Annual Report*, 1895, 10, 15.

20. Taussig to JSM Co., Jan. 21, 1891, original in JSM, copy in box 12, folder "I&SLB, 1890s," VPC.

21. The consortium included the strongest eastern trunk line (the PRR), a powerful granger road (Burlington), and an ideal western partner with connections to San Diego and Los Angeles (Santa Fe). Taussig to J. P. Morgan, Feb. 22, 1893, original in JSM, copy in box 12, folder "I&SLB, 1890s," VPC. This account of the corporate maneuvers between the TRRA and the Merchants company ca. 1893 revises long-accepted views of the episode.

22. To ensure that the bankers understood the urgency of this challenge by the consortium, Taussig added that "they have a splendid piece of property, valuable charter, and a set of rich people who are willing to advance all money necessary to develop it into a formidable competitor." Taussig to J. P. Morgan, Feb. 22, 1893, original in JSM, copy in box 12, folder "I&SLB, 1890s," VPC.

23. For the details of these takeovers, see Brown, "Unnatural."

24. In his 1868 report, Eads never explicitly advocated a bridge monopoly, while he did condemn the monopolies of the ferry and transfer company. But he intended St. Louis Bridge to take all the trans-river business, even if Taussig had to remind him of that on occasion (ch. 7).

25. Terminal Railroad Association of St. Louis, *Annual Report*, 1891, 11.

26. Even before this turn to costly stations, passenger services generated on average just two-fifths of the gross revenues derived from freight. Worse, passengers were far less profitable than freight. Hadley, *Transportation*, 109.

27. Osmund Overby, "A Place Called Union Station: An Architectural History of St. Louis Union Station," in Grant, Hofsommer, and Overby, *Union Station*, 59–90.

28. Terminal Railroad Association, *St. Louis Union Station*, 9–16.

29. Wayman, *Union*, 26–29.

30. Ciampoli, "Tornado." Other sources given different casualty figures, all terrible.

31. Curzon, *Great Cyclone*, 228, 415–16.

32. The interchange of freight cars between carriers began in a serious way after the Civil War (notwithstanding some precedents), accelerating greatly during the 1880s. One leading mechanical official dated its origin to 1866. See "The Present General Condition of Freight Cars for Interchange Traffic," *Railroad Gazette*, Mar. 27, 1885, 198.

33. White, *Freight Car*, 55–63; Hansen, Hofsommer, and Schwantes, *Crossroads*, ch. 8.

34. Goldfeder, "Railroads," 35, 38, 60.

35. Taylor and Neu, *Network*, 79–82.

36. "Interchange of Freight Cars," *Railroad Gazette*, Nov. 29, 1873, 477–78.

37. In "What the Car Service Might Be," the *Railroad Gazette* reported a range of problems constraining car interchange for carriers across the country. Jan. 19, 1883, 35.

38. Aldrich, *Death*, 109.

39. The source that best describes the railroads of St. Louis, their terminals, and their relations to the TRRA was published in 1922. See *Report of Engineers Committee*, 3, 65–66.

40. The Belt Railroad of Chicago did bear similarities. Founded in 1882 and owned by the carriers that it served, it had ninety-two miles of main line and yard tracks in 1902 but served only a portion of Chicago's railroads. Its capitalization was $1.2 million in 1902, compared with the TRRA's assets of $19.3 million that year. Data on the Belt Railway are from *Moody's Manual*, 591. See also Terminal Railroad Association of St. Louis, *Annual Report*, 1902, 29; and Young, *Iron Horse*, 112–16.

41. "Terminal Conditions in Chicago," *Railway Age Gazette*, Oct. 14, 1910, 698–700.

42. Gilbert, *Whose Fair?*, 180.

43. Clevenger, *"Indescribably Grand,"* 5–6.

44. "Handled 10,000,000 People: Business Done at St. Louis Union Station during the Fair," *Wall Street Journal*, Dec. 19, 1904, 3.

45. Public reporting by the TRRA grew sparse in this period, but data from the St. Louis Merchants Exchange show that the volume of freight moving over the river in 1905 exceeded 16 million tons.

46. *State of Missouri v. Terminal Railroad Association of St. Louis*, 182 Mo. 284 (1904). For Crow's desired remedy, see p. 290. The attorney general appealed, and lost again. See "Terminal Wins Case Again In Supreme Court," *St. Louis Post-Dispatch*, Nov. 29, 1904, 1.

47. The *Railroad Gazette*, an insightful trade journal and no crusader against injustice, agreed that TRRA freight terminals "are often badly congested and the charges for transfer across the river are high." See "East Ivory Car Ferry of the Missouri Pacific," *Railroad Gazette*, Oct. 13, 1905, 344.

48. For an account of the varied motives behind passage of the Sherman Act, see Bougette, Deschamps, and Vecchi, "Economics."

49. In the first federal antitrust case against the TRRA, Joseph Ramsay testified in July 1906 that the three terminal companies had "a general understanding" not to compete on the basis of price. Furthermore, they divided the freight transfers over the river by tonnage: 20 percent to Merchants, 25 percent to Wiggins, and 55 percent via Eads. Testimony of Joseph Ramsay Jr., July 14, 1906, 245–46, in *Transcript of Record*, Aug. 30, 1910. Ramsay had been general manager of the TRRA before taking the same post at the Wabash.

50. "*Post-Dispatch* Secures Investigation of Bridge Arbitrary by Federal Government," *St. Louis Post-Dispatch*, June 24, 1903, 1, quoting the bank's prospectus for a $50 million bond issue. Its boast centered on access to and from the East, the routes over the river.

51. "Terminal Case Argument Begun by Government," *St. Louis Post-Dispatch*, Apr. 1, 1909, 2. For a detailed account of the TRRA and antitrust law, see Brown, "Unnatural."

52. "Disagree on Terminal Case." *New York Times*, May 25, 1909, 3.

53. "Terminal Suit Is Docketed in the Supreme Court," *St. Louis Post-Dispatch*, Aug. 30, 1910, 8.

54. Because the circuit court deadlocked, this 1909 case produced no opinions. Its evidentiary record appears with the 1912 Supreme Court case. *Transcript of Record*, Aug. 30, 1910.

55. Pratt, *Supreme Court*, 37–40.

56. *United States v. Terminal Railroad Association of St. Louis*, 224 U.S. 383 (1912). The prosecution introduced ample evidence that the TRRA had charged excessive rates, fixed prices with the Merchants and Wiggins companies (before taking them over), and blocked a carrier from accessing the city (a practice known in antitrust law as foreclosure, and quite unlike a foreclosure in bankruptcy proceedings). For these offenses, the government sought far-reaching remedies: the TRRA's proprietary railroads (its owners, by then

fourteen carriers) should divest their shares in the association, and that company should divest itself of the Merchants and Wiggins properties. That result would have created three independent, competing terminal companies.

57. Despite its importance, the case has received scant attention from legal historians. Walter Pratt's book *The Supreme Court under Edward Douglass White, 1910–1921* details the *Standard Oil* and *American Tobacco* cases but omits *Terminal Railroad*. William Kolasky's article "Chief Justice Edward Douglass White and the Birth of the Rule of Reason" mirrors those choices.

58. While the opinion never used the term *natural monopoly* explicitly, it offered detailed "consideration of the natural conditions greatly affecting the railroad situation at St. Louis" (224 U.S. 395). As US law construed the term in the nineteenth century, *natural monopoly* suggested but did not explicitly reflect environmental or geographic conditions. Richard John explores its varied meanings, focusing on "technological imperatives and economic incentives," in *Network*, 157–58, 194–95. The leading scholar of monopoly in legal history, Herbert Hovenkamp, recounts the problematic efforts of legislators and judges to craft "a usable model for natural monopoly" in *Enterprise*, 110–14.

59. It is clear that the court was determined to find for the Terminal Association. Alongside their natural monopoly assertion, the justices advanced a second fallacious argument that earlier had been advanced by the association in the first Missouri trial. In this telling, without the monopoly each railroad serving St. Louis would lay its own tracks, build its own yards, and "the city would be cut to pieces . . . and thus the greatest agency of commerce would become the greatest burden." 224 U.S. 403, quoting 182 Mo. 284, 299.

60. The ITS amounted to railroad, although one powered by overhead electric wires, not steam locomotives. It ran extensive freight and passenger services over a 400-mile route across Illinois.

61. Quotation from 224 U.S. 383, 398. Inexplicably, the evidentiary record in the first federal case against the TRRA (1909) ignored that the Terminal Association had foreclosed access to St. Louis via the ITS, a fact reported in the local press and considered by local courts. The circuit court's evidentiary record and the Supreme Court's opinion also make no mention of the McKinley Bridge, which was under construction during the circuit trial. Even so, many of the justices passed through St. Louis frequently, and the new crossing was visible from any train over the Eads or the Merchants Bridge. They knew it was there.

62. The chief *economic* argument in favor of natural monopolies is that a single provider can achieve economies of scale that drive down unit costs for consumers. The evidentiary record in *Terminal Railroad* provided no evidence supporting that conclusion, and there was ample testimony against it. The justices may have concluded that dissolution of the TRRA into its component parts (the prosecution's goal) was a radical step rife with unknowns for the commerce of St. Louis—and for the court's reputation. For detailed analysis of this case, see Brown, "Unnatural."

63. "St. Louis Terminal Loses," *New York Times*, Apr. 23, 1912, 16. Journalists presented the case as a loss for the TRRA because the court forced changes to its rates and billing practices, paramount issues for St. Louisans.

64. In a 1945 case, *Associated Press v. United States*, 321 U.S. 1, the Supreme Court built on the precedent of *Terminal Railroad* to conclude that the AP provided an essential service to the newspaper industry and its readers, a service that it could not unilaterally deny to nonmembers.

65. Brinsmead, *Essential*, ch. 5; Guggenberger, "Essential," 339.

66. In the original 1904 Missouri antitrust trial, three judges rejected the company's argument (which did prevail among their four brethren) that it provided efficient service. The minority opinion noted with scorn that this was "the argument advanced in favor of all monopolies." 182 Mo. 316 (1904).

67. Eads, "Report," 1868, 532–35, quotation from 535.

68. Wiggins operated its railroad car ferries into the 1930s.

69. In choosing to preserve the TRRA, the Supreme Court decided that coordination by the monopoly served the St. Louis region better than did the results achievable by three competing terminal operators. In practical terms, this may have been true. As a question for economic analysis, it simply cannot be answered with any authority, even today. As a matter of historical fact, St. Louis had *never* known such competition.

70. Miller, "Truth." By 1900, railroads serving other gateways across the country had generally agreed to these priorities, but St. Louis alone had a single entity with the responsibility and power to enforce the rules.

71. Miller, "Truth"; Pierce, "Plan."

Epilogue

1. Reuss, "Humphreys," 18–22; Barry, *Rising*, 79–87.

2. Gould, *Fifty*, 318–20. The Corps of Engineers estimated the cost of a canal at $8 million.

3. How, *Eads*, 99.

4. "Personal," *Engineering News*, Sept. 20, 1879, 300; How, *Eads*, 110–11.

5. Eads (New York) to James Andrews, n.d., folder 1, JA.

6. Eads first consulted on the Bosporus project during an earlier trip to Turkey in 1874. See "The Bosphorus Bridge," *Railroad Gazette*, Sept. 14, 1877, 420. An 1877 profile sketched Eads's plans at that time: fifteen arches with a center span of 750 feet. "Bridging the Bosphorus," *New York Times*, Aug. 25, 1877, 5.

7. "The Proposed Isthmus Ship Railroad," *Railroad Gazette*, Sept. 5, 1879, 469–70; cover story, *Scientific American*, Dec. 27, 1884.

8. "The Chignecto Ship Railway," *Railroad Gazette*, Nov. 21, 1890, 798–99.

9. "The Manchester Ship Canal," *Engineering* (London), Aug. 29, 1884, B35/25.

10. "The Manchester Ship Canal," *Engineering* (London), Jan. 26, 1894, 98. A year later, the canal promoters offered a revised plan, Eads approved of the new approach, and Parliament authorized this massive project to make Manchester an inland port for oceangoing shipping.

11. "The Society of Arts Albert Medal," *Engineering* (London), July 18, 1884, 66.
12. "Albert Medal for 1971," *Journal of the Royal Society of Arts*, Feb. 1971, 130–31, K1.2/100.
13. John Ericsson, quoted in Church, *Ericsson*, 2:199.
14. Sellers, "Memoir," 78, K1/27.
15. "James B. Eads," *Engineering* (London), Mar. 18, 1887, 257–58.
16. "James B. Eads," *Railroad Gazette*, Mar. 18, 1887, 176.
17. White, *Railroaded*, 106–7, 127, 332.
18. Churella, *Pennsylvania*, 491; Klein, *Gould*, 250.
19. Richard John, "Gilders," argues for balanced evaluations of men like Scott, who altered much in the American political economy during these decades.
20. Gould's most recent biographer disinters the old dark robber baron. Steinmetz, *Rascal*. For a fair and authoritative account with plenty of drama, see Klein, *Gould*.
21. The TRRA's leases of the bridge and the tunnel continued well into the twentieth century, with annual payments of $664,000 to the stock- and bondholders of the companies established by the Morgans in 1879. Those payments ended when the St. Louis Bridge and the Tunnel Railroad companies were dissolved in 1944.
22. Correspondence regarding note renewal, May 3, 1876, listed in the collection guide (p. 263) to JSM.
23. Correspondence regarding payment for stock premium, Dec. 10, 1879, listed in the collection guide (p. 41) to JSM. After a radical design change and exhaustive reviews by the Board of Trade, that massive cantilevered rail crossing over the Firth of Forth finally entered service in March 1890. In light of those delays and design challenges, the St. Louis Bridge did not look so troublesome. The main structural members of the Forth Bridge are solid drawn steel tubes. It remains in daily use.
24. Carosso, *Morgans*, 275–76.
25. Writing in 1877, Carnegie detailed the stock and bond transactions by which bridge insiders tried to take their proceeds from the failing company. Although it reflected the standards of that era, today the document reads as an unapologetic outline of fraud. Carnegie to Wm. Spackman, Feb. 23, 1877, USS, K11/78.
26. Man & Parsons to Edward A. Quintard et al., June 25, 1879, in box 51, folder 8, AC, B40/6.
27. Nasaw, *Carnegie*, 173.
28. Sellers, "Memoir," 78.
29. ASCE, *Biographical*, 45.
30. JAK, biographical notes for Onward Bates (1850–1936), K12.1/7.
31. Middleton, *Québec*.
32. Petroski, *Dreams*, ch. 3.
33. "American Bridge Company," *Railroad Gazette*, July 6, 1872, 288; Darnell, *Directory*, 83–84.
34. Brown, "Not the Eads," 530.

35. Voorhees, *Alton*. The bridge used Hay process steel, not the Bessemer product typical of rails. Open-hearth steel displaced iron for bridging during the 1880s.

36. "Opening of the Glasgow Bridge," *Engineering News*, July 5, 1879, 209.

37. Designers of metal arch bridges in the Gilded Age learned to dissipate those stresses with *hinges*, expansion joints that allowed some movement in the chords.

38. "Profiles and Alignments," in *Report of Engineers Committee*, 1921, B34/6; Gilbert and Billington, "River Politics," 92; Hardesty & Hanover, "Report," 10.

39. "The East River Bridge," *Engineering* (London), June 3, 1881, 580.

40. Carnegie, *Autobiography*, 119 (emphasis in original).

41. Sellers, "Memoir," 79.

42. John K. Brown, "Heuristics, Specifications, and Routines in Building Long-Span Railway Bridges on the Western Rivers, 1865–1880," in Raff and Scranton, *Emergence*, 194–97.

43. Sullivan, *Autobiography*, 247. I am indebted to David Billington for this source.

44. Sullivan, *Autobiography*, 257.

45. Ada Louise Huxtable, "Eads Bridge—Engineering Miracle and Work of Art," *New York Times*, July 21, 1974, D-19.

46. Walt Whitman, *Specimen Days*, quoted in Nye, *Sublime*, 83.

47. The Terminal Railroad Association does much the same work today as it has since 1889. Thanks to mergers, five railroads are its proprietary owners at this writing (2023).

48. "Rehab of Eads Bridge helps extend its life beyond 2 centuries," *All Things Considered*, St. Louis Public Radio, Oct. 6, 2016, accessed Mar. 9, 2021, https://news.stlpublicradio.org/economy-business/2016-10-06/rehab-of-eads-bridge-helps-extend-its-life-beyond-2-centuries.

49. Eads, *Addresses*, 44.

Citations for most sources given in the text are included here. Materials and articles drawn from the serials listed immediately below are cited exclusively in the notes.

Congressional Globe
East St. Louis Gazette
Engineering (London)
Engineering News
New York Times, 1861–1915
Railroad Gazette, 1873–1915 (title varies)
St. Louis Post-Dispatch, 1867–68, 1874, 1893, 1902–12 (title varies)
Westliche Post (St. Louis)

Case Law
Associated Press v. United States, 321 U.S. 1 (1945).
Eads et al. v. The H. D. Bacon, 1 Newberry 274, reprinted in 8 Federal Cases 224, Case 4,232.
State of Missouri v. Terminal Railroad Association of St. Louis, 182 Mo. 284 (1904).
The St. Louis Transfer Railway Company v. The St. Louis Merchants Bridge Terminal Railway Company, 111 Supreme Court of Missouri Reports (1892) 666.
Transcript of Record. Aug. 30, 1910. *United States v. Terminal Railroad Association of St Louis*, 224 U.S. 383 (1912). 3,079-page trial record in *The Making of Modern Law: U.S. Supreme Court Records and Briefs, 1832–1978*, online resource.
United States v. Terminal Railroad Association of St. Louis, 224 U.S. 383 (1912).
United States v. Terminal Railroad Association of St. Louis, 236 U.S. 194 (1915).
Wiggins Ferry Co. v. City of East St. Louis, 107 U.S. 365.

Statutes

Act Authorizing Construction of Certain bridges and establishing them as Post Roads (Omnibus Bridge Act). Approved July 25, 1866. *Appendix to the Congressional Globe*, 39th Cong., 1st Sess., 390–91.

Act Authorizing the Construction of a Bridge over the Mississippi River at Saint Louis, Missouri (Merchants Bridge Statute). Approved Feb. 3, 1887. *US Statutes at Large*, 49th Cong., 2nd Sess., chap. 91, 375–77.

Addresses, Papers, Speeches, Websites

American Society of Civil Engineers. "James Laurie." Accessed June 29, 2023. https://www.asce.org/about-civil-engineering/history-and-heritage/notable-civil-engineers/james-laurie.

Benson, T. Lloyd, and Trina Rossman. "Re-Assessing Tom Scott, the 'Railroad Prince.'" Paper presented at the Mid-America Conference on History, Sept. 16, 1995.

Gaylor, Carl. Speech to the Engineers Club of St. Louis, Feb. 20, 1929. MHS. K12.1/40.

Gilbert, Ralph W., Jr., and David Billington. "The Eads Bridge and Nineteenth Century River Politics." Paper presented at First National Conference on Civil Engineering: History, Heritage and the Humanities, Princeton University, Oct. 14–16, 1970.

Goldfeder, Ron. "Railroads of St. Louis: When & How They Got There." B35/5.

Hetherington, James R. "The History of Union Station." Accessed Feb. 2, 2022. pdf document from www.indianahistory.org.

Kouwenhoven, John A. "The Bridge." Typescript of speech to Missouri Historical Society, Oct. 25, 1957. K27/137.

Leighton, George Bridge. "Notes on St. Louis Railway Passenger Traffic." Paper presented to the Commercial Club of St. Louis, 1894.

Miller, Henry. "The Truth about the Terminal." 1927. MHS. B37/3.

Pierce, Thomas M. "The St. Louis Plan of Unified Railway Terminals." 1928. MHS. B37/4.

Purdy, Harry L. "An Historical Analysis of the Economic Growth of St. Louis, 1840–1945." Unpublished paper, Missouri Pacific Lines. Accessed Aug. 11, 2020. https://fraser.stlouisfed.org/files/docs/publications/books/econgrowthstl_purdy_1945.pdf.

Rosenbloom, Joshua L., and Gregory W. Stutes. "Reexamining the Distribution of Wealth in 1870." Working paper 11,482, National Bureau of Economic Research, June 2005.

Taussig, William. "Development of St. Louis Terminals: History and Reminiscences." Paper presented at the Commercial Club of St. Louis, 1894. WT.

Taussig, William. "Origin and Development of St. Louis Terminals." Address to the St. Louis Railway Club, ca. 1910. WT.

Trask, Ruth. "St. Louis Bridge Company Corporate History." Typescript, ca. 1955. Author's collection.

Yenawine, Wayne. "A History of the Holdings of the Terminal Railroad Association of St. Louis." Typescript, ca. 1950. Box 1, TRRA/B.

Reports

Allen, Gerard B. "Illinois and St. Louis Bridge Company. President's Report to the Stockholders." St. Louis, 1873.

Alton & St. Charles County and the St. Louis & Madison County Bridge Companies Consolidated. Untitled, undated pamphlet [1867], with report by Latrobe and Smith. Author's collection.

"Annual Report of the Secretary of War, 1887, Vol. 2, 4 parts." US Congressional Serial Set, 1887, [i]–14.

[Boomer, Lucius]. *Proceedings and Report of the Board of Civil Engineers*. St. Louis, 1867.

Description and Plans of the Proposed Grand Union Passenger Depot in Saint Louis. St. Louis, 1870. B38/1.

Eads, James B. "Report of the Chief Engineer." Annual, 1868–71. In Eads, *Addresses and Papers*.

Eads, James B. "Review of the US Engineers' Report on the St. Louis Bridge." Oct. 1873. In Eads, *Addresses and Papers*, 77–88.

Hardesty, Shortridge. "Report on Inspection of the Eads Bridge over the Mississippi River, 1940." Author's collection.

Hardesty & Hanover. "Report on Stress Measurements and Analysis of Arch Ribs of the Eads Bridge over the Mississippi River." Typescript. Sept. 28, 1950. Author's collection.

Homer, Truman J. *Reports of the City Engineer*. St. Louis, 1865.

Illinois and St. Louis Bridge and St. Louis Tunnel Railroad. *Annual Report* (various years).

Reavis, Logan Uriah. "History of the Illinois and St. Louis Bridge." St. Louis, 1874.

Report of Engineers Committee [on the] St. Louis–East St. Louis Railroad Terminals. St. Louis: C. E. Smith & Co., 1922.

St. Louis Bridge and Tunnel R.R. *Annual Report* (various years).

St. Louis Merchants Exchange. *Annual Statement of the Trade and Commerce of St. Louis*. St. Louis, 1865–1916.

Switzer, Wm. F. *Report on the Internal Commerce of the United States for the Fiscal Year 1889*. Pt. 2. Washington, DC: GPO, 1889.

Taussig, William. "Report to the Board of Directors re the US Board of Engineers." Nov. 5, 1873. In Stockholders Minute Book, 1873–75. TRRA.

Terminal Railroad Association of Saint Louis. *Report of the Chief Engineer on Improvements Made in 1902-3-4*. St. Louis, 1905.

Terminal Railroad Association of St. Louis. *Annual Report* (various years).

Trask, Ruth. "Corporate Data Respecting [the] Terminal Railroad Association of St. Louis, Preceded by Short Tables and Summaries." St. Louis Mercantile Library.

US Congress. House. *House Report No. 2, Government Contracts*. 37th Cong., 2nd Sess. (1861).

US Congress. House. "Report to Accompany Bill H.R. 11249: Bridge over the Mississippi River at Saint Louis, Missouri." Aug. 23, 1888. H.R. Rep. No. 3307, 50th Cong., 1st Sess.

US Congress. House. "Reports on the Construction of the St. Louis and Illinois Bridge across the Mississippi River." Mar. 31, 1874. H.R. Doc. No. 194, 43rd Cong., 1st Sess.

Vernon, Edward. "Report of the Transportation Bureau of the Merchant's Exchange of St. Louis." St. Louis, April 1876.

"Wages in the United States and Europe, 1870–1898." *Bulletin of the Department of Labor* 18 (Sept. 1898): 17.

Warren, Brevet Major General G. K. *Report on Bridging the Mississippi River*. Washington, DC: GPO, 1878.

Theses and Dissertations

Imberman, Eli Wood. "The Formative Years of Chicago Bridge and Iron Company." PhD diss., University of Chicago, 1973.

Kamm, Samuel Richey. "The Civil War Career of Thomas A. Scott." PhD diss., University of Pennsylvania, 1940.

McConachie, Alexander Scot. "The 'Big Cinch': A Business Elite in the Life of a City, St. Louis, 1895–1915." PhD diss., Washington University, 1976.

Polinsky, Gerard R. "The Construction of the Illinois and St. Louis Bridge (Eads Bridge) at St. Louis, 1867–1874." MA thesis, Washington University, 1954.

Storey, Britt Allan. "William Jackson Palmer: A Biography." PhD diss., University of Kentucky, 1968.

Timmerman, Kurt. "Bridging the Mississippi: A History of the Rivalry between Suspension and Arch Modes of Engineering." MA thesis, University of Missouri, St. Louis, 2014.

Articles

Abbott, Carl. "The Location of Railroad Passenger Depots in Chicago and St. Louis, 1850–1900." *Railway and Locomotive Historical Society Bulletin* 120 (Apr. 1969): 31–47. B34/19.

Beitz, Ruth S. "Nantucket of the Middle West." *Iowan Magazine*, Winter 1964, 2–6.

Bishop, William Henry. "St. Louis." *Harper's New Monthly Magazine*, Mar. 1884, 497–517.

Bougette, Patrice, Marc Deschamps, and Frederic Vecchi. "When Economics Met Antitrust: The Second Chicago School and the Economization of Antitrust Law." *Enterprise and Society* 16, no. 2 (June 2015): 317–22.

Brown, John K. "Not the Eads Bridge: An Exploration of Counterfactual History of Technology." *Technology and Culture* 55, no. 3 (July 2014): 521–59.

Brown, John K. "The 'Rule of Reason' and an Unnatural Monopoly: *United States v. Terminal Railroad*." *Tocqueville Review* 43, no. 2 (2022): 145–72.

Brown, John K. "When Machines Became Gray and Drawings Black and White: William Sellers and the Rationalization of Mechanical Engineering." *IA: The Journal of the Society for Industrial Archaeology* 25, no. 2 (1999): 29–54.

Butler, W. P. "Caisson Disease during the Construction of the Eads and Brooklyn Bridges: A Review." *Undersea and Hyperbaric Medicine* 31, no. 4 (Winter 2004): 445–59.

Castagna, Richard. "Wiggins Ferry Company." *Magazine of the Terminal Railroad Association of St. Louis Historical and Technical Society*, Summer 2004, 110.

Chanute, Octave. "Pneumatic Bridge Foundations." *Journal of the Franklin Institute* 55 (June 1868): 387–91 (pt. 1); 56 (July 1868): 17–28 (pt. 2); 56 (Aug. 1868): 89–99 (pt. 3).

Ciampoli, Judith. "The St. Louis Tornado of 1896." *Gateway Heritage* 2, no. 4 (Spring 1982): 24–31.

Coase, Ronald. "The Nature of the Firm." *Economica* 4, no. 16 (Nov. 1937): 368–405.

Cooper, Theodore. "Notes on the Erection of the Illinois and St. Louis Bridge." *Transactions of the American Society of Civil Engineers* 3 (1875): 239–54.

Cooper, Theodore. "The Use of Steel for Bridges." *Transactions of the American Society of Civil Engineers* 8 (1879): 263–94.

Cox, John L., and F. B. Foley. "Pioneering at Midvale." *Baldwin Locomotives* 13, no. 1 (Apr.–July 1934): 36.

DeLony, Eric. "The Golden Age of the Iron Bridge." *American Heritage of Invention and Technology* 10, no. 2 (Fall 1994): 8–22.

Eads, James B. "Upright Arched Bridges." *Transactions of the American Society of Mechanical Engineers* 3 (1875): 195–238, 319–35; 4 (1875): 81–85, 162–84, 201–3.

Ferguson, Eugene F. "The American-ness of American Technology." *Technology and Culture* 20, no. 1 (Jan. 1979): 3–24.

Gow, David. "Dynasty: A Tale of Power and Money." *Guardian*, Jan. 18, 2000. https://www.theguardian.com/business/2000/jan/19/4.

Grant, H. Roger. "The North Missouri: A St. Louis Railroad." *Railroad History* 213 (Fall–Winter 2015): 91–101.

Griffin, John, and Thomas C. Clarke. "Loads and Strains of Bridges." *Transactions of the American Society of Civil Engineers* 2 (1874): 93–106.

Griggs, Frank E., Jr. "John A. Roebling's Niagara River Railroad Suspension Bridge 1855." *Structure*, June 2016. https://www.structuremag.org/?p=9982.

Griggs, Frank E., Jr., and David Biggs. "O. Chanute, C.E." *Journal of Bridge Engineering* 14, no. 5 (Sept.–Oct. 2009): 374–87.

Guggenberger, Nikolas. "Essential Platforms." *Stanford Technology Law Review* 24, no. 2 (Spring 2021): 237–343.

"Hiram Hill." *American Artisan* 6, no. 14 (Apr. 15, 1868).

Howland, Edward. "Iron Bridges and Their Construction." *Lippincott's Magazine*, Jan. 1873, 9–26.

"Ice Bridge over the Mississippi at St. Louis." *Harper's Weekly*, Jan. 18, 1873, 52.

"The Illinois and St. Louis Bridge." *Century Magazine* 2 (1871): 172–73.

John, Richard R. "Who Were the Gilders? And Other Seldom-Asked Questions about Business, Technology, and Political Economy in the United States, 1877–1900." *Journal of the Gilded Age and the Progressive Era* 8, no. 4 (Oct. 2009): 474–80.

"The Kansas Question." *Putnam's Magazine*, Oct. 1855, 427–28.

Katte, Walter. "A Description of the Proposed Plan for Erecting the Superstructure of the Illinois and St. Louis Bridge." *Transactions of the American Society of Civil Engineers* 2 (1874): 132–44.

Kolasky, William. "Chief Justice Edward Douglass White and the Birth of the Rule of Reason." *Antitrust* 24, no. 3 (June 2010): 77–83.

Kouwenhoven, John A. "The Designing of the Eads Bridge." *Technology and Culture* 23, no. 4 (Oct. 1982): 535–68.

Kouwenhoven, John A. "Eads Bridge: The Celebration." *Bulletin of the Missouri Historical Society* 30, no. 3 (Apr. 1974): 159–80.

Kouwenhoven, John A. "St. Louis as Eads Knew It." *Bulletin of the Missouri Historical Society* 30, no. 3 (Apr. 1974): 181–95.

Marianos, W. N., Jr. "George Shattuck Morison and the Development of Bridge Engineering." *Journal of Bridge Engineering* 13, no. 3 (May–June 2008): 291–98.

"Memoir of Henry Flad." *Transactions of the American Society of Civil Engineers* 42 (Dec. 1899): 561–66.

National Park Service. "James B. Eads and His Amazing Bridge." *Museum Gazette* 2 (Jan. 2001). http://www.nps.gov/jeff/historyculture/upload/eads.pdf.

Oliver, Robert E. "East St. Louis: Crossroads of the GM&O, Part 2." *GM&O Historical Society News* 156 (2021): 4–17.

"Our First Number." *American Gas-Light Journal* 1 (July 1, 1859): 1–3.

Parton, James. "The City of St. Louis." *Atlantic Monthly*, June 1867, 655–71. B34/17.

Perkins, Albert T. "Bridge and Ferry Conditions at St. Louis." *Official Proceedings St. Louis Railway Club* 11, no. 7 (Nov. 9, 1906): 14–30. B37/6.

Perkins, Albert T. "The Municipal Bridge and Terminals Commission of St. Louis." *Journal of Political Economy* 15, no. 7 (July 1907): 412–20.

Petrowski, William R. "Kansas City to Denver to Cheyenne: Pacific Railroad Construction Costs and Profits." *Business History Review* 48, no. 2 (Summer 1974): 206–24.

Pfeiffer, David A. "Bridging the Mississippi: The Railroads and Steamboats Clash at the Rock Island Bridge." *Prologue* 36, no. 2 (Summer 2004). https://www.archives.gov/publications/prologue/2004/summer/bridge.html.

Quin, C.W. "France—Mining and Metallurgical Products." *Laboratory: A Weekly Record of Scientific Research* 1 (Aug. 3, 1867): 313–14.

"Railway-Engineering in the United States." *Atlantic Monthly*, Nov. 1858, 642. B34/25.

Reuss, Martin. "Andrew A. Humphreys and the Development of Hydraulic Engineering." *Technology and Culture* 26, no. 1 (Jan. 1985): 1–33.

Roberts, W. Milnor. "Description of the 'Plenum Pneumatic Process' as Applied in Founding the Piers of the Illinois and St. Louis Bridge, at St. Louis, MO." *Transactions of the American Society of Civil Engineers* 1 (1870): 259–71.

Roediger, David R. "Not Only the Ruling Classes to Overcome, but Also the So-Called Mob." *Journal of Social History* 19, no. 2 (Winter 1985): 213–39.

"The St. Louis Merchants Bridge Terminal Railway." *Poor's Manual of Railroads* 26 (1893): 579–80.

Sarno, Don, and Norbert Shacklette. "St. Louis—Part 1." *Passenger Train Journal* 21, no. 6 (June 1990): 16–35.

Sellers, William. "Memoir of James Buchanan Eads." *National Academy of Science Biographical Memoirs* 3 (1895): 61–79.

Sinclair, Bruce. "At the Turn of a Screw: William Sellers, the Franklin Institute, and a Standard American Thread." *Technology and Culture* 10, no. 1 (Jan. 1969): 20–34.

Smith, William Sooy. "Pneumatic Foundations." *Transactions of the American Society of Civil Engineers* 2 (1874): 411–26.

Snyder, Charles E. "The Eads of Argyle." *Iowa Journal*, Jan. 1944, 73–90.

"Special Union Station Issue." *Magazine of the Terminal Railroad Association of St. Louis Historical and Technical Society*, June–July 2001, 6–20.

Taber, Thomas T. "Celebrating 115 Years of the Terminal Railroad." *Magazine of the Terminal Railroad Association of St. Louis Historical and Technical Society*, Summer 2004.

Taussig, Frank. "My Father's Business Career." *Harvard Business Review* 19, no. 2 (Winter 1941): 177–84. K7/190.

Taussig, William. "Personal Recollections of General Grant." *Missouri Historical Society Publications* 2, no. 3 (1903): 10–13.

Thomas, Larry. "East St. Louis Connecting Railway." *Magazine of the Terminal Railroad Association of St. Louis Historical and Technical Society*, Summer 2004, 67–73.

Thomas, Larry. "How the Terminal Railroad Acquired the Wiggins Ferry Company." *Magazine of the Terminal Railroad of St. Louis Historical and Technical Society*, Summer 2004, 112.

Thomas, Larry. "Special Union Station Issue." *St. Louis Union Station News*, June–July 2001.

Twombly, A. S. "The Illinois and St. Louis Bridge." *Scribners Monthly* (June 1871): 173.

Wallace, Agnes. "The Wiggins Ferry Monopoly." *Missouri Historical Review* 42, no. 1 (Oct. 1947): 1–19.

White, Richard. "Information, Markets, and Corruption: Transcontinental Railroads in the Gilded Age." *Journal of American History* 90, no. 1 (June 2003): 19–43. B34/21.

Books

Adler, Dorothy R., and Muriel E. Hidy, eds. *British Investment in American Railways, 1834–1898*. Charlottesville: University Press of Virginia, 1970.

Adler, Jeffrey S. *Yankee Merchants and the Making of the Urban West: The Rise and Fall of Antebellum St. Louis*. Cambridge: Cambridge University Press, 1991.

Aldrich, Mark. *Death Rode the Rails: American Railroad Accidents and Safety, 1828–1965.* Baltimore: Johns Hopkins University Press, 2009.

American Society of Civil Engineers (ASCE). *A Biographical Dictionary of American Civil Engineers.* New York, 1972.

Annales des Ponts et Chaussées. Vol. 7. Paris, 1874. B35/23.

Arenson, Adam. *The Great Heart of the Republic: St. Louis and the Cultural Civil War.* Cambridge, MA: Harvard University Press, 2011.

Baedeker, Karl. *The United States.* New York: Charles Scribner's Sons, 1909.

Bain, David Haward. *Empire Express: Building the First Transcontinental Railroad.* New York: Viking, 1999.

Barry, John M. *Rising Tide: The Great Mississippi Flood of 1927 and How It Changed America.* New York: Simon & Schuster, 1997.

Baskin, Jonathan Barron, and Paul J. Miranti. *A History of Corporate Finance.* Cambridge: Cambridge University Press, 1997.

Bates, Edward. *The Diary of Edward Bates, 1859–1866.* Ed. Howard K. Beale. House Document 818, 71st Cong., 3rd Sess. Washington, DC: GPO, 1933.

Baughman, James P. *Charles Morgan and the Development of Southern Transportation.* Nashville: Vanderbilt University Press, 1968.

Baxter, W. E. *America and the Americans.* London: Routledge, 1855.

Bearss, Edwin C. *Hardluck Ironclad: The Sinking and Salvage of the* Cairo. Baton Rouge: Louisiana State University Press, 1980.

Belcher, Wyatt W. *The Economic Rivalry between St. Louis and Chicago, 1850–1880.* New York: Columbia University Press, 1947.

Bellesiles, Michael A. *1877: American's Year of Living Violently.* New York: New Press, 2010.

Bensel, Richard Franklin. *The Political Economy of American Industrialization, 1877–1900.* Cambridge: Cambridge University Press, 2000.

Borneman, Walter R. *Iron Horses: America's Race to Bring the Railroads West.* New York: Little, Brown, 2010.

Boynton, Charles B. *The History of the Navy during the Rebellion.* Vol. 1. New York: D. Appleton, 1867.

Bridge, James Howard. *The Inside History of the Carnegie Steel Company: A Romance of Millions.* New York: Aldine, 1903.

Brinsmead, Simon. *Essential Interoperability Standards.* Cambridge: Cambridge University Press, 2021.

Brown, John K. "Heuristics, Specifications, and Routines in Building Long-Span Railway Bridges on the Western Rivers, 1865–1880." In Raff and Scranton, *Emergence,* 171–203.

Brown, John K. *Limbs on the Levee: Steamboat Explosions and the Origins of Federal Public Welfare Regulation, 1817–1852.* Middlebourne, WV: International Steamboat Society, 1989.

Canney, Donald L. *The Old Steam Navy.* Vol. 2, *The Ironclads, 1842–1885.* Annapolis, MD: Naval Institute Press, 1993.

Carnegie, Andrew. *Autobiography of Andrew Carnegie*. Boston: Houghton Mifflin, 1920.

Carosso, Vincent P. *The Morgans: Private International Bankers, 1854–1913*. Cambridge, MA: Harvard University Press, 1987.

Chandler, Alfred D., Jr. *The Visible Hand: The Managerial Revolution in American Business*. Cambridge, MA: Belknap Press of Harvard University Press, 1977.

Chanute, Octave, and George Shattuck Morison. *The Kansas City Bridge*. New York: Van Nostrand, 1870.

Chernow, Ron. *The House of Morgan: An American Banking Dynasty and the Rise of Modern Finance*. New York: Simon & Schuster, 1990.

Church, William Conant. *The Life of John Ericsson*. 2 vols. New York: Charles Scribner's Sons, 1906.

Churella, Albert J. *The Pennsylvania Railroad*. Vol. 1, *Building an Empire, 1846–1917*. Philadelphia: University of Pennsylvania Press, 2013.

Clarke, Thomas Curtis. *An Account of the Iron Railway Bridge across the Mississippi River at Quincy Illinois*. New York: Van Nostrand, 1869.

Clevenger, Martha, ed. *"Indescribably Grand": Diaries and Letters from the 1904 World's Fair*. St. Louis: Missouri Historical Society Press, 1996.

Clough, Annie L. *Head of the Bay: Sketches and Pictures of Blue Hill, Maine*. Woodstock, VT: Elm Tree Press, 1953.

Condit, Carl W. *American Building Art: The Nineteenth Century*. New York: Oxford University Press, 1960.

Cooper, Theodore. *American Railroad Bridges*. New York: Engineering News Publishing Company, 1889.

Corey, Lewis. *The House of Morgan: A Social History of the Masters of Money*. New York: G. Howard Watt, 1930.

Cronon, William. *Nature's Metropolis: Chicago and the Great West*. New York: Norton, 1992.

Curzon, Julian. *The Great Cyclone*. St. Louis, 1896.

Dacus, J. A., and James W. Buel. *A Tour of St. Louis*. St. Louis, 1878.

Darnell, Victor C. *Directory of American Bridge-Building Companies, 1840–1900*. Washington, DC: Society for Industrial Archaeology, 1984.

Daunton, M. J. *Wealth and Welfare: An Economic and Social History of Britain, 1851–1951*. New York: Oxford University Press, 2007.

Detroit Bridge and Iron Works. *Memoir of the Iron Bridge over the Missouri River at St. Joseph, Mo*. Detroit, 1873.

DeVoto, Bernard. *The Course of Empire*. New York: Sentry, 1952.

Dickens, Charles. *American Notes for General Circulation*. London: Everyman's Library, 1997.

Dorsey, Florence. *Road to the Sea: The Story of James B. Eads and the Mississippi River*. New York: Rinehart, 1947.

Doyle, Joseph B. *20th Century History of Steubenville*. Chicago, 1910.

Eads, James B. *Addresses and Papers of James B. Eads*. Ed. Estill McHenry. St. Louis, 1884.

Edwards, Rebecca. *New Spirits: Americans in the Gilded Age, 1865–1905*. New York: Oxford University Press, 2006.

Einhorn, Robin L. *Property Rules: Political Economy in Chicago, 1833–1872*. Chicago: University of Chicago Press, 1991.

Elliott, Richard Smith. *Notes Taken in Sixty Years*. St. Louis, 1883.

Emerson, Ralph Waldo. *The Journals and Miscellaneous Notebooks of Ralph Waldo Emerson*, vol. 11. Ed. William H. Gilman, Alfred R. Ferguson, George P. Clark, and Merrell R. Davis. Cambridge, MA: Harvard University Press, 1960.

Foner, Philip S. *The Workingmen's Party of the United States*. Minneapolis: MEP, 1984.

Franch, John. *Robber Baron: The Life of Charles Tyson Yerkes*. Urbana: University of Illinois Press, 2006.

Francis, David R. *The Universal Exposition of 1904*. St. Louis: Louisiana Purchase Exposition Company, 1905.

Gilbert, James. *Whose Fair? Experience, Memory, and the History of the Great St. Louis Exposition*. Chicago: University of Chicago Press, 2009.

Glaab, Charles N. *Kansas City and the Railroads*. Madison: State Historical Society of Wisconsin, 1962.

Glendenning, Gene V. *The Chicago & Alton Railroad: The Only Way*. DeKalb: Northern Illinois University Press, 2002.

Gordon, Christopher Alan. *Fire, Pestilence, and Death: St. Louis, 1849*. St. Louis: Missouri Historical Society Press, 2018.

Gould, Emerson E. *Fifty Years on the Mississippi; or, Gould's History of River Navigation*. St. Louis, 1889.

Grant, H. Roger. *"Follow the Flag": A History of the Wabash Railroad Company*. DeKalb: Northern Illinois University Press, 2004.

Grant, H. Roger, Don L. Hofsommer, and Osmund Overby. *St. Louis Union Station: A Place for People, A Place for Trains*. St. Louis: St. Louis Mercantile Library, 1994.

Grant, Ulysses S. *The Papers of Ulysses S. Grant, Volume 25: 1874*. Ed. John Y. Simon. Carbondale: Southern Illinois University Press, 2003.

Greenberg, Dolores. *Financiers and Railroads, 1869–1889: A Study of Morton, Bliss & Company*. Newark: University of Delaware Press, 1980.

Greenhill, Ralph. *Engineer's Witness*. Boston: David R. Godine, 1985.

Grodinsky, Julius. *Jay Gould: His Business Career, 1867–1892*. Philadelphia: University of Pennsylvania Press, 1957.

Grodinsky, Julius. *Transcontinental Railroad Strategy, 1869–1893: A Study of Businessmen*. Philadelphia: University of Pennsylvania Press, 1962.

Hadley, Arthur T. *Railroad Transportation: Its History and Its Laws*. New York, 1885.

Hansen, Peter A., Don I. Hofsommer, and Carlos Arnaldo Schwantes. *Crossroads of a Continent: Missouri Railroads, 1851–1921*. Bloomington: Indiana University Press, 2022.

Hatton, Joseph. *Henry Irving's Impressions of America*. Boston, 1884.

Haupt, Herman. *General Theory of Bridge Construction*. New York: D. Appleton, 1851.

Heathcott, Joseph, and Angela Dietz. *Capturing the City: Photographs from the Streets of St. Louis, 1900–1930.* St. Louis: Missouri Historical Society Press, 2016.

Hendrick, Burton J. *Life of Andrew Carnegie.* 2 vols. New York: Doubleday, Doran, 1932.

Hodes, Frederick A. *A Divided City: A History of St. Louis, 1851–1876.* N.p.: Bluebird, 2015.

Hovenkamp, Herbert. *Enterprise and American Law, 1836–1937.* Cambridge, MA: Harvard University Press, 1991.

How, Louis. *James B. Eads.* Cambridge, MA: Riverside Press, 1900.

Hubbard, Walter, and Richard F. Winter. *North Atlantic Mail Sailings, 1840–1875.* New York: US Philatelic Classics Society, 1988.

Hunter, Louis C. *A History of Industrial Power in the United States.* Vol. 2, *Steam Power.* Wilmington, DE: Eleutherian Mills—Hagley Foundation, 1985.

Hunter, Louis C. *Steamboats on the Western Rivers: An Economic and Technological History.* Cambridge, MA: Harvard University Press, 1949.

Hyde, William, and Howard L. Conard, eds. *Encyclopedia of the History of St. Louis.* 4 vols. New York: Southern History Company, 1899.

Jackson, Robert W. *Rails across the Mississippi: A History of the St. Louis Bridge.* Urbana: University of Illinois Press, 2001.

Jaminet, Alphonse. *Physical Effects of Compressed Air.* St. Louis, 1871.

John, Richard R. *Network Nation: Inventing American Telecommunications.* Cambridge: Belknap Press of Harvard University Press, 2010.

John, Richard R., ed. *Ruling Passions: Political Economy in Nineteenth-Century America.* University Park: Pennsylvania State University Press, 2006.

Kane, Thomas P. *The Romance and Tragedy of Banking.* New York: Bankers Publishing Co., 1923.

Kanigel, Robert. *The One Best Way: Frederick Winslow Taylor and the Enigma of Efficiency.* New York: Viking, 1997.

Karabell, Zachary. *Inside Money: Brown Brothers Harriman and the American Way of Power.* New York: Penguin, 2021.

Keystone Bridge Company. *Descriptive Catalogue of Wrought Iron Bridges.* Philadelphia, 1874.

King, Edward. *The Great South: A Record of Journeys.* Hartford, CT: American Publishing Co., 1875.

Klaus, Ian. *Forging Capitalism: Rogues, Swindlers, Frauds, and the Rise of Modern Finance.* New Haven, CT: Yale University Press, 2014.

Klein, Maury. *The Life and Legend of Jay Gould.* Baltimore: Johns Hopkins University Press, 1986.

Klein, Maury. *Union Pacific: The Birth of a Railroad, 1862–1893.* Garden City, NY: Doubleday, 1987.

Kouwenhoven, John A. *Made in America: The Arts in Modern Civilization.* 1948. Reprint, New York: Doubleday, 1962.

Krass, Peter. *Carnegie.* New York: Wiley, 2002.

Larson, John Lauritz. *Internal Improvements: National Public Works and the Promise of Popular Government in the Early United States*. Chapel Hill: University of North Carolina Press, 2001.

Laurie, Bruce. *Artisans into Workers: Labor in Nineteenth-Century America*. New York: Noonday, 1989.

Lazonick, William. *Business Organization and the Myth of the Market Economy*. Cambridge: Cambridge University Press, 1991.

Leland, Charles Godfrey. *Memoirs*. Vol. 2. London, 1893.

Leland, Charles Godfrey. *The Union Pacific Railway, Eastern Division, or Three Thousand Miles in a Railway Car*. Philadelphia, 1867.

Lewis, Gene D. *Charles Ellet, Jr.: The Engineer as Individualist, 1810–1862*. Urbana: University of Illinois Press, 1968.

MacGill, Caroline E. *History of Transportation in the United States before 1860*. Washington, DC: Carnegie Institution, 1917.

Manders, Damon, and Brian Rentfro. *Engineers Far from Ordinary: The US Army Corps of Engineers in St. Louis*. St. Louis: US Corps of Engineers, 2011.

McCullough, David. *The Great Bridge*. New York: Simon & Schuster, 1972.

McGinty, Brian. *Lincoln's Greatest Case: The River, the Bridge, and the Making of America*. New York: Liveright, 2015.

Medbery, James K. *Men and Mysteries of Wall Street*. Boston: Fields, Osgood, 1870.

Medley, Julius George. *An Autumn Tour in the United States and Canada*. London, 1873.

Middleton, William D. *The Bridge at Québec*. Bloomington: Indiana University Press, 2001.

Miller, Howard S., and Quinta Scott. *The Eads Bridge*. Columbia: University of Missouri Press, 1979.

Miner, H. Craig. *The St. Louis–San Francisco Transcontinental Railroad*. Lawrence: University Press of Kansas, 1972.

Misa, Thomas J. *A Nation of Steel: The Making of Modern America, 1865–1925*. Baltimore: Johns Hopkins University Press, 1995.

Moody's Manual of Railroad and Corporation Securities. New York: Moody Manual Co., 1902.

Morris, Charles R. *The Tycoons*. New York: Henry Holt, 2005.

Morris, James McGrath. *Pulitzer: A Life in Politics, Print, and Power*. New York: Harper, 2010.

Mortimer, John. *Zerah Colburn: The Spirit of Darkness*. Suffolk, UK: Arima, 2005.

Nasaw, David. *Andrew Carnegie*. New York: Penguin, 2006.

National Geographic Society. *Historical Atlas of the United States*. Washington, DC, 1993.

Neuzil, Mark. *Views on the Mississippi: The Photographs of Henry Peter Bosse*. Minneapolis: University of Minnesota Press, 2001.

Nye, David E. *American Technological Sublime*. Cambridge, MA: MIT Press, 1994.

Pak, Susie J. *Gentlemen Bankers: The World of J. P. Morgan*. Cambridge, MA: Harvard University Press, 2013.

Parker, N. Howe. *Iowa as It Is in 1855*. Chicago, 1855.

Paskoff, Paul. *Troubled Waters: Steamboat Disasters, River Improvements, and American Public Policy, 1821–1860*. Baton Rouge: Louisiana State University Press, 2007.

Perkins, Edwin J. *Financing Anglo-American Trade: The House of Brown, 1800–1880*. Cambridge, MA: Harvard University Press, 1975.

Peterson, Norma L. *Freedom and Franchise: The Political Career of B. Gratz Brown*. Columbia: University of Missouri Press, 1965.

Petroski, Henry. *Engineers of Dreams: Great Bridge Builders and the Spanning of America*. New York: Knopf, 1995.

Philips, John L. *The Bends: Compressed Air in the History of Science, Diving, and Engineering*. New Haven, CT: Yale University Press, 1998.

Pratt, Walter F., Jr. *The Supreme Court under Edward Douglass White, 1910–1921*. Columbia: University of South Carolina Press, 1999.

Primm, James Neal. *Lion of the Valley: St Louis Missouri*. Boulder, CO: Pruitt, 1981.

Raff, Daniel M. G., and Philip Scranton. *The Emergence of Routines: Entrepreneurship, Organization, and Business History*. New York: Oxford University Press, 2017.

Renehan, Edward J., Jr. *Commodore: The Life of Cornelius Vanderbilt*. New York: Basic Books, 2007.

Richardson, Heather Cox. *West from Appomattox: The Reconstruction of America after the Civil War*. New Haven, CT: Yale University Press, 2007.

Riegel, Robert E. *America Moves West*. New York: Henry Holt, 1947.

Ringwalt, J. L. *Development of Transportation Systems in the United States*. Philadelphia: Railway World, 1888.

Roebling, John Augustus. *Long and Short Span Railway Bridges*. New York: Van Nostrand, 1869.

Rust, Daniel L. *The Aerial Crossroads of America: St. Louis's Lambert Airport*. St. Louis: Missouri Historical Society Press, 2016.

Sale, Kirkpatrick. *The Fire of His Genius: Robert Fulton and the American Dream*. New York: Free Press, 2001.

Sander, Kathleen Waters. *John W. Garrett and the Baltimore & Ohio Railroad*. Baltimore: Johns Hopkins University Press, 2017.

Sandlin, Lee. *Wicked River: The Mississippi When It Last Ran Wild*. New York: Pantheon, 2010.

Satterlee, Herbert L. *J. Pierpont Morgan*. New York: Macmillan, 1939.

Scharf, J. Thomas. *History of St. Louis City and County*. 2 vols. Philadelphia: Louis H. Everts, 1883.

Schodek, Daniel L. *Landmarks in American Civil Engineering*. Cambridge, MA: MIT Press, 1987.

Shallatt, Todd. *Structures in the Stream: Water, Science, and the Rise of the Army Corps of Engineers*. Austin: University of Texas Press, 1994.

Short, Simine. *Locomotive to Aeromotive: Octave Chanute and the Transportation Revolution*. Champaign: University of Illinois Press, 2011.

Slap, Andrew L. *The Doom of Reconstruction: The Liberal Republicans in the Civil War Era*. New York: Fordham University Press, 2006.

Steinmetz, Greg. *American Rascal: How Jay Gould Built Wall Street's Biggest Fortune*. New York: Simon & Schuster, 2022.

Stone, Mary Amelia Boomer. *Memoir of General George Boardman Boomer: Bridge Builder and Soldier*. Boston: Rand & Avery, 1864.

Strouse, Jean. *Morgan, American Financier*. New York: Random House, 1999.

Sullivan, Louis. *The Autobiography of an Idea*. New York: Dover, 1956.

Taliaferro, John. *All the Great Prizes: The Life of John Hay, from Lincoln to Roosevelt*. New York: Simon & Schuster, 2013.

Taylor, George Rogers. *The Transportation Revolution, 1815–1860*. 1951. Reprint, Armonk, NY: M. E. Sharpe, 1977. Originally published as volume 4 of *The Economic History of the United States* by Holt, Rinehart & Winston.

Taylor, George Rogers, and Irene D. Neu. *The American Railroad Network, 1861–1890*. Urbana: University of Illinois Press, 2003.

Taylor, Jacob N., and M. O. Crooks. *Sketch Book of Saint Louis*. St. Louis, 1858.

Terminal Railroad Association. *The St. Louis Union Station: A Monograph*. St. Louis: National Chemigraph Company, 1895.

Theising, Andrew J. *Made in USA: East St. Louis*. St. Louis: Virginia Publishing, 2003.

Thienel, Philip M. *Mr. Lincoln's Bridge Builders: The Right Hand of American Genius*. Shippensburg, PA: White Mane Books, 2000.

Turner, George Edgar. *Victory Rode the Rails: The Strategic Place of the Railroads in the Civil War*. New York: Bobbs-Merrill, 1953.

Twain, Mark. *Life on the Mississippi*. New York: Library of America, 2009.

Twain, Mark. *Mark Twain in Eruption*. Ed. Bernard DeVoto. New York: Harper & Bros., 1940.

Tweedale, Geoffrey. *Sheffield Steel and America: A Century of Commercial and Technological Interdependence, 1830–1930*. Cambridge: Cambridge University Press, 1987.

Tyson, Robert A. *History of East St. Louis*. East St. Louis: Haps, 1875.

Unger, Irwin. *The Greenback Era: A Social and Political History of American Finance, 1865–1879*. Princeton, NJ: Princeton University Press, 1985.

US Department of Commerce, Bureau of the Census. *Historical Statistics of the United States*. 2 vols. Washington, DC: GPO, 1976.

Veenendaal, Augustus J. *Slow Train to Paradise: How Dutch Investment Helped Build American Railroads*. Stanford, CA: Stanford University Press, 1996.

Vitiello, Domenic. *Engineering Philadelphia: The Sellers Family and the Industrial Metropolis*. Ithaca, NY: Cornell University Press, 2013.

Voorhees. H. B. *The Alton Railroad Bridge at Glasgow Missouri, 1879*. New York: Newcomen Society, 1944.

Wagner, Erica. *Chief Engineer Washington Roebling: The Man Who Built the Brooklyn Bridge*. New York: Bloomsbury, 2017.

Wall, Joseph Frazier, *Andrew Carnegie*. New York: Oxford University Press, 1980.

Wallis, Richard T. *The Pennsylvania Railroad at Bay*. Bloomington: Indiana University Press, 2001.

Wanko, Andrew. *Great River City: How the Mississippi Shaped St. Louis*. St. Louis: Missouri Historical Society Press, 2019.

Ward, James A. *J. Edgar Thomson: Master of the Pennsylvania*. Westport, CT: Greenwood, 1980.

Wayman, Norbury L. *St. Louis Union Station and Its Railroads*. St. Louis: Evelyn E. Newman Group, 1987.

White, John H., Jr. *The American Railroad Freight Car*. Baltimore: Johns Hopkins University Press, 1993.

White, Richard. *Railroaded: The Transcontinentals and the Making of Modern America*. New York: Norton, 2011.

White, Richard. *The Republic for Which It Stands: The United States during Reconstruction and the Gilded Age, 1865–1896*. New York: Oxford University Press, 2017.

Winpenny, Thomas R. *Without Fitting, Filing, or Chipping: An Illustrated History of Phoenix Bridge Company*. Easton, PA: Canal History and Technology Press, 1996.

Woodward, Calvin M. *A History of the St. Louis Bridge*. St. Louis, 1881.

Young, David M. *The Iron Horse and the Windy City: How Railroads Shaped Chicago*. DeKalb: Northern Illinois University Press, 2005.

INDEX

Page numbers in *italic* refer to figures; those in **bold** refer to tables.